THE ELEMENTARY FORMS OF STATISTICAL REASON

THE ELEMENTARY FORMS OF STATISTICAL REASON

R. P. Cuzzort
University of Colorado at Boulder

James S. Vrettos
The City University of New York, John Jay College of Criminal Justice

ST. MARTIN'S PRESS
New York

Editor: Sabra Scribner
Managing editor: Patricia Mansfield Phelan
Project editor: Diana Puglisi
Production supervisor: Scott Lavelle
Art director: Lucy Krikorian
Text design: Dorothy Bungert, EriBen Graphics
Graphics: MacArt Design
Cover design: John Callahan

Library of Congress Catalog Card Number: 94-74792

Manufactured in the United States of America.

0 9 8 7 6
f e d c b a

For information, write:
St. Martin's Press, Inc.
175 Fifth Avenue
New York, NY 10010

ISBN: 0-312-08946-5

PREFACE

The orthodox approach to statistics sets the realm of the statistician apart from other, less "pure," unscientific descriptions of our world. We think this has generated a sense of statistics not only as an alien study, but also as something removed from the affairs of "real" people. If this assessment is even partially valid, it is to be most profoundly regretted, for there is no study more important, more universally applicable, and more enlightening with respect to the difficulties of achieving accurate accounts of our lives than the formal study of statistics.

The proper study of statistics begins with the problem of what constitutes accurate descriptions of reality. At the same time, we understand that all intellectual activity, whether formal or informal, is concerned with description. Statistical reasoning, because it is another way of describing things, is not at all irrelevant. Everyone relies on it in a loose way nearly all of the time. We believe students gain a better awareness of and appreciation for the character of statistical thinking if they can be shown the connection between statistics and everyday concerns. They make the transition to formal statistical methods more easily when they can connect the logic of statistics to the logic of thought forms with which they are more familiar. They find statistical thinking is neither more nor less intimidating than poetry, philosophy, journalism, history, or any other descriptive style.

The Elementary Forms of Statistical Reason was written particularly for those nonmathematicians in the social sciences and humanities who approach statistics with trepidation. Such people, we have found, vary widely in their knowledge of and affection for mathematics—with most of them, unfortunately, looking on such studies antagonistically. Their teachers, generally more mathematically sophisticated, commonly present statistical procedures as a form of well-defined, rigorous, and logically precise exercises. This, we are convinced, has produced several generations of statisticians and researchers who have acquired a formalized knowledge of how to get things done but who, at the same time, have an open contempt for even relatively light philosophical ponderings over the "whys" of their methods. Too often, it is simply presumed that statistics is a most worthwhile endeavor. We question that presumption and challenge those beginning their study of statistics to maintain a high level of skepticism. At one level, statistical thinking simply cannot be avoided—it is, in this case, a matter of universal and constant concern. At another level, the level of formal, elaborate, academic statistical studies, it is a matter that calls for careful evaluation. In any event, statistical sophistication is a foundation stone upon which to build nearly any kind of professional sophistication.

Our approach, one that has been increasingly successful as we have refined it through the years, emphasizes why statistical studies are necessary. It also makes the basic but controversial point that any descriptive effort must be biased. Determining how much distortion or bias to leave in or remove from a descriptive effort is an arbitrary and extremely important decision. Statisticians have to rely on a kind of "license" just as the novelist relies on "artistic license." This makes statistical methods more arbitrary and less certain. A beginning student might as well see what this implies at the outset.

We have, therefore, not ended each chapter with the traditional set of purely logical problems or exercises and answers. Instead, we have introduced sections entitled "Thinking Things Through." The issues brought up in these sections seek to reveal the extent to which statisticians, like anybody else, are constantly faced with problems that have to be resolved arbitrarily. Many matters that come up in statistical studies do not have clean, clear, solid, logical answers. (If they did, society would have resolved most of the social and political turmoil in which it seems to be constantly caught up.)

One reviewer was indignant that *The Elementary Forms of Statistical Reason* is "mostly words about statistics." We liked that, because that is exactly what this book is. It is a way of getting at statistics by examining more carefully the engaging question: "Why bother with such stuff, and if you insist on bothering with it, how do you really know what you are doing?" All too often, when statisticians are asked to talk about what they've accomplished, they lack the words to express themselves. Stories are told, for example, of graduate students who, when asked why they used percentages to report their findings, could not find the words necessary to justify their actions.

Students should not be trained to be totally dependent on whatever statistical authorities assert is the appropriate way to deal with a given problem. They should develop independent problem-solving abilities. This requires a technical knowledge of statistics and a broad intellectual sense of how statistical descriptions fit into the general spectrum of human reasoning. We encourage our students to invent their own formulas, if their doing so leads to better solutions. Statistical studies should be liberating, not oppressive. Such intellectual exploration is what a liberal education is designed to achieve.

We must conclude by acknowledging our teachers and those who inspired and challenged us through hours and hours of enlightening and spirited argument. We were encouraged to develop our own appreciation for statistics as a liberating medium. Of course, it should go without saying that we relieve our teachers completely of any responsibility for our unique opinions and any errors that appear in the body of this work. We would also like to take this moment to express our appreciation to those truly fine students who brought errors to our attention—such students are rare, and like many things that are rare, extremely valuable.

First we would like to mention our indebtedness to several teachers and "masters" with whom we worked either as students, assistants, or colleagues: James Quinn, Donald Bogue, Warren Thompson, Otis Dudley Duncan, Phillip Hauser, Richard Mark Emerson, Roy Francis, Stuart Chapin, Theodore Caplow, Norman Ryder, and Jean Phillips. These mentors are removed in time but not in spirit.

So many people provided us with direct criticism and suggestions that we are certain to overlook many. We are particularly indebted to Ron Urban, Paul Melevin, Galen Switzer, Cynthia Chien, Pamela Behan, John A. Paulos, C. Dale Johnson, Florence Karlstrom, Richard Skeen, Rolf Kjolseth, and Stanley Lieberson.

We are also especially indebted to those who reviewed the manuscript for St. Martin's Press: E. Helen Berry, Diane Felmlee, William Harris, Christopher Marshall, William Protash, William Notz, and Geoffrey Watson.

At St. Martin's Press we would like to express appreciation for the excellent efforts of Louise Waller, Bob Nirkind, and Diana Puglisi. Our thanks to everyone.

R. P. Cuzzort
James Vrettos

CONTENTS

THE ELEMENTARY FORMS OF STATISTICAL REASON

INTRODUCTION: AN OVERVIEW

"Everything is simple; nothing is simple."

—R. P. CUZZORT

"Rules make the learner's path long, examples make it short and successful."

—SENECA

"If you are unwilling to experience embarrassment, you will never learn much of anything."

—ARTHUR HINMAN

The discussions in this book are intended to accomplish several things. First and foremost, they are designed to reveal the *skeletal structure* of statistical reasoning and, to some degree, the general process of reasoning itself—wherever it appears in human communication. Many of the qualifications and refinements that are part of modern statistical techniques have been bypassed here in order to concentrate on the most significant and primary processes that direct our thinking.

Second, the discussions in this book are intended to reveal the *extent* to which statistical forms appear in virtually all forms of human communication, ranging from mathematical statistics (such as might be found in quantum mechanics and nonlinear physics) to forms in which statistical thinking is presumed not to exist—such as poetry, opera, popular movies, the thought patterns of mentally disturbed people, or even the behaviors of nonhumans.

While cats or dogs or other nonhuman creatures obviously do not practice statistics the way college professors do, they nonetheless betray an underlying involvement with statistical concerns such as drawing conclusions from samples, assessing relationships, evaluating what is typical or not typical, and determining risks or probabilities.

The main point here is that statistical forms are deep patterns—possibly having their origins in the dynamics of adaptive behavior itself. Statistical reasoning is profound. It is the study of the very foundations of thinking and acting. The elements of statistical reason are found in the communications of every conscious living human being, male or female, white or black, old or young, rational or irrational.

The primary statistical forms are so deep that, in rudimentary ways, they appear in the behaviors of plants and single-celled creatures. To understand the elemental forms of statistical reason is to understand a universal basis for thinking and acting. There is more than a suggestion, when you become familiar with statistical forms, that they are a genetic structure in much the same sense that the mind is "vulnerable" to—or genetically programmed for—language, as linguist Noam Chomsky has suggested.

Third, these discussions are designed to show that statistical reasoning enhances and is dependent upon *imagination*. There are statistical studies that are delightful works of imaginative genius. The most important factors in doing statistics—as in "doing" art, music, religion, or anything else—are a keen intelligence, a constant curiosity, and a powerful, disciplined imagination. Later in the book the power of this form of imagination is illustrated through two influential statistical stories. Good statistical studies can be as mind-boggling as a clever poem or an exciting novel.

Fourth, the discussions in this book are meant to provide *a foundation for peering into the complexities of formal statistical analysis*. Statistical analysis can mystify even well-trained statisticians. However, if a person has a solid understanding of the basics of statistical analysis, complicated arguments can be reduced to their essential forms. For example, if you are aware that relationships, in general, reduce variability around some typical value, then a complex technique such as the analysis of variance becomes a particular instance of the broader problem of examining relationships. To see this, it is necessary (a) to have a good idea of what you are trying to do when you establish relationships between variables, and (b) to have a good sense of why you bother with the whole business in the first place.

Fifth, we shall occasionally suggest, in the discussions in this book, that clear reason, which to a large degree is a matter of the proper employment of statistical reason, can do as much for *promoting moral conduct* as can the study of ethics or religious thinking. If this claim is valid, then anyone interested in the moral life must become aware of the extent to which bad reasoning underlies many moments of human misunderstanding and misery. Recent statistical studies of human sexuality, for example, have had a tremendous influence on a variety of older moral notions and are, to a considerable extent, accountable for many of the changes that have taken place in modern society with respect to sexual roles.

Sixth, this book is *not a mathematical treatment of statistics*. Human collectives do not readily lend themselves to mathematical descriptions. We can describe physical entities through mathematical devices, but human conditions such as happiness, success, love, sanity, morality, progress, or any of hundreds of similar concerns, are elusive. They are inherently ambiguous abstractions of an extremely complicated nature, and they are inescapably subject to varied interpretations. We are interested, here, in a social treatment of statistics. This means being concerned, for example, with why the mass media have increasingly relied on quasi-statistical social reports (polls, surveys, and the like) and why these reports are highly popular among modern audiences. Thus, we want to consider the problems posed by the attempt to transform the stuff of human relations into statistical forms.

Seventh, the discussions in this book stress that statistical reasoning is not a matter of separate and distinct techniques, each applied properly according to the demands of the situation. Unfortunately, that was how we learned statistics as undergraduates, as graduate students and then, in one case, as a research associate at the University of Chicago. Let us hasten to say that knowing when and where to apply appropriate statistical techniques is what constitutes technical skill in any serious research. However, to see statistical procedures more broadly—from "above" as it were—enables a person to become aware of an interesting feature of such reasoning: *all of its major parts "blend" into each other.* They are not distinct or separate techniques. It is good to know how the separate parts of statistical reason—averages, variances, sampling, associations, probability, and so on—fit together and blend into each other. Actually, statistical reason is like an intricate two-note concerto, in which the two notes consist of the idea of averages and the idea of variability—with the idea of averages the dominant note. Nearly all statistical operations are some kind of elaborated development of these two extremely fundamental ideas.

Eighth, these discussions are concerned with what the mathematician John Allen Paulos calls *innumeracy.* There is great variability, among American students especially, with respect to knowledge of, and attitude toward, numerically expressed information. We have worked with students who have no trouble with complex numbers, imaginaries, functions, quadratics, integrals, the binomial expansion, and other mathematical ideas. We have also encountered students whose numerical literacy level was low. Indeed, not only were these students unfamiliar with numerical expression but furthermore, many were antagonistic to the idea of gaining any deeper awareness of what such forms of expression are capable of revealing.

Numbers have great imaginative power. Before people became aware of the number of miles that stood between them and the nearest star, they were enchanted with the heavens, but they were not overwhelmed. Now we measure interstellar distances in light years. How many miles are there in a light year? Light travels 186,000 miles per second. There are 31,536,000 seconds in a year. Therefore, there are 5,865,696,000,000 miles in a light year. That's a lot of miles. In fact, if you tried to drive the distance in your car, averaging sixty miles an hour and driving twenty-four hours a day, it would take you over eleven million years to get just one light year away from the earth. If you stopped along the way to eat and sleep, and drove only twelve hours every day, it would take you twenty-two million years to make the trip.

Ninth, this book addresses the obvious fact that social science does not have the rigor that is characteristic of the natural sciences. One of the things we shall want to consider, since this book is primarily concerned with statistics within a social context, is how we can find, through the study of statistical reasoning, *why it is so difficult to achieve precision in social knowledge.* Indeed, there may be a kind of "knowledge barrier" as we move into the social domain: a barrier that inhibits moving beyond a particular level of certainty. If this could be proved, it would, in itself, provide a basis for thinking more deeply about the social future of the people of the earth.

While the social sciences cannot be precise in the same fashion as the physical sciences, the endeavor for precision must always be maintained. The haunting problem that confronts the social scientist is how to know when this so-called knowledge barrier has been reached, and whether it is futile to try to move further. In any event, only one thing can be said with any degree of certainty about social knowledge, and that is that there is no certain, empirically established, incontestable "truth" about social actions. Instead, it seems that we are doomed, forever, to have to deal with social dynamics in terms of relatively crude probabilistic evaluations.[1]

Tenth, it is characteristic, when new and revolutionary social movements are unfolding, for people on different sides of the movements to caricature each other. Social science—combined with an emphasis on modern statistical methods—is a modern social movement. As such, it has often denigrated literature or journalism as academically unsound. In this book we would like to make the point that *literary and statistical approaches to social events are simply different "mapping" techniques.* While there are certainly obvious differences between a statistical report and, let us say, a novel on gang activities in the United States, the information that is contained in one "map" is not necessarily better or worse than the information in the other. They are complementary approaches to an extremely complex and abstract set of processes.

Eleventh, in discussions in this book we shall occasionally consider statistics in terms of *aesthetic criteria.* Good statistical work that is clean, cogent, and carefully done can have an aesthetically pleasing quality. We concede that examples of such work are not easily found in the literature of social research. But this does not mean that social research should go out of its way to present its arguments in terms that are purposely obscure and even grotesque. Can statistical work display, as does mathematics more broadly, forms that are more aesthetically appealing than others? Or is statistics doomed to be an ugly science? Our strong feeling about this is that a good statistician should have an aesthetic as well as a technical sensitivity.

Twelfth, this book seeks to provide *a sense of the "spirit" and the inherently playful nature of statistical reasoning.* It takes up matters that standard texts are not able to deal with. For example, we examine why certain statistical forms are found in virtually all communications. Standard texts concentrate on *how* to do statistics rather than *why* we do them.

In this book we are interested in broader issues, such as the problem of whether or not anything intellectually substantive has been achieved in some instances of hypothesis testing. Standard texts rarely consider hypothesis testing in any critical manner; they simply do not have time for such discussions.

This is not meant as a criticism of standard texts. One of the problems with statistical training is that there is such a great amount of material to learn; as a result, instructors in introductory classes commonly complain that a typical text leads to a superficial review in which students learn too little about too much. This leads to confusion and an eventual reliance on rote or mechanical problem-solving, prompting students to do statistics with a modest comprehension of why they are doing it or what is actually being accomplished. At worst, many students come away from statistics with profoundly negative feelings and a deep distaste for formal statistical logic. If this book can, to some degree, correct such sentiments, to the extent that they already exist, then it will have served its purpose.

The quote that opens this introduction says that everything is simple and nothing is simple. We can think of no better way to describe statistical logic. At one level it is extremely simple—and we will try to remain at a relatively simple level throughout this book. However, at another level it is impossibly complex. Few things, for example, are simpler than the arithmetic mean. However, as simple as this computation is, the ideas behind it are complicated. Similarly, some of the complicated difficulties that arise with probabilities have their source in the simple idea of the average.

The main thing, however, is to approach the study of statistical reasoning with an open mind and with a sense that the well-balanced modern person of knowledge should be literate in both the sciences and the humanities. Statistical training is simply one of several roads that can take us into greater realms of understanding and imagination.

THINKING THINGS THROUGH

Each chapter of this book will conclude with a set of problems or questions, under the heading, "Thinking Things Through." These questions will not always have clear-cut answers. You will sometimes be openly called on to use your intuition, to question the question, to ponder and think through matters, and (though some statisticians consider this a dirty word), to reach a knowledgeable "opinion." We want you to reach the point where you feel that you are the master of statistical reasoning, rather than the other way around. Approach these problems and riddles as amusements and you will find them more fulfilling and engaging than if you think of them as mechanical exercises.

The following three questions are meant to lead you into playing around with numbers. The first two questions are set up so that you can calculate a precise figure. In the third problem, however, you are called on to evaluate data; thus, your response will differ from another person's to the extent your judgments vary.

1. This question is taken from John Allen Paulos's book *Innumeracy.*[2] Assuming human hair grows an inch a month, how fast does it grow in miles per

hour? (We have experienced any number of students who respond to this question with the statement, "That's absurd, hair doesn't grow in miles per hour." If you belong to this group, start thinking things through until you come to the precise answer to the question. You will have to deal with some modestly large and small numbers to get the answer.)

2. The planet weighs roughly six septillion tons. There are approximately five and a half billion people on the planet. If each person weighed an average of 150 pounds (a generous estimate), then what would be the ratio of the weight of all humankind to the weight of the entire planet? If a string stretching from here to the moon, a mean distance of roughly 235,000 miles, represented the weight of the earth, how long a string would represent the weight of all humankind?
3. Find out by going to the library or by looking in the Statistical Abstracts of the United States, 1994, Table 1376, what proportion of the gross national product the citizens of various countries pay in taxes. How does the United States fare? Is the United States the most heavily, moderately heavily, or least heavily taxed nation in the world? What do these figures do to your thinking? Why are Americans not especially familiar with these data? How do statistics and politics mix?

Notes

1. Statisticians vary considerably in their belief that a precise form of social knowledge can eventually be reached through careful application of scientific and statistically rigorous methods. A century of effort and millions of labor-hours have not, as yet, resulted in social knowledge that does not contain large amounts of what are, essentially, error components. We shall argue that this is the major contribution of modern social science—it has replaced the "hard" but imaginary beliefs of folk knowledge with the "softer," probabilistic, and empirically grounded knowledge of our times. For example, where folk knowledge says, "Crime does not pay," social science might say, "The probability of a given crime having a cost benefit that is positive is greater than zero." In other words, sometimes crime might pay. This is a major difference in how we think about our social lives.
2. John Allen Paulos, *Innumeracy: Mathematical Illiteracy and Its Consequences* (New York: Vintage Books, 1988), pp. 8, 11.

1

COMMON SENSE AND THE ELEMENTARY FORMS OF STATISTICAL REASONING

"Common sense knowledge is prompt, categorical, and inexact."
—SUZANNE K. LANGER

Although it is not unusual for students, especially in the humanities or social sciences, to express a distaste for courses in statistics and statistics in general, the truth of the matter is that virtually all human communication relies on several logical forms that are the core concerns of formal statistical analysis. For example, an important statistical concept is the idea of *average,* or what is *typical.* Virtually all statistical work is directed toward estimating typicalities—whether in the physical, biological, or social sciences.[1] At the same time, typification lies at the heart of ordinary communication.

The central idea of this chapter is to point out that in a sense we rely on crude applications of "statistical" thinking at all times. Statistics is not something new to us. We are, for example, always *counting* things—though we might not be aware of it, and though our "counting" is generally quite primitive. Where a statistician might say that 76 percent of the students at Siwash University favored faculty evaluations, in ordinary discourse we might say "most" or "lots" of students favored such evaluations; or, more likely, we might say, "The students at Siwash favored faculty evaluations." Notice how this last statement implies a count, but rather than giving us a precise figure of 76 percent of the students, it suggests that *all* students favored faculty evaluations.

The main point of this chapter is that ordinary communications have a crude statistical form. Academic statistical studies are a refinement of a process we have been familiar with during our entire lives. The only difference between the way we communicate on a daily basis and the way statisticians carry out their operations is that the statistician tries to be more careful. This is a difference that makes a difference. The study of statistics teaches us two fundamental lessons: (1) Ordinary communication, especially in the realm of social affairs, is filled with errors, exaggerations, false claims, and stereotypes that commonly lead to misunderstandings and conflicts with serious consequences; and (2) Precise communication is extremely difficult, and within the social realm, communication can probably never achieve complete precision.

STATISTICAL REASONING IN EVERYDAY LIFE

Courses in social statistics at the university level are largely an early twentieth-century product.[2] The term "statistics" itself did not appear in English usage until the late eighteenth century, though the term "political arithmetic" was used by the Englishman John Graunt as early as the 1660s.[3] The development of quantitative procedures relevant to modern social statistics was refined by early scientists working in a variety of areas—especially in astronomy and biology. Late eighteenth-century economists, Thomas Robert Malthus among them, found statistical arguments especially persuasive.[4] Even so, it is the work of the French social scientist Emile Durkheim that is now seen as one of the most influential early uses of statistical information to reveal possible social causes of what had previously been viewed as an extremely individualistic act. Durkheim's study, titled *Le Suicide,* was published in 1897 and marked a turning point in the use of statistical methods for the examination of social phenomena.[5]

Durkheim collected massive amounts of statistical information available at the close of the nineteenth century concerning suicides throughout the provinces of Europe. What he discovered was that the incidence of suicide varied in consistent ways with the types of social structure in the various provinces; particularly relevant were the religious practices of the communities. He concluded that suicide, rather than being a uniquely individualistic event, was related to the social order. This was one of the first statistical demonstrations of the relevance of sociological factors in the examination of an action that seemed almost totally personal or psychological.

The use of formal statistical procedures in the investigation of sociological phenomena is, therefore, quite recent. Considering its recency, it is astonishing how popular it has become. Increasingly we see nearly every aspect of the human enterprise reported in statistical or quasi-statistical terminology. For example, the mass media daily provide their audiences with statistics on nearly any subject imaginable, ranging from the rate at which the ozone layer is being depleted to the number of spots used by Disney animators to create the film *101 Dalmatians* (answer: over 1,500,000). While some statistical information was reported in newspapers or magazines even a century ago, today's newspapers are loaded with statistical information that ranges from the politically and economically significant (such as health care statistics) to the trivial and the amusing (such as statistics on the prevalence of pet psychiatrists).

In one fashion or another, people have, from the beginning of human history, engaged in crude forms of statistical analysis. Early military tacticians were concerned with the numbers of soldiers in enemy armies. Ancient cultures were, of necessity, aware of the importance of counting and measuring matters as diverse as the fruitfulness of crops or the number of stones required to build a temple. People were aware of relations between the seasons and life-sustaining events; the near universality in human cultures of rituals celebrating the transformations of the seasons is evidence of this awareness. These civilizations understood the nature of risk, chance, and **probability**—though not in the formal sense in which we understand the nature of probability today. (Note: Terms that are **boldfaced** are defined in a nontechnical manner in the glossary that appears at the end of this book.)

This is an important point to make at the outset of this introduction to **statistical reasoning.** Even in its most formal and complex mathematical forms, statistical reason rests on commonsense devices that are as old as human thought. When people began talking about the world and their relationship to it in primordial times, they also began expressing their knowledge in forms similar to statistical modes of thinking. For example, people have always been concerned with what is typical as well as what differs from the typical.[6] They are interested in relationships, risks, and probabilities. It must be noted here that people have always had to make generalizations based on limited observations. These are extremely fundamental statistical concerns. It should be apparent, then, that statistical reasoning has a history dating back to when people first began thinking and communicating their thoughts to each other. The elementary forms of statistical method are also the elementary forms of communication—permeating nearly every aspect of anything we deal with or talk about. Modern analytical statistical methods are based on forms of reasoning used all the time by people throughout the world. As you become aware of this fact, formal statistical techniques begin to lose their mystifying qualities. Thus, the academic study of statistics is, among other things, an attempt to standardize commonsense forms of understanding.[7]

The teaching of statistical reason in modern university courses and texts commonly leaves us with a sense that such reasoning is radically different from ordinary thinking. It has been our experience in teaching statistical courses to social science students over more than four decades that students often conclude that statistical methods are beyond their grasp. As a matter of fact, statistical reasoning of an intuitive kind is already within their grasp. It is something people rely on daily. It is found in virtually any form of communication—from poetry to backyard gossip to the utterances of official spokespersons. It appears in conversations in a natural and casual manner. When we talk about things, we commonly rely on one or several of the elementary rational forms that are at the center of statistical work.

Anyone encountering social statistics in a university context for the first time is confronted with mathematical techniques and formulas, along with computers and calculators. Formulas and computers call to mind the apparent precise, mathematical nature of statistical thought.[8] At the same time, however, these formulas and computers can prevent us from seeing the ways in which statistical reasoning and day-to-day thinking are similar. Both are different manifestations of the same reasoning process. One (statistical reasoning) is formal, precise, and aware of its limits; the other (ordinary thinking) is informal, imprecise, and generally unaware of its limits. But they are surprisingly alike in their *general* character. The primary difference between ordinary thinking or common sense and "higher" forms of thinking such as statistics is one of refinement and care. Broadly viewed, however, common sense and academic modes of reasoning have much in common.[9]

The central theme of this book is that not only does recognizing the ordinariness of statistical reasoning make learning statistics more pleasurable, it also makes the proper application of statistical methods, even where relatively complex problems are involved, simpler and more certain. It is not uncommon for students in the social sciences to rely on complex, formal, mathematical methods despite the fact that their understanding of what the procedures accomplish is purely mechanical—if there is any understanding at all. One must always, in social statistical studies, retain a questioning attitude. Fundamental questions should include:

> Why, in straightforward terms, am I using this particular statistical technique?[10]
>
> What will the index really tell me?
>
> What distortions are at work that might invalidate my conclusions and what can I do about them?
>
> Is the statistical procedure I am using the best way to inform people about what was found in the research?

Statistical techniques are too important to be used mechanically. Although statistical analysis can be taught mechanically and might in fact seem mechanical, it requires a serious understanding of issues and a solid theoretical grounding in one's field before it can be applied imaginatively and effectively. This means two things. First, you should have a good sense of how any figures are going to add up *before* they are added up. You might guess wrong—indeed, there would be no point in making calculations if you always guessed right—but nonetheless you should have a solid sense of how you think things will go. You should not blindly rely on computers, formulas, and calculators to do your thinking. Actually, calculators and computers do a rotten job of thinking—they exist only to confirm or rebut *your* general intuitive grasp of things. Intuition and serious thinking must come first; then the drudgery can be given over to the computers.

Second, a working sense of statistical reasoning enables you, even in casual discussions, to know when you are dealing with a problem systematically and logically. If, for example, somebody tells you that homeless people *choose* to live as they do, you are hearing a statistical argument and a rigid one at that.[11] You are being told that homelessness is a condition *caused* by choice. A careful consideration of the matter forces us to ask: Do *all* homeless people *choose* to live that way? If not, then how many do? What causes people to make such a choice? (Notice the internal contradiction here: if choice is caused, then the act was not, by definition, a matter of choice but a matter of some kind of other cause.) Where are the *data* to substantiate the existence of an empirical reality—a substantive force—subject to quantification known as "choice"? What kind of a cause are we talking about when we say something is caused by "choice"?[12] Certainly just a small amount of consideration should lead us to conclude that such a statement is in need of further logical scrutiny.

Statistical arguments differ from ordinary forms of discourse; why would we take courses in statistics if they did not? However, statistical reasoning is also ordinary. *The fundamental forms of statistical reason are the fundamental forms of thinking itself.* Statistical concepts come out of common forms of thought and reflect aspects of our thinking that are nearly "instinctive" in quality. It might be easier to understand what statistics is about if we come to see that we engage in primitive kinds of statistical reasoning from that moment when we begin thinking about anything at all. The main point is that it is not possible to communicate even simple ideas without drawing on some form of statistical logic or reason. In this sense, whether we like it or not, we are all "statisticians"—albeit usually poor ones.

Though there might be exceptions, the preponderance of human communication rests on less than a dozen primary **logical forms.** These logical forms make up the central core of statistical reasoning. We are forced to think at all times in terms of these forms. Whether we are professional statisticians or not, we rely on what are, at heart, statistical concepts to adjust to the world around us. This is true whether we are poets, politicians, professors, prophets, porters, peanut vendors, or anybody else. We begin to think statistically at the time when, as children, we become aware of the world around us. Statistical thinking is not something foreign, strange, complicated, dry, or academic. It is, instead, a basic and "natural" form of human thinking. We cannot choose to use such reason or avoid it as we wish. We can only choose—if we can choose anything at all—to assure ourselves that we are using such forms of reason as correctly as possible.

COMMON SENSE AND TYPICALITY

While people commonly distrust statistical arguments, they nonetheless subscribe to what they call **common sense.** If we trust common sense, we should trust statistics more—for statistical methods are a careful form of common sense. Only when statistical concepts and facts are distorted or abused do they lead us astray. In such instances, the fault lies not with the forms of statistical reason but with those who pervert statistical reasoning through malice or ignorance or both.

We can enhance our intuitive awareness of the statistical forms of reason by considering several parallels between common sense and statistics. There exists, in any culture, a body of commonsense knowledge. How is this knowledge created? This forces us to consider the common methods we employ in our efforts to comprehend the world around us. For example, one aspect of common sense is our ability to arrange events so we see things in terms of how typical we believe they are. We think this way without thinking about it. We readily talk about *average* ability, *normal* intelligence, *typical* appearance, business as *usual, ordinary events, conventional* circumstances, or *median* income. We do this without giving it much thought.

Actually, most of our conceptions of what is typical are derived from folk ways of thinking that defy academic understanding. If you were asked to imagine, for example, a "typical building," you could probably create a picture of such a building in your mind—one sufficiently clear for you to be able to describe it if asked. However, other people would probably offer different descriptions.

What is interesting about this is that although we have a sense of what constitutes a "typical" building, there is no way in which this form of typicality can be generated by any of the statistical algorithms with which we are familiar. Yet people have a general sense of what is typical in such cases. **Typification** is a necessary condition of human communication and we rely on it constantly. We generate a sense of typicality even when formal statistical procedures are not available for attaining a precise description of the typical case. The important point here is that typification is a logical form—one that appears in both ordinary discourse and statistical work.[13]

We are concerned with what is typical, and we commonly rely on several relatively simple ways of assessing typicality. These ways are essentially the same as those relied on by statisticians in formal statistical work. (We also rely on ways of assessing typicality that transcend formal statistical procedures—but these folk devices will not be discussed here. For example, most people have a sense of what is typical of humor. However, no intellectual, whether a statistician or anyone else, has been able to resolve the problem of what humor really is.[14]) Probably the most ordinary way of determining what is normal or usual or typical is to look around and try to figure out what is predominant in a given context—what there is the most of. For example, the typical cabinet member at the top levels of governmental administration in the United States is a wealthy, white, college-trained male with a background in higher administration or law and a political party affiliation that is the same as that of the president.

People use notions similar to the statistician's for obtaining rough assessments of typicality. At the same time, they commonly go beyond statistical methods and lean on more complicated and intricate devices than those used in formal statistical analysis. In social statistics, for example, three basic procedures are employed to assess typicality:

1. What there is the most of, or the **mode.**
2. What is in the middle, or the **median.**
3. The value that any single individual in a set would be given if the sum total of values were to be distributed equally among all members of the set, or the **mean.**

Though we assess typicality in much the same fashion as statisticians, there are times when we have a strong sense of what is typical and yet our notion of typicality violates all three of the above devices used by the statistician. For example, many of us have a notion of what a typical "mad musician" is like. We rarely think of a mad musician as a woman who plays the drums for a jazz band and is a bit on the heavy side and of Asian background. The typical "mad musician" plays the organ, piano, or violin; is male, white, and thin; and has long, tapering fingers, a pale complexion, and wears a tuxedo—with a cape. Where do such typicalities come from? They do not come from statistics nor from direct observations of "mad musicians." They come from popular literature that has developed a traditional character type that we come to think of as typical.[15]

This suggests that something as elementary as the idea of **averages** demands a great deal of thought if we are to have a real understanding of its significance. On one hand, there is nothing simpler than the idea of typicality. On the other hand, typicality proves to be extremely complicated. It takes a while to see that commonsense or ordinary assessments are often more complicated than those used by formal statisticians. Indeed, they can be so complicated that the mental procedure upon which a particular "commonsense" typicality is grounded cannot be reduced to any kind of formula or standardized procedure. Or, as a mathematician might put it, these commonsense assessments are not "computable."

We mention this only to suggest that the formal study of statistics can actually be simpler than trying to comprehend the nature of folk methods. Conceptions derived through folk methods come from a variety of sources and eventually are accepted in a taken-for-granted fashion. People get ideas from literature, gossip, hearsay, the newspapers, experience, their own unique experiences and observations, and hundreds of other sources. You can ask people what a mad musician looks like and people will often have a common notion about this kind of character. However, precisely how they came to that idea is not something most people can readily put into words. Where folk methods for reaching conclusions about events in the world are ill-defined, complex, and grounded in belief, statistical methods seek clear definitions, simplicity, and a reliance on observation.

Delineating the methods by which people resolve complex problems in ordinary life is ultimately impossible. The pursuit of good statistical procedures is the quest for a simple set of devices for resolving questions with facts. *While students are sometimes inclined to believe that folk forms are simple and statistical forms are complicated, the truth of the matter is exactly the opposite.* Folk forms of reasoning are complicated, mechanical, and inexact. Formal statistical analysis tends to be simple, probabilistic, and precise.

The point has been made that whether we are acting as formal statisticians or simply participating in everyday life, we are concerned with typicality. But *why* are we interested in the typical? At first glance it seems to be "human nature." There is a better reason, however. Knowledge of what is typical serves an important purpose. Consider the situation facing you when you are about to meet someone for the first time and you have little information with regard to who or what that person might be. If you are wise, you will *assume* (but you should not do more than assume) that he or she is a more or less typical or average person. (Now pause for a moment. You *typically* do this intuitively, but can you guess why?)

As you get to know the individual better, you can reevaluate your earlier assumption. Indeed, your own *normal* (another word for *average*) character depends on your ability to do this. If you assumed that everybody you meet is inferior, you would be an egomaniac. If you looked on everyone as above average, you would suffer from an inferiority complex.

Why is it best to assume, unless you have information to the contrary, that the people you encounter are average or typical? The answer is that this assumption, over the long haul, minimizes the total sum of errors you will make in your initial efforts to assess the strengths and weaknesses (however you choose to define such qualities) of people around you. Any other assumption increases your total sum of errors over time. *If we wish to keep cumulative errors to a minimum, our ability to make judgments of one kind or another about what is average or typical is critically important in our daily lives.* This is a fundamental fact that you need to know as you enter into the study of statistical procedures. Averages are important because over a repeated series of occurrences, reliance on a well-determined average minimizes the sum of errors made in making judgments without other information. Minimizing error is a central concern of statistics.

NINE MAJOR STATISTICAL IDEAS

Statistical (and common) thinking involves nine major themes or ideas. These themes are the basic "notes" statisticians draw on to develop well-reasoned statistical arguments. At the heart of these nine themes are the ideas of typicality and variability, two "notes" that are especially basic. Nearly all statistical work is concerned, in one way or another, with assessing typicality and variability.[16]

Here are the nine primary statistical ideas or, as we like to refer to them, the elementary forms of statistical reasoning:

1. Careful counting
2. Central tendency, averages, or typical events
3. Nontypical events, or deviation
4. Probability, or chance
5. Samples, or generalizing from limited information
6. Relationships
7. Control, or standardization
8. Logical models
9. Classification or categorization

Each of the above elements of statistical reasoning will be briefly discussed in the following pages before they are more elaborately developed in later chapters.

Careful Counting

Conversations with students and others have led us to believe that people are inclined to look on statistics as highly numerical in character (which it is); on the other hand, people are inclined to see ordinary conversations as not involved to any great extent with counting or numbers (which is false). This misconception has been given further support by the fact that initially, statistics was a term used to refer to the collection of numbers by state agencies. It is true that statistics is essentially the formal analysis of numerical data. However, the important qualifier is "formal." The major distinction between what the statistician does and what we all do all the time is simply a matter of being conscientious about details and precision.

Though statistical studies depend heavily on counting, they differ from ordinary views of the world only to the extent to which they try to be *careful* with their counting. Actually, we all count, and we count nearly all the time. (This last sentence contains several counts, for example.) However, we usually do not count carefully.

Everyday conversation is loaded with hyperbolic exaggerations that result from not counting things properly. Consider the implications of saying to oneself, "I never do anything right." This statement contains a count. The count says, if you count the number of times I do things wrong, you will find it is equal to the number of times I have done anything at all. Obviously, this is a bad count. Not only that, it is a bad count that can have serious emotional consequences for the person who believes it.

Careful counting, as opposed to ordinary, imprecise counting, de-emotionalizes the way things are described. Ordinary speech is loaded with exaggeration and hyperbole. Ordinary speech distorts reality; it distorts it in ways that often lead to hysterical responses. (How many casual counts are contained in this last sentence? What would the sentence look like if all quantifiers that are implicit in the sentence were made specific or precise?)

We listen to speakers who create casual counts and measurements of all kinds and go away with a sense that the situations described by these speakers have been well-delineated. The nature of human communication is such, however, that this is rarely the case. When, for example, poet Elizabeth Barrett Browning tells us that she is going to "count the ways" of love, she is responding to a primitive urge to know things in a quasi-numerical fashion. At the same time, we are aware, as Browning makes clear, that the ways of love are beyond counting. Even so, whether we can actually count things or not, in nearly any kind of human expression some kind of enumeration is taking place. This appears in the form of soft, quasi-numerical modifiers such as *many, much, some, numerous, a little, often, a lot, a few, plenty, commonly, rarely,* and so on.

American students (and please keep in mind that our experiences are based largely on work with students in the humanities and social sciences) commonly show an aversion to numbers. As noted in the Introduction, one writer uses the term "innumeracy" to refer to the dismal extent to which Americans are more innumerate than they are illiterate.[17] What is interesting about this is that this aversion is strong in some contexts and weak in others. There are places where Americans are seriously involved with numbers—in, for example, the realms of sports and sports records, banking and money interests, politics and voter responses, dietary accounts of calories, and so on. Whether it is hard numbers or soft quantifiers, one thing can be said with relative certainty: people are always keeping some kind of count of things. Statisticians try to do so carefully; the ordinary person in the street is not as careful.

Central Tendency or Average

Our ability to see things in the world as typical comes from our ability to notice that things share properties in common. This sort of thinking holds for people who write poems and novels as much as it does for statisticians. Here is an ordinary example spoken by a character in Richard Wright's novel *Native Son:*

> We black and they white. They got things and we ain't. They do things and we can't. It's just like living in jail.

Here Wright succinctly tells us that several things are typical of white and black communities in America. He gives dramatic value to these typicalities by adding, at the end, that what is typical of life for the American black person is the feeling that one is constantly living in jail. Wright has not used specific numerical data to make his point. However, we are not interested in that at the moment. What we are interested in is that this novelist has offered a fundamental statistical or logical form—the idea of typicality—and presented it as a literary form.

We have an extensive vocabulary of terms that refer to the idea of the average or the typical. When the statistician refers to such terms as the *mode, arithmetic mean, median,* or **central tendency** (a broad term used by statisticians to refer to measures—like the mean, mode, and median—that are used to describe that which is more common or central to a distribution of values, there is often the further implication of some of the meanings conveyed by the terms listed in the box below.

A SHORT LIST OF FOLK TERMS FOR CENTRAL TENDENCY

average, accordance, bread-and-butter, commonplace, commensurate, congruent, consistent, conventional, customary, day-to-day, everyday, frequent, garden variety, general, habitual, humdrum, invariably, likeness, mean, median, medium, mediocrity, middle, middling, nondescript, normal, ordinary, popular, prevailing, regular, the same, standard, stereotypical, stock, typical, unexceptional, uniform, usual

The Atypical

The moment we become interested in the typical we also, by implication, become concerned with the atypical. Statistical analysis concentrates on the differences and similarities between things. At the most elemental level, statistical analysis amounts to little more than carefully establishing how similar or different various events are from each other. For example, a psychologist might use statistical analysis to determine if aggressiveness scores for children who watch a lot of television are different from the scores for children who do not. Are the scores different or are they essentially the same?

Any form of comparison implies differences and is concerned with the presence of atypical events. If we believe we are superior when we compare ourselves to others, then we look upon ourselves as different. If we believe we are inferior, then we also look upon ourselves as different. Such evaluations have great significance for human beings. Even small differences—slight hints of atypicality—can precipitate fights, mental depression, arrogance, haughtiness, humility, or any of a variety of emotions. In this sense, statistical thinking is at the heart of human concerns.

Considerable debate has taken place over whether Americans are different from the Japanese. The Japanese, according to some reports, are innately more intelligent. There is an ongoing heated discussion over which of these two peoples works harder, which is more effective at producing goods and services, and which is better educated. In order to make such assessments you have to come up with some idea of what constitutes a typical American and a typical Japanese. This, in itself, is difficult. Then you have to see if there is a difference. Once you have established a difference, you have to assess its implications. Complicated as this is, people nonetheless discuss such matters with great vehemence and a sense of certainty.

Political polemics are not appropriate here. The point we are trying to make is that people are constantly arguing highly complicated ideas like these, and they do not find such argument especially difficult—in fact, they appear to argue all too easily. However, if we are not careful, we come to see differences where there are none, and no differences where there are great disparities. (Later we shall see that statisticians are more careful. They seek to find out exactly what is being compared and to what extent differences actually do or do not exist.)

People are almost neurotically interested in differences. They casually observe and act upon differences believed to be real. They are interested in differences between individuals and between groups of individuals. Much can be and has been made of absurdly small, capricious, and trivial differences. At its heart, this interest in differences is an interest in statistical matters. Some idea of how interested we are in differences can be seen in the great variety of terms we have to indicate comparisons. The box at the top of page 18 lists a number of common words that refer to differences.

A SHORT LIST OF FOLK TERMS FOR DIFFERENTNESS

alien, atypicality, antithesis, assortment, contrast, deviant, difference, discrepancy, disparity, dissimilar, distinct, distinctiveness, divergent, diversity, heterogeneity, incomparable, individuality, mismatched, modified, originality, peculiar, singular, specialness, strange, unequal, uniqueness, unlike, unmatched, unusual, variance, variant, variation, variegation

It should be noted that the enthusiasm Americans bring to small differences in the performance of athletes is an example of carrying statistical thinking to pathological extremes. Our athletic programs, along with business and economic matters, are among the most highly statistically reported events in the country. For the moment we might muse over the possibility that no other creature on earth makes so much of so little. For example, in downhill skiing it is possible to beat an opponent in a race by a mere one-hundredth of a second or less. This is what is meant by making much of a small difference.

Probability

Probabilities are a unique use of the idea of average. A probability is what you might expect, *on the average,* in some situation. (An *average,* incidentally, is also what is *probable.*) For example, in determining the likelihood of getting a good grade in a particular course, students review grade lists to find out how common A's are. The more commonly A's appear on a list, the more *likely* (that is, probable) it is, one concludes, that he or she will get a good grade in the course.

Any number of social issues are influenced by how people assess probabilities. For example, the argument is often made that drug use should be curtailed because it carries a high risk of genetic or psychological damage. Nuclear weapons are argued against because they carry with them the risk of total annihilation of the world's biosphere. On the other hand, we are presented with the argument that the United States should arm itself heavily with the most advanced weapons modern technology can provide because not to do so entails the risk of invasion from foreign forces. *Risk* is another word for probability.

The box below lists terms we commonly use to refer to probability.

A SHORT LIST OF FOLK TERMS FOR PROBABILITY

accident, chance, fortuitous, fortune, haphazard, how the ball bounces, liable, likelihood, luck, odds, percentages, possibility, probability, random, risk, tendency, toss-up, trend

The formal study of probability is complex and elaborate, and we cannot move that far in our discussion at this point. What is important to keep in mind is that probability is a familiar concept, similar in its basic commonsense applications to the idea of probability used in more formal statistical analysis.

Sampling

If anything has given statistical analysis something of a bad name, it has been the fact that much of what the statistician does relies on **sampling** and on *inferences* drawn from samples. Let it be understood immediately that there is no need to apologize for the use of samples in statistics. To focus on the limitations of sampling as a criticism of statistical procedures is absurd. The reason is evident. *All* human knowledge, in one way or another, is knowledge derived from a sampling of the world around us.

Now let there be no doubt about it, the sample is a treacherous device for drawing conclusions about how the world works. Statisticians may have problems with it, sometimes drawing broad conclusions on the basis of a very limited sample. Ordinary people have lots of problems with it—though they are often unaware of the fact that they may be overgeneralizing. The major difference between the statistician and an untrained person is that the statistician knows more clearly what these problems are and how to cope with them. Nonetheless, the appeal to generalize from limited experience is nearly irresistible. Even statisticians succumb to the urge to generalize beyond the limits set by their samples. However, it is the folk statistician, the ordinary person, who is more likely to make wild generalizations on the basis of astonishingly restricted information.

Consider how deeply sampling is part of our ordinary lives. It does not take long to see that sampling is essential and that we rely on sampling as a matter of our daily routines. For example, sampling occurs in each of the following incidents.

1. You meet a person and, after three minutes of conversation, you decide you do not like him. This is called a "first impression." However, first impressions are nothing more than samplings of a person's actions. (We often make broad, and usually unfair, generalizations about a person's character on the basis of a brief contact.)
2. You are preparing fettuccini Alfredo for a group of friends and you take a taste of the dish to determine if it is seasoned properly. You conclude that it is. (Anyone against sampling would have to eat the entire dish to see whether it was satisfactory.)
3. You look out your west window and it is raining. You conclude it is a rainy day. (A true skeptic would go to the other windows facing east, south, and north before reaching a more definite conclusion. The sun might be shining in the other windows. If this sounds silly, remember that a few decades ago a philosophical school known as logical positivism actually advocated such restrictions on the extent to which we might generalize.)

4. You pick up a book and leaf through it to see if you want to buy it. You put it back. You have just engaged in sampling activity. (If you were completely opposed to sampling, you would have to read the entire book before you could decide whether you wanted to read the entire book—and that, obviously, makes little sense.)
5. You go to the doctor's office for a physical examination. The doctor draws a specimen of blood. The doctor is using a sample to reach some serious conclusions about your vital fluids. (If you do not like sampling, you might suggest that the doctor take all of your blood before coming to any serious conclusions about it.)
6. You are bitten by a dog when very young. You are afterwards frightened whenever any dog comes near you. Your action is a generalized response to a single incident or, as a statistician might put it, a sample of one. (It is possible, by the way, that we are genetically "wired" to draw general conclusions from specific instances in this way.)
7. You are brought up in a middle-class, American, white, Protestant home. You come to the conclusion that all "normal" people are like those with whom you have associated all your life. Anyone who does not act the same way is "bad." (You have made a generalization, with moral implications, based on a limited sample.)
8. You go to a garage to have your car repaired. The mechanics at this garage bungle the job. You comment that you will never go there again. (You have made a generalization about the quality of the garage on the basis of a sample of one experience. As it turns out, this is the best garage in the world. They make only one error out of ten thousand repairs, but you happened to be the one in ten thousand. Here is a case where the connection between sampling and probability is obvious.)

It should be evident by now that we all engage in sampling and we all draw conclusions from samples. (This generalization is, incidentally, based on a sample.) Indeed, as these examples suggest, we do not hesitate to make generalizations based on small samples or even samples of just a single case—what the statistician would refer to as a situation where the sample $N = 1$ (where N refers to the Number of cases in the sample). We do this constantly. Even so, when a statistician draws conclusions based on a sample, critics often throw up their hands in dismay. When a good statistician relies on a sample, he or she is generally operating with a sophisticated awareness of what can and cannot be reliably based on such information. But most of us do not make a serious effort to investigate the adequacy of the samples we rely on daily. Possibly no other logical blunder causes as much human confusion and distress as this one.

Actually, human life can only sample, in the most finite manner, the infinite variety of the universe around us. We are, by virtue of our fates, forced to sample life. We cannot know it in its entirety. It behooves us, then, to try to comprehend the implications of the fact that our experiences are merely samplings of the world. The following box provides a short list of commonly used terms that refer to samples.

A SHORT LIST OF FOLK TERMS FOR SAMPLES

case, cross section, embodiment, experiments, illustration, instance, inspections, observation, polls, sample, specimen, studies

Later we shall see that some of the most sophisticated developments in statistics came out of an intensive attack on the problems that result from the use of samples. This realm of statistics is commonly known as **inductive statistics.** Inductive statistics has to do with the **reliability** of conclusions drawn from samples. How much **confidence** can we put in generalizations drawn from samples? Inductive statistics is commonly contrasted with a different realm of statistical analysis that is not interested in generalizing. This area of statistics is called **descriptive statistics.** For example, you may want to know only the mean for the grades in a given class without being interested in generalizing to some larger population; in this case you are only interested in describing what is typical for this particular class—hence the term *descriptive* statistics.

To anticipate a topic that we will address later, it should be mentioned that one of the basic conclusions of inductive statistics is that a big sample is not necessarily a good one. At the same time, a small sample is not necessarily a bad one. Remember, a doctor can take one relatively small drop of blood and make reliable generalizations about the other thirty thousand or so drops of blood that make up the typical blood supply of a human being.

Relationships

The human mind is not content merely to observe the typical and the atypical and make generalizations from samples. People also look for *patterns* in the world. They are interested in the ways events are related to each other. This quest for the discovery and description of related events has led to what we now refer to as science. The major difference, and it is a real difference, between commonsense **relationships** and those of science is that the latter are more carefully tested, more subtle, and more elaborate in character. Otherwise, the same basic notion is involved—the idea of this being related to that. The importance of relationships and our understanding of them is revealed in the following instances having to do with ordinary, day-to-day types of activities.

1. You catch a cold and wonder if it happened because you stayed out too late and were chilled. (Are cold symptoms related to being chilled? Modern research says they are not. Rather, colds are related to unclean hands that transmit the cold to vulnerable areas such as the mouth or nose.)

2. You wonder if your girlfriend or boyfriend will appreciate you more if you buy her or him a gift. (Gifts are thought to be related to positive feelings.)
3. You want to entertain and you decide to take your friends to the most expensive restaurant in town—even though you know nothing else about the place. (You believe price is associated with quality. You also expect it to be related to the impression you will make on your friends.)

It is not necessary to go further. It should be evident that we constantly think about how various parts of the world are related to each other. Our comprehension of relationships is basically the same, whether it takes the form of common sense or appears in the more refined forms that characterize statistical analysis and other kinds of scientific work. This is important to keep in mind because the formulas and apparatus that are so much a part of formal statistical reason can sometimes cause us to lose sight of the fact that we are dealing with the more fundamental idea of relationships.

The idea of relationship can be summed up in this way: where you find *X*, you find *Y*; where you do not find *X*, you do not find *Y*. This summary describes a *positive* relationship. There is also a *negative* form that occurs whenever *X* is present but *Y* is not present, or when *X* is not present but *Y* is present. The following table presents these various possible relationships graphically (here a plus sign indicates a positive association and a minus sign indicates a negative association).

		Y	
		Not Present	***Present***
X	***Present***	–	+
	Not Present	+	–

The idea of relationship has a rich variety of ordinary terms associated with it. The box below lists some of the words we commonly use when talking about relationships.

A SHORT LIST OF FOLK TERMS FOR RELATIONSHIPS

affiliation, affinity, agreement, association, bearing, bear upon, belonging to, bonds, causal relation, communion, comparable, coherence, congruence, concur, conformity, conjunction, connection, correlation, consociation, contingency, dependence, effect, fall in with, fit, grouping, interconnection, interdependence, interrelationship, link, linkage, mutuality, pattern, proportionate, reciprocity, relation, with respect to

Our constant reliance on presumed or real relationships is seen in other forms of linguistic expressions. For example, when we say that our team would have won the championship *if* it had been playing at home, we are asserting an implied relationship. *Any statement that sets forth conditions is also setting forth relationships.* If you tell someone you will go to the movies if you can get your work done, you have established a relationship between getting the job done and going to the movie. That is:

> If work is done (*X*), then we will go to the movie (*Y*); if work is not done (non-*X*), then we will not go to the movie (non-*Y*).

Relationships also appear in moral assertions in an implicit fashion. For example, the commandment to honor your mother and father implies a conditional to the effect that if you do not honor them, then bad consequences will ensue. Moral arguments generally have an implicit statistical argument underlying them.

Formal statistical analysis has many ways of investigating the properties of relationships. These techniques are quite important and a few of the more common devices used in modern statistical work for determining degrees of relationship will be discussed later in this book.

Control and Standardization

The topic of **statistical control,** which includes techniques for achieving rigorous control, is one of the most complicated concerns in statistical analysis. Despite its complexity and difficulty, the idea of statistical control also appears in ordinary discourse. As is the case with sampling, not understanding or ignoring the problem of what the statistician means by control can create problems.

Here is an example of an error we can make if we ignore control. Say we compare two groups, the Blues and the Grays, and we find the Grays are terribly deficient. However, we did not control the amount of training both groups received. As it turns out, the Grays received no training—so, the contrast between the two groups is not fair because training was not controlled or *standardized.* If training had been controlled, then perhaps the two groups would have performed equally well. **Control** or **standardization** is, therefore, extremely important in making comparisons between individuals and groups. Though standardization and control are technical problems, they also have humanistic connotations. If we believe in fairness, then we must be interested in problems of standardization. For example, we could not justly evaluate two runners who were forced to race on tracks that were not standardized for length. So it is, then, that standardization becomes a most critical concern in statistical work.

Control and standardization are closely related concepts. Most of the statistical measures one works with in statistical analysis involve some kind of standardization. The idea behind control and standardization is to simplify complex matters in a sufficiently reasonable manner to allow comparisons to be made. Psychologists seek measures, or standards, whereby the intelligence of one person can be reasonably compared with that of another person, for example. In sports, rules are highly standardized in order that the relative performances of teams can be compared—and they are compared constantly in sports statistics that appear in the daily newspapers of our society. Standardization appears throughout the entire domain of statistical work.

The issue of control and standardization also appears in common discourse. Here is an ordinary kind of conversation in which control is introduced as a concern. Two students are talking about race and class, in connection with life expectancy. One of them says, "Whites live longer than blacks." The second student replies, "Yes, but . . . ." (When people say, "Yes, but . . ." or "Have you taken such and such into consideration . . ." they are introducing, in an informal manner, the problem of control and standardization.) The second student goes on to say, ". . . blacks receive lower incomes in our exploitive and racist system. If they were not so poor, they would live longer. Being poor they cannot get the kind of medical attention they need." The second student has introduced a further complication, an extraneous influence, into the discussion of the relationship between race and life expectancy—an interesting problem that has been explored statistically in elaborate ways in the literature of Western social science.

We do not need to become further involved in a discussion of the relationship between racial difference and life expectancy. What we want to point to here is the form of reason that is implied when the second student says, "*Yes, but* it is necessary to take economic conditions into account as well as race." This student suggests that if you were to control the effects of wealth by using common standards—comparing wealthy, educated blacks with wealthy, educated whites, for example—the difference between the whites and blacks with respect to longevity would be changed. Perhaps it would completely disappear. One could do this, of course, by studying the life expectancies of wealthy black people and wealthy white people. In this instance, wealth would be controlled or standardized. Or, you could study poor whites and poor blacks and see if they both tend to die at an earlier age. Again, wealth would have been removed from the picture, so to speak, as a factor interfering with the differences in the life expectancies of the racial groups.[18]

In one study of the mortality rates of men and women, the researcher examined mortality rates for nuns and monks—groups that share similar or standardized environmental conditions. Men still died earlier, lending greater credence to the possibility that men are biologically destined to die at earlier ages, on the average (there's that word), than women. The environment could not be blamed for the difference in mortality because it had been held constant. (If you are about to say "Yes, but . . ." perhaps you have thought of other factors that were possibly *not* controlled. Perhaps, for example, the men were more involved in the stress of trying to be leaders in the church hierarchy. This might cause them to die at an earlier age.)

Here are some common statements that bring the problem of control and standardization into focus. See if you can sort out what factors should be controlled in each example.

1. *Helen:* "They say that if you eat less, you lose weight. I am eating less, but I'm still gaining."
   Gladys: "*Yes, but* how much exercise are you getting?"
2. *Margaret:* "If poor people would develop some ambition, they could have anything they want. America is a land of equal opportunity for all."
   Egbert: "*Yes, but* have you considered the fact that successful people almost always have affiliations with powerful organizations and poor people don't?"
3. *Foster:* "If we can get better gas mileage from our cars, we will be going a long way toward improving the environment."
   Jeanette: "*Yes. However,* you're assuming we're not going to have an increase in the number of cars in the world that will exceed the percentage of efficiency achieved by the lower mileage rates."
4. *Morris:* "I figure I have a good lawyer in my case because he said he won every trial he was in."
   Eloise: "*That's right, but* did he tell you he has only gone to court once?"

Typically, in any discussion of statistics, control has to do with determining whether *X really* is related to *Y*. It might be that *X* and *Y* appear to be related, but this is because both are related to a third factor, *Z*. For example, at one time it was believed that logical skill was associated with gender in such a way that men had a lot of it and women had a small or negligible amount. When early childhood training is controlled, however, such supposed differences become less marked or disappear altogether. Logical skill might not be related to gender but to a third variable, training, that needs to be controlled before the relationship between the two primary variables can be seen more clearly.

This book will not go deeply into formal problems of statistical control, although the reader will be constantly reminded of the fact that it is a pervasive concern wherever precise knowledge is sought. Statistical controls are much more limited than the careful experimental controls available under laboratory conditions. For example, despite the mountains of evidence compiled with respect to nicotine ingestion and lung cancer, there is still controversy over whether nicotine per se actually *causes* lung cancer. The real villain might not be nicotine but radioactive traces common to any kind of smoking of dried vegetative matter. Unless we can control the radioactivity, along with nicotine consumption, we cannot precisely evaluate the relationship between what is going on with respect to smoking and lung cancer.

In sum, control tries to obliterate the "messiness" or, as physicists might put it, the "noise" or "static" that we have to deal with when we are examining relationships. Even though it is not dealt with in great technical detail in this introduction to statistical reasoning, the problem of control is extremely important. Attempts to deal with it, either in formal statistical research or in informal and casual conversations, are essential in the quest for better forms of understanding and ways of thinking about human problems.

We do not know of a list of folk terms that specifically refer to control as that term has come to be used by statisticians. However, the phrases in the following list often imply that the person using them wants the discussion to move into a more complicated consideration of possible contaminating influences that might be distorting some relationship.

"Yes, but you have to consider . . ."

"You are leaving something out . . ."

"If you take so and so into account, it changes things . . ."

"Yes, but that could also be caused by a lot of other things . . ."

Whenever someone begins a remark in such a way, you should notice the argument has moved from a simple two-factor discussion into the deeper forests of three, four, or more factors.

Models

So far we have mentioned seven basic forms of thinking that are a part of both commonsense understandings and the elemental features of statistical analysis: counting, the typical, the atypical, probability, sampling, relationships, and control. An eighth basic form that plays a role in commonsense and statistical thought is the idea of **models.**

Essentially, a model is a pattern of thinking, usually complex (though it can also be extremely simple), that leads us to expect something. For example, if you believe some individual is a moron because he is a football player, you are thinking in terms of a simple and moronic model (a mental "map") of football players. Like any other group, football players display considerable variability with respect to intelligence. Such a model is much too simple.

We establish a model and then look around to see whether what it leads us to expect is actually a part of what is going on in the world. A model is a kind of fantasy or "toy" image of how things work. The fantasy might be based on careful logic and rigorous reasoning, but it is still a fantasy until it is matched against the reality it pertains to. Commonsense models of the world usually do not call on the careful construction of ideas. Instead, common sense tends to take its models, ready-made, from the rack of those that already exist at a particular time.

For example, one kind of commonsense model consists of what we refer to as *stereotypes*. A stereotype leads us to expect certain actions that we can test against reality if we want to. However, so powerful is the stereotype that when expectations are frustrated by contrary observations, the model is maintained and the observation rejected. For example, we might read in the newspapers of an adventurous librarian who scaled a Himalayan peak. This violates our stereotype of the librarian as a mousy, timid kind of person, and we might conclude that this particular mountain-climbing adventure was an "overcompensation" of some kind. This concept of so-called overcompensation permits us to retain our stereotype of the librarian as a mousy person in the face of contrary evidence.

There are a few words in general use that approximate the notion of a model, but only a few. The box below lists some of the words we commonly rely on to refer to the idea of model.

A SHORT LIST OF FOLK TERMS FOR MODELS

archetype, dummy, exemplar, ideal, image, map, paradigm, paragon, pattern, picture, portrayal, prefiguration, presentation, stereotype

We encounter all kinds of devices that serve as models to help us anticipate complex events or determine what we should do given a particular set of circumstances. The following are just a few of the devices that serve as models for people: religion, science, averages, maps, novels, political slogans, stereotypes, political cartoons, advertisements, pictures, diagrams, working models (such as model planes or model buildings). There are countless others. Later, in chapter 15, we will examine models that are created through the use of statistical concepts and techniques.

Categorization

Statistical reasoning, like any other form of human thought, is closely bound to language and symbols. This means, at the very least, that we somehow "capture" the world around us in words and in expression. The poet does this, for example, with lines like the following: "Time turns the old days to derision, / Our loves into corpses or wives. . . ." With these lines Algernon Charles Swinburne tries to "capture" the nature of growing old in Western culture.

The scientist is almost morbidly concerned with **categorization.** Well the scientist might be, for categorization, extremely precise and careful categorization, is fundamental to the scientific enterprise. It would not be too radical to suggest that science is basically the attainment of categories that enable fairly exact communication to take place between those who are familiar with the language of science. Where the poet never really tells us what he means by "the old days," the scientist might, more coldly, define them as a matter of being sixty-five years of age or older. The latter statement makes up in precision what it loses in rhetorical or dramatic appeal.

The importance of categorization can be illustrated in the following two statements. Both statements are similar in form. Both also appear to provide information. However, though they are formally similar, one statement tells us something, while the other tells us nothing.

Statement 1. The estimated average density of the known universe is equal to one hydrogen atom per ten cubic meters.[19]

Statement 2. The average classroom contains twenty students.

The first statement contains well-defined categories. The second statement contains categories that are not well defined. A scientist, for example, who uses the term *hydrogen atom* does not have to ponder the question of whether the atom is little or big, sick or healthy, rich or poor, old or young, religious or nonreligious, male or female, attractive or unattractive, heterosexual or homosexual, and so forth. Every significant feature relevant to the term is contained within the term itself.[20] The category *student,* on the other hand, offers no clues, in and of itself, as to what it pertains to—save in the most general way. We could be talking about older students, special education students, graduate students, foreign students, social scientists, children in grade school, military cadets, and so on. In other words, the term *student* is highly ambiguous. The term *classroom* is similarly ambiguous.

Whether we are talking about commonsense understandings or statistical understandings we must rely on various categories that are necessary for conducting conversation and otherwise communicating with those around us. We must assume, then, that when categories are precise and correct, our understandings will be precise and correct. But if our categories make no sense or are otherwise confusing, then we are in trouble. (One of the primary difficulties in establishing good statistical information with regard to human social affairs is that it is difficult—in fact we would say it is impossible—to find categories that are precise in the same way that the concepts of the natural sciences are precise.)

Categorization is the beginning point and end point of statistical analysis. Good typification is essential to having good categories. (Once more we are brought back to that all-important concept of "the typical.") Curiously enough, good categories both generate and are a result of good typification. Physical science is powerful because its categories offer precise typifications. Moreover, observations of physical phenomena reveal many phenomena that never deviate from their average or "typical" states. Such phenomena are called *constants* and are invaluable to physical science. Social science does not have any established constants. There are conditions that are ubiquitous, or universally found in all communities, such as conflict, but conflict is not constant. While it is a part of all communities, it varies in degree from one place to another.

SUMMARY

Both ordinary discourse and formal statistical analysis rely on basically the same set of broad logical forms that are necessary for coping with the infinite complexities of the world around us. The primary concern of this chapter and, more broadly, of this book as a whole, is to get the reader to see that he or she already makes use of a kind of "folk" statistical reasoning in everyday life. Once you see the extent to which everyday reasoning is similar to that of the statistician, the transition to formal statistical studies should be easier.

Whether it is everyday conversation or a formal research report, there are nine major logical devices that determine the extent to which the conversation or the report will accurately portray its subject matter. These devices or logical forms are outlined in the following questions:

1. Are things being counted properly?
2. Are your notions of what is typical of some category valid or false? Moreover, are you incorrectly applying what is typical for a group to an individual who, very likely, is deviant from what is typical for the group?
3. Are you aware of the extent to which there is variation in the subject you are concerned with? (Variability is usually greater than you think.)
4. Are you describing things as certain or are you aware of the extent to which probability is at work? For example, it is one thing to say that if you work hard you will succeed, and another to say that if you work hard you might succeed.
5. When you generalize are you aware of the sample from which your generalization comes? Is it a good sample or a bad one? How do you know if it is a good or bad sample?
6. How do you present relationships? How do you know whether events are related, not related, or only partially related?
7. When you are comparing individuals or groups, have you concerned yourself with problems of control or standardization? For example, Harvard is an excellent university, but its student population is unique. We cannot fairly compare teaching efficiency at Harvard with, let us say, teaching efficiency at Northern Arizona University, because the student populations are not standardized.
8. What kind of model is dominating your thinking, either in everyday conversation or in more academic research? How has the model been created? What kinds of expectations does it lead to? Have you gone to the trouble to check your model against the facts? Does it provide realistic expectations?
9. Have you carefully defined the events you are concerned with? If your categories are no good, then the rest of your work is basically a waste of time. Good categorization is, in our opinion, the most difficult part of the use of statistics in the social sciences.

Because these nine forms are virtually universal in all statistical work (and in ordinary conversation as well), they provide a solid foundation for understanding any and all statistical work. Therefore, we recommend that you memorize these nine forms and have a solid sense of their basic nature before continuing with the more detailed discussions that follow in this book.

Thinking Things Through

1. This chapter suggests that statistics can be studied not only as a method but as a relatively new form of literature that has become popular in the mass media. Why do you think modern people are more subjected to statistical surveys, opinion polls, trivial statistical information, and other forms of statistical or quasi-statistical information than in the past? What has made our society more inclined to rely on, and to be entertained by, popular statistical reporting?
2. Categorize the term *student* in such a fashion that it includes all people who really are students and excludes all people who are not. (We warn you in advance, this problem is nearly impossible to resolve in an uncontroversial fashion. However, this assignment is a good exercise for developing a little humility with respect to how difficult even the simplest data collection can be.)
3. Without looking at the book, list the nine forms of reasoning discussed in this chapter, and briefly define each.
4. Relying only on the front page of your local newspaper, identify at least a dozen quasi-statistical or actual statistical arguments. How many appear valid? How can you tell? To what extent do a lot of quasi-statistical arguments appear simply as assertions without factual support?
5. When Galileo dropped two iron balls, one much heavier than the other, from the Leaning Tower of Pisa and both struck the ground at the same time, what kind of model was he trying to get people to accept? How would you put the model into a simple formula? (You can make up your own terms, as long as you identify them.)
6. If stereotypes are models, then why have they acquired such a bad name?
7. Which is superior, knowledge based on observation or belief? Is it better not to believe anything, if you have no basis in observation? Or is it better to believe in something, even if it is bizarre, rather than accept the fact that you simply don't know what the situation is?
8. If science is precise description, then what stands in the way of precise description in the behavioral sciences? How can we work against imprecise description?

Notes

1. One reviewer of this text castigated us for using the term *typification*. Unfortunately, the reviewer did not tell us what was wrong with this term. The truth is that whatever is meant by such words as *average, typical,* and *common* is, especially with respect to human affairs, somewhat ambiguous. This should be kept in mind whenever statistical indexes are being calculated. The use of formulas to create a value does not necessarily mean the ambiguity of the value has been done away with.
2. Although the term *statistics* is associated with the collection of information for the state, the term is more broadly used in modern times to refer to the collection and

analysis of numerical information. Late eighteenth-century writers were using demographic (human population) data both to support and refute the idea that social reforms are good. However, it was not until the middle of the twentieth century that analytic statistics acquired a general and more systematic mathematical form.

3. Herman J. Loether and Donald G. McTavish, *Descriptive and Inferential Statistics: An Introduction,* 3rd ed. (Boston: Allyn and Bacon, 1988), 8.
4. Thomas R. Malthus's arguments were extremely influential and were among the first in Western literature to rely heavily on statistical information. His work first appeared in 1798. It is discussed in more detail in Chapter 3 of this book.
5. The important thing about Durkheim's work was, first of all, that it was a movement away from the earlier European tradition of "armchair" philosophizing about social issues by bringing available statistical information into the discussion. Durkheim is not simply important because he used statistics, however, but because he attempted a tour de force, using the most apparently individualistic of human actions—suicide—to demonstrate the existence and power of "invisible" social forces. In other words, statistics for the sake of statistics does not attain much. Statistical knowledge and techniques must be combined with a powerful and imaginative argument to attain their full use. Durkheim's work is still generating research and controversy. See Emile Durkheim, *Suicide,* trans. E. K. Wilson and H. Schnurer (New York: Free Press of Glencoe, 1951). The original work was published in 1897.
6. The quest for typicality virtually dominates human thought. It is not merely a statistical exercise. Underlying the arguments of this book is the authors' deep conviction that all realms of human discourse are grounded in some sense, either assumed or somehow empirically assessed, of the typicality of various situations.
7. Just as there is folk knowledge, there are also folk methods for dealing with the problems of day-to-day living. Folk knowledge and academic knowledge are not the same. Nor are folk methods and academic methods the same. A sociologist named Harold Garfinkel established a new approach to social theory by turning to an examination of the methods people commonly use to solve problems that academic theorists find impossible to solve using more careful, rational methods. He coined the term *ethnomethods* to refer to the ways in which people handle the nearly infinite complexities that threaten to overwhelm them at any particular point in time.

   The last part of *Studies in Ethnomethodology* discusses differences between scientific rationality and folk rationality. Though the two forms are different, we should never lose sight of the fact that folk rationality is the prior form out of which the academic forms emerged. See Harold Garfinkel, *Studies in Ethnomethodology* (Englewood Cliffs, N.J.: Prentice-Hall, 1965).
8. There is a story about a study of young engineers who were given "gimmicked" computers to help them solve a bridge-building problem. The computers kept getting more and more erratic. Researchers wanted to see when the young engineers would realize that the computers' solutions were getting wacky. Few ever saw what was going on. The bridges designed by the young engineers would never have held up.

   A group of old, "wet-thumb-in-the-breeze," slide rule engineers with lots of experience were given the same computers. They spotted the glitches immediately. The moral of this story is that a blind reliance on computers is no substitute for a well-honed intuitive intelligence and lots of experience.

9. Most texts in statistics that we have reviewed give the impression that there is a great division between statistics and ordinary forms of inquiry and discussion, or what might be called folk reasoning. One of the primary concerns of this book is to develop the position that the choice between folk reasoning or statistical reasoning is not an "either-or" proposition but is, instead, a matter of degree.
10. There is much faddishness or false formalism in the use of statistics in the social sciences. Pitirim Sorokin, who was aware of this problem many years ago, wrote a book with the provocative title *Fads and Foibles in Modern Sociology and Related Sciences* (Chicago: H. Regnery Co., 1956). What is *la mode* this year can be *passé* next year. Leading journals, particularly, are sensitive to whether a researcher's article relies on "state of the art" statistical devices—even when such devices do as much to obscure what was found as to reveal it. When this happens statistics is made to serve ritualistic rather than rational functions.
11. Statistically put, the argument would be expressed by saying that homelessness is associated with a person's wishes or choice. If we do not choose to be homeless, we will not be. If we do, we will. What makes this argument so rigid is that no matter how much people might protest that their homeless condition is beyond their control, someone can always say that they are simply letting themselves (that is, choosing) to live that way. If they would take responsibility for their affairs (the conditional implies a causal relation), then things would be different. This is another example of stereotypical logic and its powers.
12. The concept of "choice" is never, as far as we know, used as an analytic term in the physical sciences. It is only used in the social sciences. We do not say that water chooses to run downhill nor do we say that a rose chose to have yellow blossoms. When choice is used in social arguments a powerful sense of understanding is gained because it seems obvious that a given action could not have taken place unless a person chose to do it. Therefore, any action is a matter of choice, and choice is always perfectly associated with any action. But how do we know the person *chose* to act in a particular way? The answer: because that's what the person did. In other words, we are correlating something with itself. This, in formal logic, is called a *tautology* or circular argument.
13. Communication cannot take place without typifications being presumed. It is almost a kind of a paradox that in order to do statistical work—such as obtaining a mean value for some quality for a given group of individuals—it is necessary to presume typification at the outset of the research. For example, if we make a statistical assertion about the differences between the incomes of men and women working at the same jobs in the United States, we have to have some sense of what is meant by terms such as *men, women,* and *jobs* in order to make use of the statistic. When we talk about what is meant by *jobs,* for example, we are required to have a sense of what is typical of the events that are referred to by this term.
14. The work of Harold Garfinkel in *Studies in Ethnomethodology* (see note 7) addresses some of the peculiar ways in which people come to create novel and idiosyncratic conceptions (typifications) of the world around them. Much of the work on cognitive processes in psychology and modern phenomenological theory in the social sciences reveals that there is much more to the process of gaining a sense of what is typical than the devices used by statisticians would indicate.
15. Not only are folk forms of reasoning complex in nature and difficult to systematize, but they are also extremely influential in determining how people think. Early twentieth-century sociology was dominated by the dream of enhancing social reform by introducing rational social knowledge. It is a moot issue as to whether the

social sciences have improved society by providing it with more carefully assessed knowledge (the authors of this book are inclined to think that they have). However, despite the best intentions of the social scientist, the social problems facing us today appear to be as overwhelming as they have always been. This might be due in part to the fact that people, when given statistical evidence repugnant to their beliefs, simply reject the evidence. It might also be due in part to the fact that social scientists themselves cannot establish a consistent, noncontradictory body of findings with regard to a large body of social policies and programs.

16. The nine themes listed here actually underlie all intellectual work, though they have come to be identified specifically with statistical research. Literature, for example, is extremely interested in creating typifications, drawing relationships, and making generalizations—several of the basic concerns of statisticians. What makes statisticians and literary writers different is not the broad logical forms on which they rely, but how they go about placing their observations within those forms.
17. John Allen Paulos, *Innumeracy: Mathematical Illiteracy and Its Consequences* (New York: Vintage Books, 1988).
18. Professor Richard Rogers, using statistically controlled data, presents a strong case for concluding that black and white mortality rates in the United States are socioeconomically based rather than genetic in nature. See Richard G. Rogers, "Living and Dying in the U.S.A.: Sociodemographic Determinants of Death Among Blacks and Whites," *Demography* 29.2 (May 1992): 287–303.
19. This is the well-known Omega state, the point at which the density of the universe is sufficient to prevent it from eventual dissolution. So far, this state has not been observed.
20. See Thomas Kuhn, *The Nature of Scientific Revolutions* (Chicago: University of Chicago Press, 1970). Kuhn offers a nice example of categorical confusion in the physical sciences in an anecdote about a physicist and a chemist responding to the question, "Is a hydrogen atom a molecule?" The chemist says it is; the physicist says it is not. Interestingly enough, both are able to give solid experimental or observational reasons for their claim. Such categorical confusion is more common in the social sciences. For example, two people can be doing exactly the same activity and in one instance it is called "work" and in the other "play." On the other hand, two people can be engaged in extremely diverse activities, and in both instances what they are doing is referred to by the same term—"work." The ambiguities in a popular term such as *work* reveal deep categorical problems. Social categories are extremely confusing and difficult to define in any precise manner other than by arbitrary devices.

2

CATEGORIES, CLASSIFICATION, MEASUREMENTS, AND SCALES

"Oh, call it by some better name . . . ."

—Sir Thomas Moore

"The shepherd drives the wolf from the sheep's throat, for which the sheep thanks the shepherd as his liberator, while the wolf denounces him for the same act. . . . Plainly the sheep and the wolf are not agreed upon a definition of liberty."

—Abraham Lincoln

Most introductions to statistics focus on logical procedures and assume that the categories or measures or data to which those techniques are being applied are ideal.[1] With respect to social data, conditions are never ideal. To overlook this problem by pretending to deal with idealized data gives statistical procedures in the humanities and social sciences an appearance of precision that can be misleading. Such an approach also may predispose the statistical researcher to dwell on logical or technical procedures while ignoring profoundly frustrating problems that come from another area—the area of categorization. Without good categories we cannot perform good statistical analysis. Therefore, this book includes categorization as an essential part of statistical analysis.[2] The task of statistics is to describe the world more precisely, and in order to do so it must rely on precise language.

THE IMPORTANCE OF PROPER CATEGORIZATION IN STATISTICAL ANALYSIS

In ordinary conversation we discuss things in ways that seem almost instinctive. A conversation works because most of what is being said is accepted in a taken for granted manner. At the same time, common conversation, in its language, is always imprecise. It is important at the outset of this examination of the forms of reason to acquire an awareness of the extent to which ordinary language is vague, imprecise, misleading, exaggerated, and ambiguous.

For example, we might overhear a woman telling a friend, "My husband Ralph is good enough, I suppose, but when I add everything up, I can't help thinking he's a failure." The category "failure" tells us something, though exactly what it tells us is impossible to say. About all it tells us is that Ralph's wife has a negative opinion of her husband's efforts. In what way and to what extent is Ralph a "failure"? By eavesdropping a little longer on the conversation, we might get a better sense of what the woman is talking about. The deeper meaning of "failure" in the conversation can only be established through a lengthy story or narration about Ralph. The story, fitted to the particular case, provides a broader, more detailed definition of a complex quality.

Thus we can begin to see how **social categories** are typically defined by elaborate stories that keep filling in the particulars as we seek to comprehend what is meant by a given reality. The social category of "love," for instance, is defined in terms of a variety of stories that tell us how people experienced or gave form to this quality. Defining social forms is one of the functions of good literature; it is, specifically, one of the social functions of literature. (There is more to literature than merely the intent to amuse or entertain.)[3]

In statistical analysis we do not have time to tell elaborate stories about each particular case in a study unless we are examining small samples. The larger the sample, the more necessary it is to simplify things. When you want to convey information about millions of people, as is the case with census data, for example, the categories have to be especially simple. If you gave each person even a single page to tell his or her "story" you would wind up with a research volume millions of pages thick. A single page for a person to tell his or her life story is not much—but the more pages you add the bigger the volume gets.

To bring this point home, suppose each of us could tell our life story in a volume of a mere 500 pages. If this were published as a census of the population of the United States, we would have a book more than 125 billion pages long! If each life volume were only one inch thick, the total census would be 4,000 miles thick. We mention this because there are those who think statistics is a damnable activity because of the information it excludes. However, such exclusion is unavoidable. Unless we perform operations of essentially a statistical nature, we become swamped by all the information available to us. Statistical work is, among other things, the reasoned simplification of events.[4] But making things simple can be complicated.

So, statistical procedures are essential. We cannot function within the complexities of the real world without simplification. The question is, shall we simplify methodically or whimsically? Either way, we must necessarily distort reality. The trick is to minimize distortion. Another definition of statistical analysis is that it is an effort to achieve comprehensive, simplified descriptions of the world that minimize distortion.[5]

This is a basic point. We are now touching on one of the primary differences between literary and statistical approaches to describing human events. Literature tells highly elaborate stories about relatively few individuals. Statistics tells extremely superficial stories about relatively large numbers of individuals. Literature and statistics are polarities along a continuum. As we move in one direction we abandon the advantages that come from moving in the other. Whether we rely on stories or statistics, we need to compromise to get on with things.

Good literature tries to represent grand human qualities by searching for literary typifications that "capture" universal experiences. A great novel or story defines an experience, an emotion, or a moral quality. Because these qualities are complex, varied, and abstract, the stories take on an abstracted, varied, and complex character. A single story, such as that of Shakespeare's *Hamlet,* can be so complicated and involved and its definition of its subject so elaborate that scholars spend lifetimes interpreting its meaning.[6]

In contrast, statistical studies dealing with human experience try to summarize the lives of hundreds, thousands, millions, or even billions of people. Obviously, statistical reporting cannot assume the same form as literary reporting. Something has to give. In order to summarize vast amounts of experience, individual idiosyncrasies have to be eliminated. Just as literature must make compromises (often referred to as exercising "artistic license") in order to handle the complexities of human experience, statistical studies make compromises as well.[7] The extent to which compromise takes place is most easily seen in the ways in which statistical studies deal with categorization.[8] The demand for large samples forces us to move toward highly simplified categories. The idea behind statistical studies, especially in any kind of social research, is to collapse the stories that might be told about any of the individual people involved in a given study in order to describe what is happening at the aggregate, or combined, level.[9]

In the following discussion of the problem of categorization in statistical analysis in its human applications we stress two fundamental points:

1. There is no way to avoid arbitrarily simplified categorization for purposes of statistical analysis.
2. Arbitrarily simplified categories create serious problems in the quest for precision.

Statisticians must rely on highly simplified and abstracted descriptors. Consider the categories commonly used by state and federal agencies for obtaining statistics on racial and ethnic minorities. *The Statistical Abstract*[10] uses the categories "minority" and "white" to divide populations into ethnic or racial categories. Obviously these are abstracted terms. A Korean running a shop in south central Los Angeles and an Alaskan Eskimo building an igloo on arctic ice will both be placed in the category "minority" even though their individual stories differ radically. An entertainer such as Bill Cosby, whose wealth is measured in the millions of dollars, would be included in this category along with a Navaho Indian living at a subsistence level in an isolated desert region of the American Southwest. These are "minority" people in terms of these categories, but they are not really similar people.

This is troubling. The whole idea behind categorization is that the members of a category should be alike. In the physical sciences, all of the members of the category "hydrogen atom" are precisely alike—*no* measurable differences whatsoever exist between its individual members.[11] A serious problem exists in the social sciences because the similarities we rely on to categorize events are superficial. We might, for example, obtain a sample of one thousand "women." The most reasonable way to do this is simply to define a woman as any person of a certain age who possesses specific secondary sexual characteristics.

This approach has the advantage of being logically consistent, but it is a superficial way of defining sex. While it is superficial to assert that the category "sex" is defined by physical sexual characteristics, if we attempt to move beyond this superficial device, we will be trapped in the problem of establishing what is really meant by "femaleness" or "maleness." This brings us back to the impossibility of telling each individual's story.

In sum, social categorization is haunted by the problem of superficiality; physical categories have less difficulty with this. It is probably for this reason, more than any other, that the social sciences have not developed precise, stable, and uncontroversial arguments and observations despite a century or more of extensive research efforts.[12] As we shall see, the social sciences rely almost exclusively on **probability models.** Probability models are less satisfying than more rigorously deterministic or **mechanical models.** The reasons behind the dependency of social science on probability models are closely connected to the problems social scientists have with establishing precise categories.[13]

Observation begins with categorization. In turn, the adequacy of our categories is established through our observations. Categorization and observation go hand in hand; each is integral to the other. As we focus on our observations, we can lose sight of the fact that underlying our observations are the categories we were forced to rely on to bring order to what we observed. If our categories are good, we have a fair chance of making good observations. If our categories are weak, we cannot hope to move beyond limited forms of observation. Statisticians, like anyone else, are forced to deal with the problem of what constitutes good categorization.

The traditional or orthodox solution that social statisticians came up with to this problem is a simple, interesting, and, ultimately, shaky one.[14] Basically it consists of a categorization of how we categorize things, with great emphasis placed on four categories of categories:

1. *Nominal categories:* names of things
2. *Ordinal categories:* things ranked according to some order
3. *Interval categories:* things ordered according to equal intervals
4. *Ratio Scales:* things ordered with a fixed zero point and equal intervals

Of these categories of categories, ratio scales, from a statistical viewpoint, are an ideal way of categorizing events. Let's find out why ratio scales are especially appealing when we want to be precise.

IDEAL CATEGORIES

What are the properties of an ideal category? This is a tough and profoundly significant question; after all, we are forced to rely on language. If our categories are confused, so are we. Good communication cannot take place when our terms are vague, ambiguous, ill-defined, or otherwise imprecise. Precision is an inherent demand of human communication, though it is seldom, if ever, attained. Only within a small envelope of physical science research is precision raised to near perfection. What are the characteristics, then, of an ideal way of categorizing events and happenings in the world? It will be sufficient to raise the question and outline the way statisticians have dealt with it. Keep in mind that a rigorous solution to this problem has yet to be established in such fields as social psychology and sociology.[15]

Formal statistical analysis is regimented by the categories used in the analysis. If simple, imprecise terms are all we can rely on, then the statistical devices we can employ for analysis are seriously limited by this fact. On the other hand, if we have precise categories, we can use more sophisticated logical procedures. The first question anyone should ask in doing statistical work is: What is the nature of the categories involved and how are they defined? Once this is settled, we can think about specific techniques. Crude categories can only rely on crude statistical procedures.

Because it is easy to get into abstract discussions when talking about categorization, a single example should clarify the matter. One category or quality Americans are fond of is "success." We differentiate people into groups of those who are "successful" and those who are "unsuccessful." We admire the former and pity the latter.

If we want to be more rigorous in our thinking, the question to be answered is this: How do we go about precisely identifying success? The first thing we want our category to do is to differentiate correctly between those who are successful and those who are not. Our category should include all members who belong to the class or category of "success" and exclude all who are not members of that class. How might this be done?

Americans characteristically define success in terms of money. Most of us know this definition has severe limitations because we like to believe it is possible to be successful though poor, and a failure though rich. (In other words, money as a criterion of success is superficial.) Despite this belief, we continue to look on money as a basic way to evaluate success. Why is this reliance on money so popular? In seeking a general answer to this question we begin to see some of the power of the ratio scale.

An ideal way of categorizing success would be to set up distinctions in such a way that people who are highly successful would be ranked above those who are less successful. Ideally we would like to see how much more successful one person is than another—that is, we would like standard units of success. This brings us to still one more ingredient in an ideal system for categorizing successful people: we would like to know what constitutes "zero success."

The best measures in the sciences have established **zero points**—of temperature, mass, weight, velocity, gravitational attraction, and so on. Having a zero point enables us to say things such as "This is twice as hot as that," "Charles is three times as heavy as Charlene," or "A jet plane is ten times faster than an automobile." If we had a notion of what constitutes "zero success," we could talk about one person being twice or three times or one-fiftieth as successful as another.[16]

This sounds like nit-picking. Nonetheless, we are forced back to an awareness that if our categories are imprecise, so are our communications. We commonly say that someone is "ten times smarter" or "five times prettier" than someone else. Are we ever aware as we utter such statements that they are pure fictions? When we talk this way in ordinary conversations we endow our concepts with false precision. If delusion is to be avoided in communication, then this kind of false precision must be eliminated as much as possible.[17]

FOUR TYPES OF CATEGORIZATION USED IN STATISTICS

As mentioned previously, statisticians deal primarily with four types of categorization ranging from crude *nominal* classes to *ratio scales*. Now we will discuss these categories in greater detail.

Nominal Categories

Nominal categories are names or, essentially, nouns. They are surely the most common form of classification used in human communication. However, they are also quite vague. It follows, then, that *any statistical analysis that relies on purely nominal categories will seriously lack precision, no matter how elaborate our formulas for making the analysis might be.*

The problems we have in communication commonly come out of the vagueness of such concepts. At the same time, we rarely think of these concepts as vague; they always seem to have rich meanings for us. If someone mentions seeing a horse, we presume we know what he or she saw. We are likely to presume it was a live horse, healthy, standing up, with four legs, a long tail, and so on. We presume meanings. Presumption, of course, is often a prelude to trouble—in research as well as in human affairs. It is what good categorization is designed to remove, if possible. Good categories do not presume that appropriate divisions are being made; they try to *make certain* that such divisions are being made. The following list provides examples of some familiar nominal categories.

Examples of Nominal Categories

Protestant	criminal	success	professor	sunset
sweetheart	apple	airplane	dead	horse
mountain	marvelous	telephone	book	painter

Ordinal Categories

At the nominal level we can say someone is Muslim or not Muslim; a success or not a success; dead or not dead. However, we might want to know if the members of the category *Muslim* differ in terms of how Muslim they are. We can name people successes, but we know some are more successful than others. What is called for, here, is *order* within the category. So it is that we come to establish a second class of categories that *order* events. These are referred to, appropriately enough, as **ordinal categories** or scales. If we rate five employees from best to worst, we would have an ordinal categorization:

Mary is most successful.

Carlotta is second.

Jose is third.

Okuma is fourth.

Ivan is fifth.

We have put these five individuals within the category of "success" and gone further, ranking them within the category. This is an *ordinal,* or *ordering,* categorization. It is an improvement because it gives us more information than we had before, when we simply lumped everybody together under the same heading of success. With ordinal categories we both name *and* order things.

However, the *ordinal* category only tells us that Mary is the most successful of the five employees. It does not tell us how much more successful she is than Carlotta. It might be that the difference between each is minuscule. Or, it could be the case that Mary is astonishingly more successful than Carlotta, who is a modest success, as are Jose and the rest. We are ignorant of how these people should be placed along a continuum of success.

Interval Categories

If we could find some kind of constant unit of success, we could develop a way of categorizing success that would solve this problem. Let us pretend to have such a unit and call it an S-interval. We can now put our employees on a continuum that distinguishes them by showing where they fall along a continuum of equal

Figure 2.1. Individuals Ranked along an Equal Interval Scale

```
Ivan    Okuma    Jose Carlotta              Mary
S-^---S-----S-^---S-----S-^---S---^-S-----S-----S-----S----^S-----
```

intervals (see Figure 2.1). For this kind of categorization to work properly, the intervals *must* be equal to each other; this is a very important qualification. It also happens to be a tough one for social-psychological forms of research. **Equal interval scales** are commonly assumed in various tests and measures used in the behavioral sciences.

The development of equal intervals provides us with still more information. We now know more about our five people than we knew before with *nominal* and *ordinal* categories. Note that all we are seeking is a better way of describing a quality. It is accepted as self-evident that better categories provide greater clarity of thought. So it is that we move beyond nominal and ordinal categories, where possible, and seek categories that enable distinctions to be made in terms of equal intervals. Such categorization is known as an *interval scale.* With an interval scale we *name, order,* and *ascertain how far along the scale things fall.* Interval scales provide more precise categorical information than nominal or ordinal scales. A fundamental requirement in such scales is that each interval be the equivalent of each other interval. For example, in a common yardstick, an inch is always the same no matter where it appears on the yardstick.

With physical scales an inch is an inch, a pound a pound, a kilogram a kilogram, or a degree of Centigrade a degree of Centigrade, regardless of context. When we move into the domain of human affairs it is difficult to ignore the extent to which various contexts can alter the "meaning" of a given degree of some quality. For example, a pound of flesh means quite different things depending on whether it is a part of a man's bicep or a part of his nose. Should we take the subjective into consideration in human measures? Is the difference between ten and twenty dollars the same thing as the difference between one million dollars and one million and ten dollars? In a sense, yes. In another, no. There is also the nagging problem of whether our scales are precisely interval measures. Is the difference between 120 and 125 IQ points the same as the difference between 80 and 85 IQ points? Does the term "fifty percent" mean the same thing when we talk about fifty percent employment and fifty percent illiteracy? Half, mathematically, is always half. In human affairs mathematical precision quickly becomes elusive.

Ratio Categories

There is one final form of category that offers still more information than the interval scale. Although we can see that Jose is halfway between Ivan and Mary, we cannot say that Jose is half as successful as Mary. Nor can we say he is twice as successful as Ivan. We can see why when we place these employees on another scale, as shown in Figure 2.2.

Figure 2.2. Individuals Ranked along a Ratio Scale

```
Ivan         Okuma     Jose   Carlotta                  Mary
0-^--5----10----15^---20----25----30--^-35----40----45----50---^55
```

We can now see that Ivan has only two units of success and is only one twenty-fifth as successful as Mary, who is about twice as successful as Jose. We can now use *ratios* to evaluate the positions of our employees along a scale of success. This adds still further information to our categorization. In order to accomplish this not only must we have equal intervals, but we must also be able to establish a zero point for the scale. This type of categorization is known as a **ratio scale,** a powerful form of descriptive category.[18] A ratio scale does what all the other scales do, and also provides comparisons in the form of ratios.

The popularity of money as a measure of many qualities in our society must derive, at least in part, from the fact that it is one of the few social measures that is a true ratio scale. Its utility as a ratio scale makes it unique in the realm of human affairs. Each dollar (or yen, or mark) is equal to another dollar (or yen, or mark); such units are equivalent, at least during the brief periods of time before their value is affected by inflation. There is also a condition in which a person can have zero sums of money,[19] and indebtedness is easily stated as negative money.

Therefore, we can readily state that Marguerita, who has $10,000, is twice as wealthy as Oscar, who has $5,000. From this kind of ratio measure it is easy, but erroneous, to conclude that Marguerita is twice as successful, important, worthwhile, bright, powerful, and so on, as Oscar. Clearly money is not actually a measure of spiritual worth or value; but as a simple ratio measure it has few, if any, equals in human affairs.[20] Money is a grand simplifier of human complexities. It is also a superficial category.

There are, then, four basic forms of categories relevant to statistical work—and, more generally, to human communication: the *nominal, ordinal, interval,* and *ratio* scales. Nominal categories are less informative than ordinal; ordinal scales are less informative than interval; and interval scales are less informative than ratio scales. (This last sentence, incidentally, is an example of ordinal categorization.)

ARE OBJECTIVE AND PRECISE SOCIAL CATEGORIES POSSIBLE?

The statistician's basic solution to the problem of categorization in the form of nominal, ordinal, interval, and ratio scales works well in some contexts and runs into problems in others. First there is the problem of whether we can realistically scale purely social qualities. Social events are complex, abstracted, structural entities and *processes.* You cannot, for instance, take a photograph of a social event—it is, by definition, something that requires time to take place. To clarify, consider the social category "student." While we are aware, in a commonsensical fashion, that some individuals are more "student-ish" than others, whatever is meant by the term "student" is complicated, structured or ordered, and abstract; it is also an action. By using the term *student* we apply a noun to what is actually an extended verb. A student is not a *thing* but an *action*.

Transforming complex, abstracted activities into nouns is done all the time in ordinary discourse. We are forced, as well, to do it in social statistical studies. Nonetheless, using the term *student* to categorize a set of activities is not the same thing as using the term *triangle* to categorize a set of Euclidean forms or the term *planet* to categorize a set of bodies in orbit about stars. The categorization of social reality, no matter how elementary our concern might be, cannot be accomplished with any great rigor.

We can determine physical properties with scalar accuracy. We can classify people relatively precisely with respect to such things as age, height, weight, physical strength, skin pigmentation, and secondary sexual characteristics. However, we cannot classify people precisely with respect to social statuses (such as student or worker) or qualities (such as success, loyalty, morality, and so on). When we attempt to count, or measure, qualities such as these, we have to be especially careful.[21]

We bring up the problem of scaling social qualities because it is a real problem. We cannot pretend it does not exist and go on under the assumption that our categories have the same logical qualities as those used in the natural sciences. They do not. At the same time, it should be obvious that even though social categories are difficult to define for precise work, statistical research in the social sciences, properly done, is necessary and valuable. We do need to know how many people there are in the world—regardless of their individuality. We do need to know what proportion of the population is female or nonwhite or going to college. In modern society we cannot make decent social policy without statistical information. We must never, however, allow ourselves to come to the belief that more refined mathematical techniques will overcome problems that are inherent in social categorization itself. Social statistics can be invaluable. But a good statistician remains constantly aware of the strengths and limitations of the categories used in statistical reporting.

A second problem must be mentioned. The trend, in modern social science, has been toward "objectification" of data. This means that whatever categories are involved in one's research, they should be free of value implications. They should not explicitly or implicitly suggest superiority or inferiority, or goodness or badness. At the same time, it is a basic quality of social categories to establish superiority or inferiority with respect to human actions. Social categories make social evaluations—they rank and socially order people.[22]

The social statistician must, therefore, always accept the possibility that any statistical finding, regardless of how carefully objective the attempt, will have controversial implications. One reason for this is that social categories are always subject to varied interpretations and definitions. The statistician can only try to be as careful as possible in categorizing events and to make clear all procedures used in a given study.[23]

SUMMARY

Statistical studies seek precision. They rely on counting, and counting relies on being able to classify or categorize events. Social categories are characterized by abstractness and complexity. Statisticians have dealt with this problem by relying on arbitrarily simple categorical devices. The simplest form of category consists of *nominal* classifications, or simply "naming" things.

At the next level is the *ordinal* scale or ordered categorization, in which events are not only named but ranked from lower to higher (or higher to lower) degrees of some quality. Ordinal scales provide more information than simple nominal scales.

A still better form of categorization is the *interval* scale, which names, orders, and provides equal intervals along which observations may be placed. While simply ranking or ordering observations does not let us see how close or far apart they might be, the interval scale gives us information with regard to how much more of a quality one case might have than another.

The best form of categorization for purposes of statistical analysis is the *ratio* scale. Such scales name, order, rank along equal intervals, and provide a given zero point, thereby allowing us to determine whether one event is two, three or "n" times greater than another. It is worth mentioning, again, that only when we can establish ratio scales can we rely on the most informative and precise statistical techniques.

Typification as a central concern in statistical research appears in the problem of what constitutes a good category. A good category, by definition, should typify the members contained in the set. With respect to social categories two major problems exist in attaining precise typification:

1. No matter what dimension we might select to establish typicality, there is variability within the category. For example, if we divide people into two classes, male and female, there is within-category variability with respect to these qualities: some males are "more male" and some females "more female" than others contained within the same set.
2. Even if we establish ratio scales, we are still faced with the problem of superficially or arbitrarily selecting some single dimension to represent an event that is, in fact, a complex of dimensions. For example, does selecting cases according to physical sexual attributes define sexual differences in any but the most superficial manner? Is sexuality only, and perfectly exactly, a matter of type of sex organ? Or is it more complicated?

The problems that exist with respect to categorization should never be ignored by the good statistician. At the same time, one cannot afford to be overwhelmed by them. Just as it is possible to bring an ordinary conversation to a halt by constantly demanding more precise definitions of terms, it is possible to pick to pieces nearly any statistical work in the social sciences by attacking its categorical base—especially when social categories are being used. The purpose of this discussion has been to point out that statistical categories are necessarily simple in the social sciences. There is no way out of the matter. If you want comprehensive data, you have to make compromises with your categories. Good research makes good compromises. Unfortunately, the art of statistical compromise and arbitrariness cannot be easily taught—it tends to come with experience.

Thinking Things Through

1. What makes categorization in mathematics and the physical sciences simpler than in the social sciences? Or is it simpler?
2. Why is a ratio scale superior to nominal forms of categorization? Why is it difficult to establish a real or true zero point for social qualities?
3. You want to do a study of successful people. Describe five different ways you might go about categorizing "success." What are the advantages and disadvantages of your definitions of success? To what extent was your response to the problem "culture-bound"? That is, to what extent did you arbitrarily decide to deal with the notion of success from an American vantage point? Do you think men and women have the same or different conceptions of success? Is it possible to be objective about something as culturally valued as is success in America?
4. What kinds of problems are involved in giving precision to the data in Table 2.1, with respect to the categories of "vehicles" and "cars"? Remember, we are concerned with real precision. Since most people would see the following table as having no categorical problems, we would like you to consider whether the "car population" of Japan, let us say, is the same as the "car population" of the United States. In what ways might these two populations differ so that a perfect comparison is compromised?

Table 2.1. World Motor Vehicle Registrations: Six Selected Countries, 1991 (registrations in millions)

Country	*Total Vehicles*	*Cars*	*Persons per Car*
Bulgaria	1.5	1.3	6.8
China (mainland)*	4.1	1.0	1,075.0
France	28.8	23.0	2.4
India*	2.9	1.5	566.0
Japan	59.9	37.1	3.3
United States	188.4	143.0	1.7

*Data for 1989.

Source: Statistical Abstract of the United States, *1994, Table 1374.*

5. This exercise explores a serious categorical issue. It is not only a statistical problem but a broader ethical and moral issue. Give it some thought. A University of Massachusetts sociologist named Paul Hollander wrote about anti-American writers and thinkers in the twentieth century.[23] How would you go about carefully defining the category "anti-American"? Can you think of some way of arranging individuals so that they would fit along a scale of anti-Americanism? Is there such a thing as "zero" Americanism? Such a zero point is implied here, because if there is pro- and anti-Americanism, then a continuum is created that should have, at some point, a zero value. Discuss this category as a problem with respect to collecting objective data—that is, data without value implications.
6. You are an American environmentalist who would never think of using a plastic foam cup at the office for drinking coffee. On the other hand, you drive to the big football game that is being played one thousand miles from your home town. The round-trip consumes 65 gallons of gas. Let's assume it took 100 gallons of oil to distill the 65 gallons of gas. Let's also assume that it takes one ounce of oil to make a plastic foam cup, and that an ounce of oil is roughly one four-hundredth of a gallon. How many plastic foam cups did you use up on your holiday car trip to the football game? If you used two plastic foam cups a day for your coffee, how many years would it take you to use up all the plastic foam cups that could have been made from the oil you consumed on your football venture?
7. Using the same values as those presented in problem 6, assume that five thousand alumni take the car trip to the game. How many plastic foam cups could have been created from all the oil consumed by this population? If each cup were four inches high and all of the cups were laid end to end, how long a line of cups would you have?

Notes

1. This is commonly noted in statistical texts. Herman J. Loether and Donald G. McTavish note that " . . . the derivation of formulas tends to be based upon sets of ideal conditions such as scores free of measurement problems and error or 'a normally randomly distributed random variable' " (*Descriptive and Inferential Statistics* [Boston: Allyn and Bacon, 1988], xii). This sweeps a lot of dirt under the rug. In real research, ideal conditions are never encountered—leaving us to wonder how one trained to deal only with ideal conditions is going to react when faced with reality.
2. Traditionally, sociology divides its methodology into two parts: one deals primarily with the problem of how data are to be organized and simplified, which is essentially the task of statistical analysis; the other deals with how data are to be defined and counted, which is essentially the task of measurement. However, this division is awkward insofar as each division, at least in part, determines how one will make decisions with regard to the other.
3. People in the fields of communications, linguistics, sociolinguistics, and literature have given a lot of serious thought to the problems posed by social categories. A good statistician can benefit greatly from reading what literary theorists have to say about the problem. While it is impossible to summarize the character of his books

in a sentence or two, the works of Kenneth Burke are especially insightful. Burke argues that literature is, in part, a response to the problem of defining the social terminology that makes up our day-to-day existence. A love story, for example, deals with the complex problem of defining love by offering paradigmatic case studies of relationships. Stories are elaborated definitions of concepts. See Kenneth Burke, *A Grammar of Motives* (New York: Prentice Hall, 1945); *On Symbols and Society* (Chicago: University of Chicago Press, 1989); *The Philosophy of Literary Form: Studies in Symbolic Action* (Baton Rouge: Louisiana State University Press, 1967), and *A Rhetoric of Motives* (Berkeley: University of California Press, 1969). These books might appear remote to a statistically oriented professional. However, for the social scientist, few writers provide more insight into the categorical problems involved in social communications than does Kenneth Burke.

4. Curiously enough, good literature is also the reasoned simplification of complex events—a well-crafted poem is designed to be informative in just a few lines. Literature and statistics differ in what they are trying to simplify. Literature simplifies an event or social quality, such as courage or love or loyalty, in the form of a condensed story. Statistics simplifies and orders large masses of information by giving it some kind of numerical form.
5. Notice that we can reduce distortion but never entirely eliminate it. Any statistician should remain aware of the essential distortions inherent in statistical operations.
6. This book often refers to the differences and similarities that hold for literary and statistical descriptions of the world around us. In order to get a better awareness of what statistical methods can achieve and what they cannot achieve, we need to compare them with nonstatistical methods. In all of our discussions there has been no overt attempt to "stack the deck" in favor of either literary methods or statistical methods. Each approach has its own strengths and its limitations. We are convinced that it is both wiser and fairer to avoid the older notion held by some statisticians, that literature is simply the entertaining and ingenious lies of ingenious liars. A good indication of how pervasive such a notion once was can be seen in the great popularity of a book, during the middle of the twentieth century, that was vehemently critical of literature: George A. Lundberg's *Can Science Save Us?* (New York: Longmans, Green, 1961). Lundberg actually proposed, at one point, that children be protected from literature until they are old enough to comprehend that it is fiction and, therefore, misleading. It is, of course, also unfortunate to notice that some literary figures think of statistical methods as a form of barbarism. In this book we try to avoid these extremes and move instead toward a point of view that suggests we can learn a great deal from both camps.
7. It is interesting to notice that people are generally aware of the fact that writers lean on "artistic license" in order to get on with the problems of completing a novel or drama. They are not as generally aware of the fact that statisticians must make use of a kind of "statistical license" in order to get on with the problems of completing a piece of statistical research. At no other point in statistical procedures in sociopsychological research is the necessity for license as obvious as it is with respect to the problem of how events are going to be categorized. This is especially the case with social statistics. Purely mathematical approaches to social statistics tend to gloss over the problem of categorization. Herbert Blalock's attempt at a grand theory of social conflict is a case in point. While Blalock's logic is meticulous, the categories involved in his study are imprecise. See Herbert Blalock, *Power and Conflict: Toward a General Theory* (Newbury Park: Sage Publications, 1989). In any event, the point here is that one cannot escape the problem of "license" in the

investigation of human affairs, regardless of whether a literary or statistical approach is taken.

8. What is especially nasty about these compromises is that, like nearly any compromise, an element of arbitrariness is unavoidable. We like to think that science or the truth or "hard" research does not deal arbitrarily with reality. Unfortunately, there is no way of avoiding arbitrariness. Some element of arbitrariness is found in both literature and in statistics. If you ask how the arbitrariness is resolved, the answer is that it is generally resolved politically. The literature pertaining to political resolutions of research issues is scattered, but large enough in scope to provide a unique study in its own right. For some examples see Daniel M. Fox, "Health Policy and the Politics of Research in the United States," *Journal of Health Politics, Policy and Law* 15.3 (Fall 1990): 481–499; Charles W. McCutchen, "Peer Review: Treacherous Servant, Disastrous Master," *Technology Review* 94.7 (October 1991): 28–40; Jeffrey L. Rogers, "Missed Opportunities: Politics, Research, and Public Policy," *International Journal of Offender Therapy* 35.4 (Winter 1991): 279; Elli Lester, "Manufactured Silence and the Politics of Media Research: A Consideration of the 'Propaganda Model,' " *The Journal of Communication Inquiry* 16.1 (Winter 1992): 45–55.
9. Actually, the problem of categorization is a troublesome one for any intellectual discipline—and the physical sciences are not an exception. Physical science is constantly revising its categories and finding fault with them. At one time, for example, one of the qualities attributed to the category "space" was that it was absolute. Now the absoluteness of space depends upon which kind of space you might be talking about. Several popular books deal with some of the problems that scientists have had with categorization. Especially recommended is Thomas Kuhn, *The Nature of Scientific Revolutions* (Chicago: University of Chicago Press, 1970), which was also cited in Chapter 1. Also readable and at the same time provocative are Roger Penrose's *The Emperor's New Mind: Concerning Computers, Minds, and the Laws of Physics* (Oxford: Oxford University Press, 1989) and Douglas Hofstadter's *Godel, Escher, Bach: An Eternal Golden Braid* (New York: Vintage Books, 1980).
10. *The Statistical Abstract of the United States* is a collection of roughly 1,500 tables that give statistical information about conditions in the United States ranging from abortion rates to energy consumption. It is updated annually. We use it as essential reading in introductory statistics courses. There are numerous other statistical sources, but this one is excellent for its conciseness and its comprehensiveness. We consider it an essential reference book for any college student (or professor, for that matter).
11. Physical scientists can talk about precision of such a high degree that they can say there are *absolutely no differences* between particles of a certain order. If one particle of a particular type is replaced by another, the situation remains identical to what it was before. See Penrose, *The Emperor's New Mind* (cited in note 9). In a social unit, on the other hand, if one person replaces another, the situation is never what it was before. The reason is obvious—no human being, including an identical twin—is *exactly* like any other human being.
12. In the social sciences, the best models are usually physical ones. For example, Zipf's gravity models, an early statistical finding, achieved correlations greater than 0.90, an unusually high degree of precision in social science. However, the factors correlated were distances between cities and amount of physical interactions between the cities (for example, phone calls per week, and so on). See George

Kingsley Zipf, *Human Behavior and the Principle of Least Effort: An Introduction to Human Ecology* (Cambridge, Mass.: Addison-Wesley Press, 1949). Zipf developed so-called gravity models to account for activity between human settlements. Although this work was done nearly fifty years ago, it still provides some of the finest approximations to mechanical models to be found in social science literature.

13. The postmodernist and deconstructionist schools of thought have emphasized the frailty of social constructions without turning around and looking at the other side of things—the unavoidability of simplification. Simplification always leaves an observation open to criticism.
14. Statistically relevant categories are tied in with the problem of counting things. Statistical categories must have some kind of relevance with regard to transforming information into a numerical form. The question that is being raised here is, How far should you go in trying to transform some quality in such a way that you can grant it numerical form? Some things are easily dealt with in terms of numbers: a car's gas mileage, an elephant's weight, the tonnage of an ocean liner. Other things resist numerical transformations: beauty, manliness, courage, morality, wisdom. In this sense, the categorical scheme that is so popular among statisticians must always be seen as a limited scheme—applicable only where there is a quest for a count of things. The quest for numerical information is motivated, in turn, by the desire for precision. There is sufficient difference in the meaning of quantitative work with respect to physical studies and humanistic ones to suggest that in humanistic studies numerical descriptions take on the quality of an analogy. Some statisticians will find this assertion extremely distasteful. It is not meant to be—analogy is not necessarily a bad way to go about describing things.
15. It would help the social sciences greatly if this contention could be mathematically demonstrated. There are still social scientists who treat social categories as though they were logical or mathematical categories. The problem ultimately has to do with whether or not some condition is "computable." Is there a consistent algorithmic process that consistently provides exactly the same parameters each time the category is used? While this can be done in some areas of the physical sciences, we do not know of any purely social category that meets these criteria in a precise manner.
16. The invention of the zero was one of the major inventions of mathematical terminology. The zero appeared around the ninth century in India and also in Mayan culture. See C. J. Brainerd, *The Origins of the Number Concept* (New York: Praeger, 1979), Byrne J. Richard, *Number Systems: An Elementary Approach* (New York: McGraw-Hill, 1967), and Georges Ifrah, *From One to Zero: A Universal History of Numbers* (New York: Viking, 1985). It is a peculiarity of social qualities that it is difficult to imagine them being possessed of a zero point. Any kind of zero, for qualities such as success, knowledge, intelligence, morality, dependability, and so on, has to be established arbitrarily. Arbitrary zero points do not provide true ratio scales.
17. The elimination of "false precision" in human communication cannot be achieved in any total manner. Oddly enough, false or presumed precision is necessary in the greater part of human communication. Constant efforts to bring greater precision into a conversation tend to destroy the conversation. The rationalist's dream of a precise language and a scientifically established body of knowledge directing all forms of human conduct is a utopian vision. Social language is inherently imprecise, and attempts to generate greater precision commonly disrupt relations rather than improve them. If you want to disrupt a conversation, simply keep demanding

that people define their terms—whatever the terms might be. No one has demonstrated this better, perhaps, than Harold Garfinkel. See Garfinkel, *Studies in Ethnomethodology* (Englewood Cliffs, N.J.: Prentice-Hall, 1967).

18. Rational numbers come from the Latin *ratio,* meaning to account, reckon, or calculate. Rational numbers are numbers that permit fractions, hence the use of the term *ratio* to refer to the value obtained from dividing one number by another.
19. This is true only in a formal sense because even though a person might not possess a single penny, he or she remains worth something and, in that sense, no individual achieves a true zero point with respect to wealth.

    There is also a problem with respect to money inflating or changing value over time—this makes comparisons with earlier periods more difficult. Standardizing for inflation is one of the most common "control" or "standardizing" procedures used in statistical work.
20. It perhaps says something about the power of money as a measure that economists are the only social scientists recognized by the Nobel Prize committee.
21. Social categories are socio-dramatic categories designed to rank people and induce control over actions. They have inherent rhetorical properties that make them difficult to use as purely objective scientific terms. This is the argument of Kenneth Burke (see note 3).
22. The unique nature of social categories, as opposed to purely physical or logically descriptive categories is, in our opinion, the major barrier to achieving a mathematical sociology. The very nature of social categories precludes a mathematical form of social description.
23. Paul Hollander, *Anti-Americanism: Critiques at Home and Abroad, 1965–1990* (Oxford: Oxford University Press, 1992).

3

TWO STATISTICAL SHORT STORIES

"Population, when unchecked, increases in a geometrical ratio. Subsistence increases only in an arithmetical ratio. A slight acquaintance with numbers will show the immensity of the first power in comparison with the second."

—THOMAS ROBERT MALTHUS

Statistical works vary both in their complexity and in the costs of gathering the data they analyze. Some people erroneously believe that unless they have a wealth of expertise and a million-dollar grant for data collection, they cannot conduct a worthwhile study. This is not so. Statistical work is still a place where an imaginative individual can accomplish a great deal with a modest amount of material and by relying on simple techniques. The statistical "short stories" presented in this chapter illustrate this point. Statistical analysis can be elemental and still be extremely effective. Perhaps it is most effective when it is simple (we will not get into this argument here). Our two tales make clear that it is certainly possible to do ingenious work with simple data and a few elementary statistical techniques.

These stories have the basic qualities of story telling—they are imaginative, evocative, creative, and simple. They also represent good statistical, quantitative argumentation. Neither study portrayed in these stories was expensive. Each brought recognition to its author. One shook all of Europe and Western civilization, and its power is still great, even though it was first published in 1798.

Many scientists or researchers have an interest in the beauty or aesthetics of research. They talk, in mathematics or in science more generally, of a concern with "elegance." Bertrand Russell, for example, says, "Mathematics . . . possesses . . . supreme beauty . . . sublimely pure, and capable of a stern perfection such as only the greatest art can show."[1] However, we do not know, offhand, of social statisticians who talk about the *beauty* of statistics. This might be one of the reasons students commonly dislike statistics—because it does not seem to offer anything attractive. Sometimes it appears as though statisticians go out of their way to make their arguments in the most ungainly manner possible. Social statistics are, for the most part, clumsy and awkward in appearance, lacking the grace of literature and the subtlety of physical science. Yet this does not need to be the case. Good statistical work can also have an aesthetic appeal. A good statistician should never be disdainful of the idea that intellectual activity—statistical work included—can be aesthetically pleasing as well as informative.

The two statistical stories in this chapter were chosen because they present tidy pieces of work. Part of their appeal comes from the fact that they reveal how statisticians can produce research that is well-crafted, powerful, and a stimulus to the imagination.[2] These stories are intrinsically interesting. More importantly, these stories could not have been written in any other than a statistical form. They are introduced here to show that statistical stories, like any other kind of story, can be provocative and, in a sense, charming.

There is one other reason for presenting the stories we have included here. It has been our experience that students (especially in the social sciences), when asked to identify favorite studies in their field, are often unable to do so. Any knowledgeable person should be able to identify at least five favorite studies. In one test we found that a class of thirty social science students, when asked to identify, by name and title, their favorite social science works, averaged less than 2.6 identifications per individual. Many students could not list a single book. They said they did not have to know books, they only needed to know "great ideas." When asked to list great ideas, they averaged zero. We therefore enjoin people who are embarking on a career in statistical research to have at least ten or twelve favorite studies that they can refer to instantly and with which they are deeply familiar. Here we have time to describe briefly only two of our own favorites.

THE BIRTH OF THE GODS

Our first story involves a group of graduate students at the University of Michigan who, in the late 1950s, examined the problem of the origins of religious belief. The instructor for their graduate seminar, Professor Guy E. Swanson, was interested in the ideas of a brilliant turn-of-the-century social theorist named Emile Durkheim. It was Durkheim's contention that religious ideas are associated with the structural character of the societies in which they historically develop. Durkheim maintained that early or "primitive" forms of religious expression are actually ways of talking about abstract and invisible, though powerful, social forces. He saw religion as a "personification" of the abstract forces of the social order. (The term *Lord,* for example, is used to refer to the creator of the cosmos and also to the lord of the manor. In many societies, including early Western culture, the most powerful leaders were accepted as gods in their own right. Durkheim merely developed this obvious parallel between human social organization and the characterization of religious forces into an elaborately orchestrated interpretation of religious expression.)

Durkheim was not merely saying that the gods are an invention of human beings; he was saying that the way people talk about their gods is influenced by the nature of the communities in which they live. Durkheim presented his arguments in a book entitled *The Elementary Forms of the Religious Life,* one of the most influential statements in Western literature concerning the relationship between social forces and religious practice. It was the arguments in this book that interested Swanson and his students.[3]

In essence, the participants in the seminar began to ask a special kind of question: was it possible to make some kind of observation about societies and religious belief such that an observable relationship could be established between the two? Swanson was aware that a considerable amount of data about so-called primitive societies and cultures had been accumulated over the years by the University and was kept in a large file known as the *Human Relations Area Files.*[4] Any moderately large university or college has such a file, usually in the library, and its records are readily available. This file provides records of various cultures and their kinship systems, forms of religious belief and practice, economic systems, war activities, child care practices, and other pertinent social and cultural information.

Two pieces of information were of special interest to Swanson and the students in the seminar. First, the area files contained information about religious practices. Second, the area files contained information about how the societies included in the file were organized. Given this situation, all that was called for was

1. A careful procedure for categorizing societies by the way they were ordered, organized, or structured.
2. A careful procedure for categorizing the kinds of religious belief each society endorsed.

This is an important point. The statistical work—techniques, calculations, tabulations, and so on—was simple. But the categorization was difficult. It should be emphasized, once again, that good statistical work begins and ends with good categorization. Good categorization is at the heart of good reasoning, whether it is a matter of doing statistical research or any other kind of thoughtful activity.

There is space to discuss only one of Swanson's findings here. Imagine a simple social system in which there are no sub-units. Next, imagine a social system in which there are several sub-units—units that have to be kept together, so to speak. The latter form of society would have more organizational problems than the simpler form. It would require centralized leadership. It is in the second type of society, Swanson speculated, that more clearly defined conceptions of a controlling or regulating god appear. Such conceptions will not develop in simpler societies with no sub-units. At least, this is what we might expect if we have correctly understood the arguments of Durkheim.

It happens to be the case that in some societies there is no well-defined conception of a controlling or regulating god. Other societies, such as our own, have an elaborate notion of a god who is omnipotent and controls everything, and yet, paradoxically (despite this power), has trouble with forces that rose in rebellion against his authority. Why is it, then, that some societies follow what Swanson called a "high god" religion and others believe in a more diffused, vaguely defined force that is sometimes referred to as *mana*? How do we account for this difference? Is it something random? Is it psychological? (If we presume that people are essentially alike in terms of their neuro-physiological characteristics, then we have to presume further that such beliefs are not the product of individual qualities or psychological factors because these conditions would be the same from society to society.) The answer must lie in something other than psychological factors. The most obvious route to pursue is the character of the society itself. It was this line of reasoning that led the group at the University of Michigan to the finding that is the "plot" of our first statistical short story.

The problem of religious complexity was crudely resolved by making a distinction between cultures that believed in mana and those that believed in high gods. The problem of complex and simple social systems was solved by identifying the number of sovereign groups in the system. For example, a society that had only a kinship system (that is, one organized only in terms of family relationships) would have just one sovereign group. A society that had a kinship system, plus tribal organization, would have two sovereign groups. A society that grouped its tribes into a larger organization, or "nation," of tribes would have three sovereign groups.

The issue of categorizing what was meant by complex religious belief and complex social organization was dealt with. Now the researchers were able to code data taken from the human relations files for thirty-nine cultures. Each culture was coded as having a high god religion or one based on the idea of mana. Each culture was also coded as having a social system made up of one, two, or three sovereign groups. Once this task was accomplished, the statistical analysis was a matter of sorting data into tables. Table 3.1 is a slightly modified version of one presented in Swanson's book *The Birth of the Gods*.[5]

Table 3.1. The Observed Relationship between Complexity of Social Structure and the Existence of a High God in the Beliefs of Thirty-Nine "Primitive" Cultures*

High God Concept	*Complexity of Social Structure*	
	Simple	***Complex***
Present	2	17
Absent	17	3

**Data from Guy E. Swanson,* The Birth of the Gods *(Ann Arbor: University of Michigan Press), 65. The above table is a slightly modified version of Swanson's presentation.*

Although the table is simple it is worth considering, because it illustrates a basic form of statistical presentation common to many fields—from physics to the social sciences—that draw on evidence to support an argument. The table involves two categories, each divided into two parts. The result is a fourfold table sometimes referred to as a *two-by-two contingency table.*

Even without extensive training in statistical analysis, we can tell that something is going on here. Out of the 19 cultures that are "simple" in organization, 17 do not have a high god concept. They have simpler religious symbolism ranging from a belief in mana up to, but not including, a belief in high gods. Out of the 20 cultures that were coded "complex" in social structure, 17 had a high god concept. Whether or not a high god notion occurs in the religious beliefs of a people appears to be contingent on whether a society is simple or complicated in structure. Religious ideology, at least in some of its manifestations, is associated with social structure. But the association is not perfect. There are five exceptions, and Swanson attempts to account for them; his accounting is too elaborate to deal with here.

The same data, presented graphically, make the point of the table even more dramatically.

Figure 3.1. A Graphic Representation of Swanson's Findings as Presented in Table 3.1

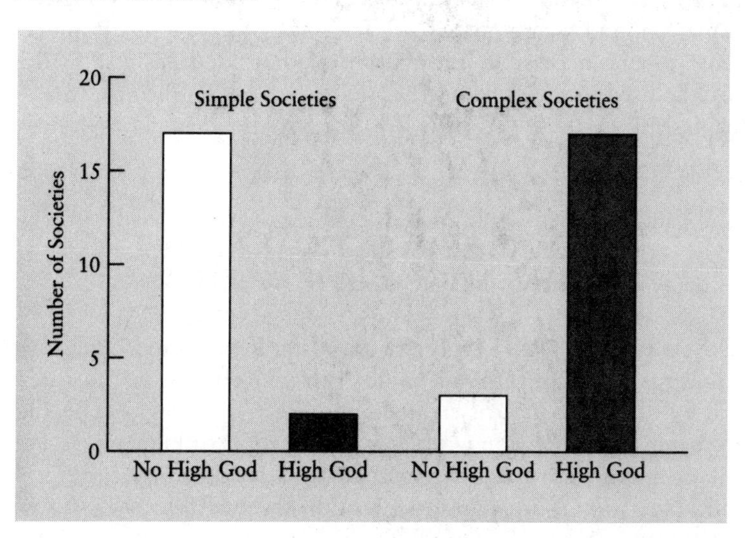

This story illustrates further basic aspects of statistical work. First of all, the researcher made use of readily available information to test an argument that has profound implications. The primary concern was to see if there is some kind of association between two conditions. Such a concern takes a common form: *If X, then Y; if non-X, then non-*Y. Or, in the case of Swanson's data: If the society is complex, then there is a high-god ideology; if the society is not complex, there is no high-god ideology. This finding does not differ from commonsense ways of thinking. The extent to which there is a difference between Swanson's work and ordinary approaches to resolving problems lies not in the form of the logic but in the care shown in thinking about the problem theoretically, the care given to collecting information, the care taken in coding data, and then the care given to tabulating information.

Note, also, that Swanson was aware of the extent to which chance might have brought about some of the differences in the data. The possible influence of chance was tested, and it was determined that the differences were so great that it was extremely improbable that they were produced by accident. Most statistical analysis in the humanities and social sciences is of this variety. It deals with whether or not what we are looking at is nothing more than the product of random fluctuations.

Undoubtedly there are those who think the gods, demons, and angels are strictly the subject matter of poets, philosophers, and theologians. However, the gods are bigger than that. In the hands of an imaginative statistical worker, the subject of the gods is as engaging as it is in literature. Before moving on to our next story, it should be mentioned that Swanson examined other social factors and their relationship to such religious conceptions as witches, a personally concerned god, the idea of a devil, the idea of heaven or life after death, and others. What he found was carefully scrutinized in terms of logic, statistical information, and techniques. Read *The Birth of the Gods* for a nice example of how theory, accessible information or data, and basic statistical techniques can be used to investigate issues that seem, on the surface, to be beyond statistical testing.

AN EARLY STATISTICAL MODEL THAT SHOOK THE WORLD

At the close of the eighteenth century, nation-states were preoccupied with two major concerns. One concern had to do with military defense, the other with wealth. Both of these concerns created an interest in numbers of people. With regard to defense the question was, Do we, the people of England (or France or Germany or Spain), have sufficient numbers of people to defend our boundaries and wage war? With regard to wealth, the question was, Do we have sufficient numbers of people to work the mines and the various growing industries whose wealth augments the power and the glory of the nation? The general sentiment, in response to both questions, was that one could hardly have enough people to deal with such pressing matters. There was, in the late eighteenth century in England and elsewhere, the feeling that if there was a problem regarding numbers of people, the problem was that there were not enough. The belief also developed that the growing strength of the large nation-states was such that it would generate new forms of human progress—that wealth, power, and civilization would move forward to new heights. The future was something to be anticipated with great hope. (These two concerns and this philosophy are still powerful today, and the issues of population, wealth, growth of industry, and the power of nations dominate a great deal of social statistical work.)

This optimism about the times became the subject of a debate between two Englishmen, a father and son, with the son arguing against liberal optimism and the father arguing for it. The son, as a result of his cordial debates with his father, won lasting fame as one of the greatest social thinkers in Western civilization. His name was Thomas Robert Malthus, and the simplicity of his argument, along with the careful attention he gave to its support, not only sustained him in his discussion with his father but eventually led to one of the most powerful critiques of reform liberalism in recent history.

After Malthus published his essays on population, it was no longer as easy to claim that the future was inevitably one of progression toward liberty, equality, and fraternity. Instead, humanity appeared to be locked in a natural and eternal embrace with war, famine, disease, pestilence, and, as Malthus put it in the terms of his day, "vice." The end toward which human destiny was directed could not have a happy ending, as the reform optimists believed, but was naturally inclined toward catastrophe and unhappiness. Where liberals were of the opinion that nature could be tamed and life could be made happy for everyone, the Malthusian argument claimed that nature cannot be tamed, and that the end result of human effort is to return once more to the understanding that nature contains within it much cruelty.[6]

Even if we happen to be familiar with the argument, it is worth looking at again because it illustrates, as well as any work we might call to mind, the tremendously persuasive power of well-developed statistical models. Malthus carried the day in the battle against liberal reformism by putting together a model of how two major forces in the world are related to each other. The term *model* was not used in his day, but what he did was basically the same thing that scientists and statisticians do today when they construct a model. Malthus's model was one that, through an act of intellect (or, if you will, pure fantasy), created a set of expectations. Malthus drew an intellectual "map" that suggested, but did not prove, that all roads lead toward eventual suffering.[7] Let us see how it worked.

Malthus knew that organic populations have a tendency to double over given periods of time. This is true of any organic population, including human populations. (Incidentally, we are beginning to see in the twentieth century that this growth problem appears to be true of mechanical populations as well—as seen, for example, in the growth of automobile populations over the past half century.) The question was, What is a reasonable time for such doubling? Malthus accepted as one possibility the interval of twenty-five years as a time within which it was reasonable to expect human populations to double. Given this arbitrary observation, the model involved little more than *playing* with the idea of what happens as a result of doubling. Still, for late eighteenth- and early nineteenth-century intellectuals, the results of this "playing" were astonishing.

Malthus set his model in motion by considering what might have happened had two people existed at the time of Christ, and their number doubled each generation up to his own time, the close of the eighteenth century (remember that he defined a generation as twenty-five years). The calculation requires more patience than anything else. At the beginning you have two people. Twenty-five years later you have four (2^2). By the end of the twelfth generation, some three hundred years later, there are only 4,096 people (2^{12}). By the end of the twenty-fourth generation, a period of six hundred years, there are almost 17,000,000 people (2^{24}). It takes about seventy-two generations (where each generation is defined as an interval of twenty-five years), from the time of Christ, to reach the era in which Malthus lived.

Malthus's model shows that with doubling taking place in each generation, we would have, at the end of seventy-two generations, 2^{72} or 4,722,366,482,880,000,000,000 people. If the current world population is set at five and a half billion people, then the number of people expected on the basis of this model, in 1800, would have been about one trillion times greater than today's population. The planet earth has a total area of about 197 million square miles. There would have been enough people, in 1798, to place about 24 billion people on every square mile of the surface of the earth. We would have over four times the current world's total population occupying each square mile of the planet, including the surface of the oceans! It would be a crowded planet indeed. This is the model from which we have come to refer to the idea of population "explosions."

Malthus was careful to make the point that such doubling was easily possible in 25-year generations. However, it makes little difference whether the interval is twenty-five years or fifty years or even a hundred or five hundred years. The same kind of doubling process leads to astronomically high numbers within relatively brief periods of time. Even today few people comprehend what a simple doubling model can lead to. For example, business and political leaders, without exception, talk about "growth" economies. However, in time a "growth" economy produces a doubling of wealth, goods, or other economic quantities. Eventually, such doubling leads to an exhaustion of space available to hold the materials produced by the "growth" economy.

Malthus made effective use of this model. Note that he set up conditions that are conservatively reasonable. These conditions lead to specific expectations—fantastic expectations, but fairly well-defined ones. If the conditions of the Malthusian model had been met, we would have expected, in 1800, an unbelievable number of people to have been living on the earth.

What Malthus did back in 1798, when he published his first essay on population, was to confront the intellectual world with the problem of **expected values.**[8] With respect to numbers of people, he pointed out, there was a considerable difference between what one might expect on the basis of simple procreation and what was actually observed. We did not then, nor do we now, have an outrageous number of people occupying every square inch of the planet. In fact, when you look around, you find the planet is surprisingly bare of people. In Malthus's time the population of the world was approximately 700 million people. (The world's population is now estimated to be 5.4 billion people, and this number is expected to increase to over 8 billion by the year 2025.) The difference between what would have been expected from simple doubling and what actually happened is obviously a huge difference. By 1798, according to Malthus's doubling model, there should have been almost five septillion people on the earth; but there were only 700 million. Somewhere along the way an astonishingly large number of people who should have showed up somehow got squeezed out of the picture.

We expect almost five septillion people. A lot of people are expected, but they don't appear. How do you account for the difference? It certainly appears too large to ascribe to chance.

Malthus had a ready answer—a gloomy one. The pressures of population growth are taken care of by natural processes (an argument that, a few years later, made a deep impression on another major Western thinker—Charles Darwin). Basic to these processes are lethal forces such as war, famine, disease, pestilence, and vice, as well as other disasters.[9] The eternal lot of living creatures, from this point of view, is to endure the brutal prunings of nature. There may be moments of peace and prosperity, but they are momentary. Sooner or later, the powerful hand of nature brings about a balance between population and the means available for supporting that population. There is no way out. No matter how hard we work toward some better future, our eventual destiny will be one in which people experience massive misery. Of course, today, we might advocate an avoidance of such a fate through controlling the birth rate and working toward the idea of zero population growth. For reasons that cannot be gone into here, Malthus did not accept this possibility of controlling the birth rate or, as it is commonly referred to today, family planning.

Malthusian thought, whether we agree or disagree with it, has compelling features. It is simple. Malthus did his calculations by hand; computers did not exist in 1798. The factors in the model are well defined: human populations and the means of sustaining such populations. It is a reasonable model of how things might be expected to work. It compares what is expected with what is observable. It relies on data to sustain the argument that optimism with regard to the future is not warranted.

Malthusian thought has implications for how people should deal with major problems of social policy. In recent times, the exploration of simple Malthusian doubling problems by ecologists has led to more complex models that are now a part of what is called *nonlinear analysis,* or the study of "chaos." Finally, the Malthusian model was influential in bringing about another model in another sphere of intellectual effort: the model of evolution. All in all, this was quite an accomplishment for a writer.[10]

A FURTHER NOTE ON THE MALTHUSIAN MODEL: THE COMPOUND INTEREST FORMULA

The Malthusian model is a simple compound interest model. It relies on repeating a formula over and over. Malthus doubled populations every twenty-five years and discovered the result of fantastically large numbers in a relatively short time.

If we wanted to solve the same problem using *annual* growth rates instead of rates for every twenty-five years, we could substitute an annual rate that would produce a doubling every twenty-five years. This turns out to be a growth rate of approximately 2.81 percent per annum (year)—a rate of growth that is considered a moderately healthy growth rate for the economy.

The Malthusian dilemma is a specific example of a set of problems that are solved by an elementary but extremely important formula. It is the formula for compound interest problems. This formula is both practical and playful. One can use it to stimulate the imagination—and that is what both art and science, at least in part, are all about. The **compound interest formula** is simple and clean. It is like a line sketch by Picasso—it reveals much in just a few strokes of the pen. It involves only nine characters, less than are found in many single words.

In terms of practical implications, few formulas have as much significance as this easily grasped model. For those who hate formulas, this set of nine characters should be given some thought. Like any good formula, it contains volumes of possibilities. More importantly, like all good formulas, it can be played with. It is meant to be played with. That's one of the things that makes it worth knowing.

When we "play" with this formula, we should keep in mind that in the real world things do not go on growing indefinitely. *Constraints are always imposed on growth.* This is not a demographic or social law or principle; it is a strictly physical principle. (For one thing, the universe itself is limited in how much growth it is capable of containing, but that carries us beyond our current discussion.)

Let's plug some real values into the compound interest formula to see where they lead us. Here is a place where numbers can boggle the imagination. Most people are quite unaware of the force of growth phenomena and their potentially catastrophic consequences. The compound interest formula tells us what happens if a given rate of growth is allowed to continue over a given number of periods of growth.

$$F = B(1 + r)^n$$

Where: F is the final number.

B is the beginning number.

r is the rate of change.*

n is the number of times you are calculating the occurrence of change.

*Note that a rate of growth of 25 percent per unit of time would be represented, within the parentheses, by the value 0.25.

Here is an example of a compound interest solution. Within the first six months of life a human infant more than doubles in weight. What would happen if a child continued to double its weight every six months until it reached the age of thirty? The growth rate, *r*, in this instance would be 100 percent every six months or, in the formula, 1.0. The beginning number, *B*, would be the child's weight at birth; let's say the child weighs seven pounds. We want to calculate the final weight at age thirty, so *n* is 60 (60 comes from a doubling every six months for 30 years, or 60 doublings). Substitution gives us:

$$F = 7 \times (1 + 1.0)^{60}$$

or 7 times (2 to the 60th power)

This means a weight of 8,070,450,532,260,000,000 pounds (over four quadrillion tons). Our seven-pound baby turned into a fair-sized whopper.

More seriously, consider the production of motor vehicles on a global level. In 1970, about 29 million vehicles were produced throughout the world. In 1986, about 44 million vehicles were produced.[11] The growth rate in vehicles for this sixteen-year period was equal to

$$\frac{(44\text{mil} - 29\text{mil})}{29\text{mil}} \times 100 = \frac{15\text{mil}}{29\text{mil}} \times 100 = \text{about 50 percent}$$

With a growth rate of 50 percent for sixteen years, let's set the annual growth in production of vehicles at 2 percent. How many motor vehicles will be in production, if a 2 percent per annum growth continues unabated, by the year 2100? Starting in 1970, the year 2100 is 130 years in the future so our period, *n*, is 130. The rate is 2 percent, or .02. The beginning value is 29 million. Our formula gives us:

$$F = 29 \text{ mil} \times (1 + .02)^{130}$$

$$= 381 \text{ million vehicles in production annually in 2100}$$

Within little more than a century, if this trend were to continue without interruption, we would be producing eight times the number of vehicles we are now producing. With the Baltic nations, China, India, and a great number of other nations seeking consumer parity with the United States, it is difficult to predict growth rates for automobile productivity on a global scale. Here we are simply playing a game of "what if." As we will see later, a great deal of fundamental statistical analysis is, in fact, a matter of "what if." Such exercises at least sensitize us to possibilities inherent in the idea of unlimited, continuous, and unabated industrial productivity and growth. Can the earth tolerate eight times the motor vehicular load it is now trying to sustain? The above figures pertain to production per year, but the number of vehicles on the road is actually far greater. From 1960 to 1990 the total number of vehicles registered in the world went from 127 million to 583 million.[12]

Malthus talked about doubling every twenty-five years, but this is a much greater rate of increase. Obviously it cannot go on without eventually coming to a halt. Here we are interested in showing how playing with formulas and figures can make one think a bit more seriously about the future consequences of growth. Will the pressures to increase motor vehicular productivity prevail over those that might seek to restrict it? Such questions cannot be readily answered, but playing with compound interest formulas can make us see their importance.[13]

If the economy were to grow, unabated, at a rate of 2.8 percent per annum, a rate that most industrial and economic leaders consider moderately "healthy," we would have a doubling of whatever we might mean by the "American economy" every twenty-five years. Well within the span of five centuries, at this rate, the world would be saddled with over one million current "American economy" equivalents comparable to the size and consumptive power of the present United States. (Keep in mind that we are not considering other huge global economies also dedicated to growth.) At this point in time, that looks like more than the planet might be able to deal with. The compound interest formula is easy to play with. Always remember, though, that we are merely playing with a model. However, such play can be instructive, as the following use of the formula and the graph in Figure 3.2 illustrate.

Number of American economies 500 years from now
$= 1 \text{ Am. Econ.} \times (1 + 0.0281)^{500}$
$= 1{,}041{,}549$ American economies

Figure 3.2. The Form of the Growth Curve If the Current American Economy Were to Grow at 2.81 Percent per Annum for the Next Five Hundred Years

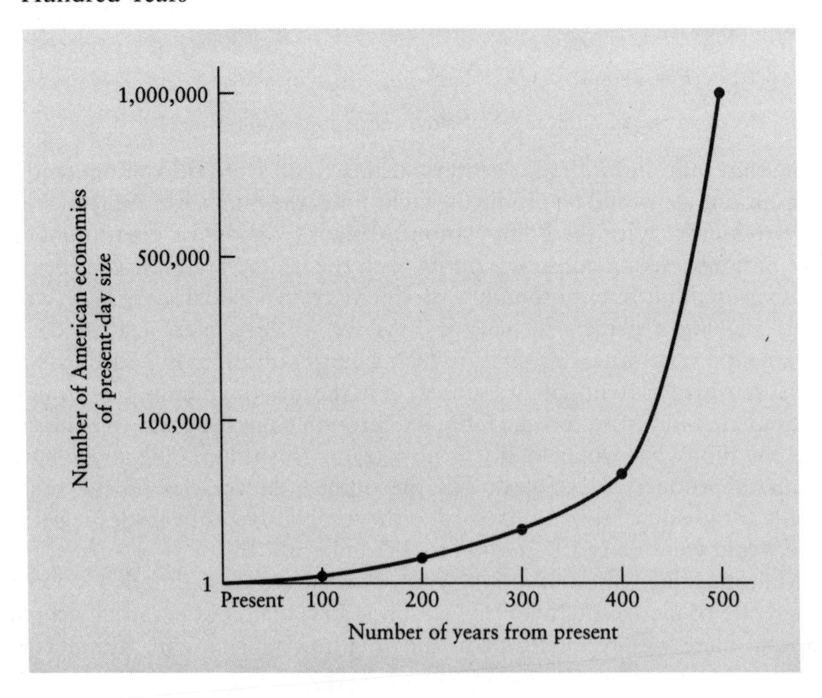

The point is not to argue that we are going to have a million economies the size of the United States's economy crushing the planet in five hundred years. Rather, the point is to show that numbers have an argumentative power of their own. They lead us somewhere even when we are playing around with them. Numbers are meant to be played with; they are powerful intellectual toys. In the example above, numbers suggest it is madness to view growth as pure progress or progress as pure growth. We must think seriously about what we mean when we talk about promoting growth. So far, there is little evidence that anyone, aside from Malthus and a few other intellectuals, has seriously bothered with the implications of even simple compound interest formulas. What will our future be if we do not control growth? On the other hand, what are the consequences of bringing to a halt that system of economic growth that has led to the present prosperity of leading industrial nations? Neither of these questions has an easy answer.[14]

SUMMARY

We have suggested that statisticians do not openly concern themselves to any great extent with aesthetics or gracefulness in their writing. Yet well-done statistical work, like work in any other discipline, can be engaging not only because it informs us of something but also because it is well-crafted.

In this chapter, two statistical reports provide examples of engaging and imaginative uses of the statistical imagination. The first study, by Professor Guy E. Swanson and his graduate students, demonstrates that statistics can be used to look at the gods—a subject that many would believe to be beyond statistical examination. The second study goes back to 1798 to show that a well-crafted statistical argument can have earth-shaking consequences. For two hundred years the writings of Thomas Malthus have remained a classical demonstration of the power of statistical reasoning to provoke thought and controversy.

Finally, we emphasize that good statistical work is simple and clear. Reliance on simple procedures at the beginning of the examination of a problem almost invariably leads into more complicated developments.

THINKING THINGS THROUGH

1. Theology and statistics appear to be miles apart in their concerns. However, if one accepts Swanson's findings, what are the theological implications? Do such studies undermine religion? Is there a "natural antagonism" between statistics, which tends to be suspicious of belief or faith, and religion, which relies on faith?
2. What do you think will be the eventual fate of nations and industrial systems that require constant growth to survive?

3. Write the compound interest formula from memory and work out a fantasy growth problem relative to some interest you might have.
4. In the past, stories developed out of events or happenings that captured people's interest. Identify a recent story that has an underlying statistical origin. For example, the tremendous power and influence of the story of the Holocaust comes, in part, out of the numbers that were involved in that chapter in twentieth-century history. Recently, the Timorese have been killed in large numbers. Why isn't this horror story being publicized? What does this suggest with respect to what gets statistically reported and what doesn't? Can social statistics and politics be separated?
5. In what ways are story telling and statistical studies different? In what ways are they similar? Why is it that stories that exaggerate and twist reality sometimes give us a better sense of the "truth" than documentaries that attempt to be realistic in every detail? Is it necessary, in order to tell a story (whether literary, folk, or statistical), to twist things? Consider some of the ways in which you are forced to distort what really happened when you describe an event or an action. What are the implications for statistical "story telling"?
6. Using the logic of exponential growth patterns and the arguments of Malthus, as well as data obtained from official sources such as *The Statistical Abstract of the United States,* prepare your own argument for or against the proposition that humankind will solve the problem of environmental contamination and pollution.
7. Get data on AIDS. Is the incidence of this illness growing or leveling off? Is this problem solvable? Why? Why not? Is AIDS a serious world problem, or is it a threat we do not need to take seriously? Use some numbers to make your case.
8. Check the figures in this chapter. We suggest this because an earlier version of this manuscript, read by hundreds of students, contained some highly inflated errors. No one bothered to check because the numbers were, we suspect, intimidating. Don't let figures in a book deceive you. Read them. Think about them. Check them where they begin to push at your imagination. Everybody makes mistakes, including your professors and the writers of books. A good statistician is always checking figures. While you cannot check everything, you should check what is important to you.

Notes

1. Bertrand Russell, *The Study of Mathematics* (Cambridge: Cambridge University Press, 1903.)
2. It more and more appears to be the case that writers cannot be classified easily as purely statistical or nonstatistical in nature. Most journalism of today, for example, mixes popular statistical information with more general story telling. There are, of course, purists who try to remain completely within a given form, statistical or narrative. However, it is our impression that such writers are increasingly becoming a minority. The point here is that if you are going to be a modern writer, either

academic or popular, you are going to have to have some familiarity with statistical procedures.

3. Guy E. Swanson, *The Birth of the Gods* (Ann Arbor: University of Michigan Press, 1964).
4. The *Human Relations Area Files* can be found in most large university libraries. They are a summarization of the ethnographic research done by anthropologists around the world, coded by categories that are of common interest in social research such as kinship systems, economic systems, political or organizational systems, and so forth. These files have also been put into computerized data bases and can be accessed through home computers.
5. Guy E. Swanson, *The Birth of the Gods* (Ann Arbor: University of Michigan Press, 1964), 65.
6. Modern times have turned this around. Now conservatives are pro-growth and liberals are pessimistic about unrestricted growth. The issue is so touchy that a teacher in Colorado was threatened with the loss of her job for teaching anti-growth philosophies. To endorse anti-growth philosophies in the United States today comes close to being subversive.
7. Thomas Robert Malthus, *An Essay on Population* (New York: Dutton, 1958). There is an extensive library on the writings of Malthus. The original essay was published in 1798. Malthus was severely criticized for not anticipating the extent to which modern technology, especially agricultural and medical, would temper the effects of population growth. However, in just the past few years researchers are becoming alarmed over the possibility of new forms of plague that threaten to devastate the world's population. At the present time, AIDS is a much more serious plague than many people are willing to concede.
8. The idea of an expected value is commonly relied on in statistical analysis. Malthus's essay provides as nice an example of the difference between expected and observed values as any we know of in the entire realm of social statistics.
9. Malthus did not predict the precise moment of disaster, nor did he really define disaster in a way that people could unquestioningly accept. Once again we are confronted with categorical problems. While millions starve on a global level, there are also millions enjoying luxurious and indulgent lives. Is Malthus right or wrong? Is disaster a problem that can be dealt with statistically?
10. Modern mathematicians have found much deeper patterns in growth phenomena than Malthus suspected. For a light treatment of how complicated growth patterns can be, see James Gleick, *Chaos* (New York: Penguin, 1987). This book is about the astonishment of physicists when they found that all natural processes have a probabilistic element in them. It is worthwhile reading for anyone going into the social sciences, because nowhere is there a larger element of probabilistic (error) factors in any descriptive effort than in the social sciences.
11. *The Statistical Abstract of the United States,* 1990, Table 1435.
12. *Ibid.,* Table 1453.
13. Several reviewers of this book argued that the compound interest formula is not really a "statistical" formula and does not belong in a statistics book. All we can say is that we vehemently disagree. The compound interest formula is more significant, in terms of its implications for social and economic policy, than any other formula we can think of. The general ignorance of the implications of this formula is possibly more widespread today than it was in 1798 when Malthus managed, for a brief period, to popularize the necessity of including it in any discussion of what might be meant by "progress."

14. See Charles C. Mann, "How Many Is Too Many?" *The Atlantic Monthly*, February 1993, 47–67. This is a nicely balanced article on the dispute now raging between those who think we do not have enough people on the planet and those who think we are in danger of the wretched anticipations of Thomas Malthus. It is also a nice example of a popular writer's reliance on statistical information to inform the general public—providing further proof that statistics is no longer the singular domain of academic purists.

4

FUNDAMENTALS: THE ART OF TABULAR DESIGN

"I drink to the general joy of the whole table."

—SHAKESPEARE

"Write the vision, and make it plain upon tables, that he may run that readeth it."

—HABAKKUK 2:2

The bane of people who have a distaste for statistical literature is the statistical table—row upon row of figures referring to this, that, and the other thing. Statistical tables appear dry, cold, and bloodless. However, like anything else, the dryness or coldness of a table is a relative matter that depends on the interests of the individual looking at it. Sports fans can be found poring over complicated tables with an enthusiasm that only comes from a deep affection for statistical information. Racing fans carefully review lists of numbers that detail the records of horses, jockeys, and other matters relevant to betting. Business people read stock listings with a devotion that can make a Nobel laureate in physics look like an intellectual laggard. College students show a deep interest in elaborate arrays of data that pertain to the evaluations and ratings of teachers and courses, grade lists, and the like.

The point is that tables are not, in and of themselves, something people shy away from. Some tables are boring; some are not. The important thing to keep in mind is that it is not the form of the table that determines interest level. What determines our eagerness is how the *content* of the table relates to our interests. As our interests broaden, it should follow that we find ourselves interested in greater varieties of tabular presentations.

The quantitative table is the heavy workhorse of statistical literature.[1] Most research begins and ends with tabular presentations of data and information. It is essential, then, to have a good grasp of what a table is, how one goes about putting material into tabular form, and, finally, how one goes about interpreting tables. This chapter is concerned with the design of tables. A brief look at their interpretation concludes this chapter's discussion.

TABULAR DESIGN

In designing any table it is important to keep in mind that its primary task is to make life easier for the reader. The **statistical table** is to statistical data or information pretty much what a poem is to literary information or what the sketch is to graphic information: its primary task is to summarize large amounts of material in a brief space. Many people think of tables as complicated forms. On some occasions this can be the case. For the most part, however, tables are simple in structure. Like writing a poem or conveying an artistic impression in just a few strokes of the pencil or pen, saying a lot in a little space is not always easy.[2]

What constitutes a good table? One rule to go by is that a well-designed table should contain *all* relevant information necessary for interpreting its contents. Another way of stating this rule is that if you put together a table and leave it lying somewhere, and someone picks it up, that person (assuming he or she possesses a modest degree of literacy) should be able to comprehend fully what the table is about. A good table should stand by itself.

Tables, though simple, take time and can be difficult to construct. Nonetheless, it is not much more difficult to do the job well than to do it poorly. The following example illustrates a situation in which just a bit of added effort would have cleared the muddy waters.

Table 4.1 is a mock version of a table that appeared in a major journal. It has been modified slightly to serve the purpose of illustration. The first thing we see in this table is that back-up information is required. By itself, the table is ambiguous. What is meant by "perceptive limits"? Where was the sample of

Table 4.1 Perceptive Limits by Understanding of Advertisements

Understanding of Advertisements	*Cognitive Ability*	
	Lower (%)	*Higher (%)*
Low	61	42
Medium	39	58
Total Percent	100	100
Total Number	190	250
TOTAL	440	

440 people obtained? Was the study done in the United States? In Japan? In Tibet? When was it done? What kinds of advertisements are being referred to? Television? Radio? What does "understanding an advertisement" mean? How low does cognitive ability have to be for a person not to comprehend an advertisement at all? Does the sample refer to children or adults? The table does not give us any clues; it is basically meaningless and requires an accompanying text for clarification. With slight modifications, this table could be much more self-sufficient.[3]

SIMPLICITY VERSUS DETAIL

There are two opposing ideals that interfere with the design of a good table. The first ideal, the very *raison d'être* of the table, is the attainment of simplicity and compactness. The attainment of this ideal, however, comes at the cost of another ideal, namely the need to provide as much information as possible. Unfortunately, the more detail we pack into a table, the more complicated it becomes. If we seek simplicity, we must give up information. In this respect, the table is no different from other forms of communication—poems, letters, articles, stories, maps, or anything else that attempts to convey information. Simplicity and complexity are polar opposites; they cannot be achieved simultaneously. Yet this is the quest of good communications, whether in art or in science: to combine in a single expression or form the opposing ideals of simplicity and fullness of information.

There are several features of a good table. Keep in mind that no hard and fast rules exist for putting together information in the form of a statistical table—just as there are no hard and fast rules for writing a good poem. Different people, working with the same data, will organize those data differently. However, if they do their work properly, the tables they construct will offer basically the same understandings of the data.

The first feature of a good table is a proper title. This cannot be overemphasized. The title should provide clear information about the table's contents. At the same time, it must be brief. Like the writer of the Japanese poetry form known as haiku, the writer of a table has a limited space in which to make clear the nature of the information being offered. A good table title is certainly not as difficult as good poetry, but it has its expressive problems. Within just three or four short lines the title should state the following:

1. To *whom* do the data refer?
2. *Where* were the data collected?
3. *When* were the data collected?
4. *What* kind of information is in the table? (Absolute frequencies? Relative frequencies? Correlation coefficients? Some other statistical index or coefficient?)

5. From what *source* were the data obtained? (For example, were they obtained from a secondary source? Were they obtained in an informal manner? Is the data source official?)
6. What *categories* are involved in the table? (Does the table simply present the frequency of incidents of driving while intoxicated, for example? Or is it an investigation of the relationship between the incidence of child abuse and socioeconomic status?) In sum, what categories are involved? What is the argument of the table?

Table 4.2 (page 73) is moderately elaborate and complex. It is also quite simple. You could pick up this table, look it over, and find a number of things to think about if you happened to be interested in the topic the table addresses (teacher evaluations). The basic point we want to make is that this table is self-contained. Its title is clear. Its information is straightforward. An individual who wanted to carry out further statistical operations on these data would have an opportunity to do so. For example, it appears to be the case that when students rank an instructor high in one area, they also rank that same instructor high in most other areas. The relationship between exams and grading ratings and instructor ratings is a positive one, as this smaller table illustrates:

		Exams and Grading Rating		
		A	*B*	C
Instructor Rating	*A*	11	4	0
	B	4	11	1
	C	0	4	0

We will not dignify this little working illustration of the relationship between instructor rating and examination rating by identifying it as a formal table. We leave that to you as a small assignment. (If you were putting together a title for this smaller table, for example, what kind of title would you write for it?) The basic point we are trying to make here is that Table 4.2 is well enough designed for one to do a lot with it. Good research includes basic data tables of this kind. It is a way of laying your statistical cards on the table. The kind of table that is represented by Table 4.1 hides its cards behind a facade. There is little or nothing to be done with the data provided there.

A second feature of a good table is that it divides its data in a way that does not overwhelm the reader. At the same time, however, it should provide as much detail as possible. As a general rule, this usually means not having more than ten or twelve rows and ten or twelve columns in a table. (*Columns* are vertical arrays of data. *Rows* are horizontal.) A table with twelve rows and twelve columns reaches the point of being cumbersome for most readers. After all, such a table will have 144 **tabular cells,** or entries.

Like all other rules in statistics, writing, art, or science, this rule of limited rows and columns can be violated and a good table can still be the result. In Table 4.2, data are presented in thirty-seven rows and nine columns. There are

Table 4.2. Student Evaluations of Teachers and Courses Offered in the School of Education, University of Colorado, Boulder, Fall Semester, 1982. (A+ = highest rating and F = lowest.)*

Instructor	*Course No.*	*No. Respondents/ Class*	*Course Rating and Effectiveness*	*Assignment Effectiveness*	*Exams and Grading*	*Communication*	*Student Contact*	*Overall Instructor Rating*
Openshaw	500	9/12	B+ & B+	A	B+	B+	B+	B+
Haas	500	12/15	B+ & A−	A	A−	B+	B+	A
Hopkins	503	27/33	C+ & B−	B	B+	B−	B−	B−
Anderson	503	20/25	B & B+	B+	B+	B+	B	B+
Smith	504	12/15	B & B+	A	B+	B	B	B
Openshaw	506	6/7	B & B	A−	B+	B−	A−	A−
Ratliff	508	5/5	B+ & A+	A	A+	A+	A	A−
McKean	509	17/19	C & C	C+	B	C+	B	C
Hodge	511	11/18	B− & B	A−	B	B−	B	B
Keenan	512	9/13	C & C+	A	B−	B	B	C+
Westerberg	516	6/7	A & A	A	A	B+	A+	A
Carline	525	15/16	C+ & B−	B+	B	B	B	C+
Olson	526	10/13	B− & B−	B	A	B	B+	B−
Distefano	527	11/11	A & A+	A+	A+	A	A	A+
Swadener	537	4/11	D+ & D	C−	C+	C−	C	D+
Olson	542	2/5	B & B+	A−	A	B+	B+	A−
Kraft	544	6/8	A & A	A−	B+	A	B+	A
Swadener	549	11/24	B & B	B+	B+	B	B	B+
Sanders	553	8/13	B & B	B+	B−	B	B+	B+
Sanders	553	16/27	B & B	B	B	B	B+	B
Roark	556	10/26	C+ & B−	B	B−	C+	B	C+
Turner	556	11/12	B+ & A−	A	A−	A−	B+	A
Haas	562	21/30	B & B+	B+	B+	B+	A−	A−
Ramos	563	5/6	D & D	C−	NA	C−	C	D+
Meyers	565	15/15	B & B	B+	C	B	B	B
Hemenway	566	19/25	A− & A	A −	B+	A−	B+	A−
Wubben	567	6/6	A+ & A+	A	A+	A	A	A+
Wubben	567	28/34	A & A	A	A	B+	A	A
Rose	569	11/11	B+ & A−	A	A	B+	A−	B+
Swadener	595	16/24	B+ & B+	B	B	B−	B+	B+
Smith	603	22/24	B & B+	A	B+	B	B	B
Hodge	605	13/15	A & A	A+	A	A−	A+	A
Hemenway	612	22/26	B & B+	A −	A−	B+	A−	B+
Mokean	618	15/23	C & C+	B+	B+	B	B	C+
Hannafin	625	10/17	B+ & B	A−	B+	B−	B+	B+
Hannafin	627	1/1	A+ & A+	A+	A+	A+	A+	A+
Kraft	629	10/10	B & B+	B	A−	B	B+	B

**Source:* Faculty Course Evaluation, *Ombudsman Office, Univ. of Colorado Student Union, Spring, 1983.*

333 cell entries of one kind or another. Yet, the table is interesting and not especially difficult to browse through. At least it is safe to say that faculty evaluation tables such as this one have considerable popularity on university campuses, where students shop around for courses that have lively or popular instructors (that is to say, instructors with good course evaluations).

Perhaps tables are frustrating to many people because they feel that they have to understand all the information presented in them. This is not so. A table permits, indeed it asks for, a kind of "browsing" as the reader looks for information. In Table 4.2, for example, there is no need to become completely familiar with all the information contained in the rows and columns. It is interesting, though, to see who got the highest and the lowest grades. (It is also amusing to notice that the only teacher who got straight A+'s across the board was also the only teacher rated by a single student.)

A third feature of a good table is proper spacing of rows and columns. When data are packed together, confusion can result. Table 4.2 violates this rule slightly. The columns had to be squeezed together to fit everything onto the page. It would have been better either to have had a bigger page or a smaller table.

A fourth feature of a good table is that each cell should provide the reader with some kind of information—including the statement that information was not available for some cells, if necessary. (This is commonly done by using the letters *NA*, for "Not Available," in cells where information is lacking.)

A fifth feature of ordinary statistical tables is that percentages should be displayed in as clear a manner as possible. For instance, in the following array of data it is difficult to determine, by inspection, what the percentages refer to:

Grades	*Men*	*Women*	*Percent*
A	33	42	17
B	78	80	33
C	105	96	39
D	43	22	9
F	12	4	2

In the next array of data, the meaning of the percentages is clarified by the use of marginal totals and subclassification. Here, the meaning of the percentages is not ambiguous; it is clearly stated. Though this set of numbers looks more complicated than the one above, it is actually easier to read and interpret.

Grades	*Men*		*Women*		*Total*	
	Number	*Percent*	*Number*	*Percent*	*Number*	*Percent*
A	33	12	42	17	75	15
B	78	29	80	33	158	31
C	105	39	96	39	201	39
D	43	16	22	9	65	13
F	12	4	4	2	16	3
TOTAL	271	100	244	100	515	101*

*Total does not equal 100 due to rounding.

A sixth feature of a good table is that, unless a **value** is precisely zero, it should not be represented by a zero. For example, if you find that a particular percentage is equal to 0.0003 percent, this should be listed as 0.0+ percent. The plus sign indicates that the zero value deviates from true zero by a small amount that could not be specified.

Finally, it is wise not to complicate a table by carrying figures out to absurd decimal places that exceed the accuracy of your measurements. There is much pressure on researchers today to make their work appear precise, which is reasonable. However, in the quest to appear precise it is easy to go overboard and make things look considerably more precise than they are. It is not uncommon to find social scientists recording correlation coefficients to four digits, such as $r = 0.7346$. The correlation is probably accurate, at best, to only one digit, which would mean reporting it simply as $r = +0.7$. It is, however, traditional to report correlation coefficients to at least two digits. This is done even when the data do not warrant such precision.[4]

In sum, the basic rule of table construction is to keep things as simple as possible while conveying as much correct or valid information as you can. How one balances the demands of simplicity and complexity is a matter of individual choice. There are no hard and fast rules.

A good table can get by with very little or no additional textual explanation. If you need a lot of text to make your point, there is not much purpose in having the table—you might as well use text alone. The *Statistical Abstract of the United States,* updated annually, consists at present of about 1,500 simple frequency tables.[5] It is an astonishingly informative collection of tables and an excellent source of examples of tabular design. No other single statistical source book contains so much information within such a small space. At the same time, it never uses complex techniques. All of its data are presented in the form of tables with little or no accompanying text. Each table is capable of telling a great deal by itself. We cannot think of a better source to turn to for models of tabular design.

THE FEATURES OF A GOOD TABLE

Table 4.3 (page 76) is an example of a well-constructed table. It is worth examining for the features that make it effective.

The title of Table 4.3 informs us that the population referred to is the civilian noninstitutional population. The information in the table also tells us that the data came from a sample. Also mentioned is how to find out more about the sample, should the reader want such information, and the fact that the data for 1985 and later are not strictly comparable with earlier data.

The meaning of *injury* is explained in a note below the title, in which we are told that injuries that did not restrict activity or lead to medical attention were excluded from the survey. The 1987 injuries are further classified by *place of occurrence.* This is made clear in the bottom four rows of the table, which

Table 4.3. Persons Injured in the United States, by Sex, 1970 to 1987, and by Age and Place, 1987*

Covers civilian noninstitutional population and comprises incidents leading to restricted activity and/or medical attention. Beginning 1984, the levels of estimates may not be comparable to estimates for 1970–1980 because the later data are based on a revised questionnaire and field procedures; for further information see source. Based on National Health Interview Survey.

	Persons Injured (000,000s)			*Rates per 100 Population*		
Year	*Total*	*Male*	*Female*	*Total*	*Male*	*Female*
1970	56.0	31.8	24.2	28.0	33.0	23.3
1975	71.9	39.4	32.5	34.4	39.1	30.0
1980	68.1	39.0	29.1	31.2	37.1	25.8
1981	70.3	40.1	30.2	31.2	36.9	25.9
1982	60.0	32.4	27.6	26.4	29.6	23.4
1983	61.1	33.0	28.1	26.6	29.8	23.7
1984	61.1	27.4	26.4	26.4	30.1	22.9
1985	62.6	34.6	26.0	26.8	30.6	23.1
1986	62.4	34.0	28.4	26.4	29.8	23.3
1987 total[1]	62.1	33.6	28.4	26.0	29.1	23.1
Under 5 years	4.8	2.7	2.1	26.1	26.9	23.2
5–17 years	14.4	8.4	6.0	31.9	36.4	27.3
18–44 years	29.0	16.7	12.2	28.3	33.5	23.5
45–64 years	8.0	4.0	4.0	17.9	18.7	17.2
65 years and over	6.0	1.9	4.1	21.1	15.9	24.8
Home	21.0	11.0	9.9	8.8	9.6	8.1
Street or highway	7.7	4.2	3.5	3.2	3.6	2.9
Industrial	7.4	5.5	1.9	3.1	4.7	1.5
Other	15.8	8.7	7.1	6.6	7.6	5.8

1. Includes unknown place of accident, not shown separately.

**The above table was obtained from Table 183 of* The United States Fact Book, The Statistical Abstract of the United States, *1994. (Any student working in the social sciences should have a personal copy of this amazingly broad and detailed source of information. It is published each year by the Government Printing Office and costs only a few dollars.)*

refer to injuries that happened in the home, on a street or highway, in an industrial setting, or in some other locale. Note the care shown in informing the reader that injuries where the place of occurrence was not known are included in the total for 1987 but, otherwise, not shown separately.

The table presents both **absolute** and **relative** values. Observe that absolute values are reported in millions, which simplifies things. Relative values are shown as rates per 100 population; this is the same as percentages. Relative values are made explicit in this table because the term *relative values* can refer to values per 1,000 population, 10,000 population, 100,000 population or even a population of 1,000,000. Most relative values use a base of 100 or 1,000, however.

Before you continue reading, take a close look at Table 4.3 and browse through it. What does it have to say? It conveys a tremendous amount of information in a very small space, telling us about injuries to people living in the United States through a period from 1970 to 1987. The first thing we might notice is that a lot of people get injured during any given year. In 1975, for example, your chances were about one in three of being injured seriously enough to be incapacitated or require medical care. In 1986, the chances went down to about one in four. What might account for this decline in the ratio of injured people? Table 4.3 also provides us with information concerning the age and the sex of the injured person. From these data, we can see immediately that men tend to have higher injury rates. However, as the table indicates, this difference declines as people get older. At advanced ages, women have a higher rate of injuries.

Note that the table makes the source of its data clear. The character of the table—that is, the nature of its subject—is also made immediately clear in the title; it is a simple presentation of the frequency of injuries. The total population of the United States is involved. If the data referred only to the adult population, then this would have been indicated in the title through a modification such as the following:

> Persons Injured, by Sex, among the Adult Population of the United States (aged eighteen or older) by Sex, 1970–87, and by Place of Occurrence.

What is most important about this table is that it stands by itself and provides volumes of information in less than a page. A detailed textual interpretation of everything this table tells us either explicitly or implicitly would take at least fifty pages of text, and possibly more.

Although census data are usually more concerned with detail than with simplicity, the tables that appear in the *Statistical Abstract of the United States,* because they are abbreviated versions of more complicated tables, have to be especially concerned with simplification. Table 4.3 is actually three simple tables combined into one. The first, which presents an absolute count of injuries, involves nine rows and three columns. This creates only 27 cells for the reader to review. The second table, which presents the information in the first table in relative terms, also contains only 27 cells. The information in this table is easily reviewed by any interested reader. The third table presents detailed data for 1986. It consists of ten rows and six columns.

With regard to design, the columns are nicely separated by vertical lines. Major separations are indicated by heavier lines. Note that the rows are separated at one point, further facilitating the reading of the table.

All cells provide information. Had information been lacking for, let us say, the males in 1975, the appropriate cell would have been filled in with *NA* (Not Available) or some other notation. Furthermore, a footnote at the bottom of the table would likely have been included to inform the reader as to why this information was lacking.

The column heading for the injury rates expressed as relative values makes it clear that each rate refers to the number per 100 population who suffered injuries. In 1970, for instance, 33 out of 100 males suffered an injury, while

23.3 females out of 100 had a similar experience. (Note, again, that in using relative figures such as these, we have "standardized" or "controlled" the size of the groups from which the information was obtained.)

We should also note that this table does not get bogged down in unnecessary precision. The rates are reported to one decimal place—a common tradition in statistical reporting of this kind. The absolute values are reported only to the nearest hundred thousand. Instead of the number of injuries for both sexes in 1975 being reported as 71,936,741 or some such number, it is reported simply as 71.9. The column heading informs us that this refers to millions, or to approximately 71,900,000 injuries.

Finally, we should develop an appreciation for the considerable amount of information this table offers. In this table we are informed about the differences between rates of injuries among men and women, about short-term trends in injuries, and about three key locations in which injuries often occur. We are also given an age breakdown in accident rates. Furthermore, there is information here that permits us to examine the interplay between these factors (for example, possible trends in the differences between men's and women's accident rates).

As is the case with other forms of writing, developing and presenting good tables is a skill that is gained through practice and through noticing, and *appreciating,* the work of those who are skilled in this art. It is not, like some of the other arts, a difficult skill to acquire. Nonetheless, it is a skill that is rather rare, as the large number of poorly designed tables that we see indicates.

PAIN, POETRY, AND COLD NUMBERS

Statistical presentations, when given more than a casual consideration, can evoke emotions. They should evoke emotions. Literary and statistical presentations differ in form, but they deal with much the same substance when involved with the human scene. The fact that injuries are being reported in tables does not make the pain any less than when it is reported in a story or poem. For the person with imagination and sensitivity, a construct such as Table 4.3 should be almost overwhelming in the amount of pain and trouble it conveys.

There are innumerable ways to talk about pain and accidents. We can try to describe such things through poetry, drama, or painting. We can also describe them by means of statistical tables. The data in Table 4.3 are as poignant as any account we might get from lines of poetry. This table tells us that the possibility of a male being injured, in some incapacitating way, in any given year, is about one out of three. For women it is about one out of four. The absolute numbers give us something to think about—more than 60 million people injured in the United States in one year. That means a lot of hurting, loss of work, demands on medical assistance, and grief for those related to the injured. In the domain of human affairs, numbers are always more than "just numbers."

Statistical tables are neither cold nor warm, feeling or unfeeling, square or hip. They are simply another way—a concise way—of presenting information.

Whether they are poignant and filled with meaning is, to a great extent, a function of the degree to which the person reading the table is sensitive and imaginative or, to the contrary, insensitive and unimaginative. Some tables should make us cry or go into a rage, just as a sad poem might. Others should make us shake our heads in amusement or admiration as we see, in the abbreviated form that tables offer, further evidence of the folly or the greatness of humankind. The difference between poetic emotionality and statistical emotionality is that poetry, according to our stereotypes, is supposed to evoke emotion and statistics are not. Statistics are commonly supposed to be objective and, therefore, beyond emotionality. This does not mean, however, that reading statistical tables should deprive us of our sentiments, be they admiration or outrage, as we gradually begin to see what really underlies the figures arrayed before us.

HOW TO READ A TABLE

Table 4.4 (page 80) is excerpted from *The Statistical Abstracts*.[6] It is moderately complicated, but certainly not "dry." We shall mention only a few suggestions for reading tables, using Table 4.4 as an example, before bringing this discussion to a close.

1. *Don't try to read the whole table at once;* there's too much information. Trying to read a table the way you would read a page of printed matter will produce frustration and an urge to put the table aside in favor of something more "interesting." Moreover, don't get tied up in how the table is defining things. Later, as you get into more serious research, you have to do this. For the moment, however, just casually glance at the table and *look for items that come closest to things that interest you.* If you are interested in skiing, you can quickly see, in Table 4.4, that downhill skiing was twice as popular as cross-country skiing, and that people who live in households with large incomes are four times more likely to indulge in downhill skiing than those with small incomes.
2. *Look for extreme values.* Swimming, camping, and walking (as an exercise or sport activity) are the three most popular activities mentioned in Table 4.4. Cross-country skiing and soccer are the least popular. If you are a soccer fan, you might be interested in seeing that softball is only twice as popular as soccer in high-income households, while it is about three times more popular in lower-income households.
3. *Draw on commonsense knowledge, where possible, to check the credibility of the table.* For example, we would expect golf to be more popular in higher-income households, and the table shows that it is, increasing regularly in popularity from a low of 4.8 percent in low-income households to over 18 percent in high-income households. If the table had shown something different, either the table would be in error or our stereotype of the golfer would be in error. In any event, the credibility of the table might be challenged.

Table 4.4. Participation in Sports Activities by Household Income (in dollars), for Persons Seven Years of Age or More, United States, 1986

Activity	*Household Income*				
(% participating in)	*Under 15,000*	*15,000 to 24,999*	*25,000 to 34,999*	*35,000 to 49,999*	*Over 50,000*
Aerobics	8.1	10.1	10.5	12.8	15.0
Backpacking	3.4	5.0	4.0	4.8	3.9
Baseball	4.4	5.4	6.9	7.9	6.9
Basketball	7.9	9.1	11.6	13.4	12.0
Bicycling	19.1	24.4	25.5	27.9	28.8
Bowling	12.8	16.3	20.7	20.5	18.5
Calisthenics	5.0	6.0	6.0	7.1	7.7
Camping	26.4	29.2	28.3	29.0	31.4
Football	4.7	5.5	5.6	6.9	6.1
Golf	4.8	8.1	9.7	12.5	18.5
Hunting/guns	6.6	9.5	9.0	8.4	6.3
Racquetball	2.9	3.0	4.2	5.1	6.6
Running/Jogging	8.0	9.7	10.9	12.4	12.5
Skiing—downhill	2.8	3.1	5.3	5.7	12.5
Skiing—cross-country	1.3	1.8	3.5	2.8	4.2
Soccer	2.3	2.7	4.9	4.7	5.8
Softball	7.0	8.4	10.3	12.4	10.1
Swimming	24.1	30.4	35.4	36.8	39.5
Target shooting	4.0	5.4	6.0	5.0	3.1
Tennis	5.3	5.8	8.3	8.9	12.3
Volleyball	8.0	9.3	11.2	12.4	10.2
Walking	26.4	29.2	28.3	29.0	31.4

Note: This table was slightly amended to facilitate its use in this context. The original table appears in the Statistical Abstract for the United States, *1990, Table 392.*

4. *Look for information that contradicts popular knowledge.* For example, target shooting as a sport tends to be more popular in middle-income households. However, there is not much of a difference across income levels. This violates the stereotypical notion that people who enjoy guns and target shooting tend to be lower-class, "redneck" types. Guns and target shooting involve a modest percentage of people participating in sports, and they appear across all income levels. (Look at the category of bowling for a similar statistical refutation of popular ideas; bowling tends to be more popular in higher-income households.)
5. *Gain a sense of what the relative values are saying.* About five percent of all the households, on the average, engaged in target shooting—that is, five percent of the American population over the age of seven. That is five percent of more than 200 million people, or 10 million target shooters. Notice that relative values tend to "diminish" the size of things while absolute values commonly give an impression of greatness. Five percent, for example, does not sound like much, but 10 million is a big number. Be careful in reading tables, because propagandists are aware of this aspect of relative and absolute values and frequently use it to manipulate the sentiments of naive readers.

6. *Examine the table long enough to gain an overview of some of its basic findings.* For example, we can quickly see in Table 4.4 that higher-income households participate in nearly all sports in greater numbers than people at the lowest income levels. (Perhaps this was, in part, what Thorstein Veblen meant when he wrote about the "leisure class."[7]) It is also worth noting that while there are differences, they are relatively small; the lowest-income households still participate in all recreational activities at relatively high rates.
7. *If you become more seriously interested in the data, begin to check its figures against tables from other sources and tables that deal with the same subject but provide different comparative categories* (such as age, sex of respondent, education, region or state of residence, and so forth).
8. *Is each case counted once or more than once?* In this instance, a single individual could participate in five or six different recreational activities and be included in each recreational category. This might mean that some of the differences found between household income categories could result from the simple fact that people in higher-income households have greater opportunity to participate in a variety of recreational activities.
9. *If you have a choice, it is better to remain conservative in your speculations over why some differences exist.* For example, upper-income households are three times more likely to play golf than hunt with guns. However, this does not necessarily mean that higher-income people are more inclined to indulge in "gentle" sports. The proportion that does hunt is just about as high as that for the lower-income group. Possibly all that is going on here is that higher-income people have greater opportunity to golf *and* to hunt, while lower-income people must choose between one or the other. Don't place too much interpretive weight on statistics such as these.
10. *Be willing to spend a bit of time simply browsing around in the figures and letting them suggest things to you.* Keep in mind that the table is only suggesting possibilities. Use it as a stimulus to your imagination. Note, also, how the table contradicts images that come from the mass media. The most popular recreational activities are swimming, camping, and walking. How often do you see people engaged in recreational walking depicted in printed advertisements or on television commercials?

A final suggestion: Don't read too much precision into these figures. They are approximations, which means they are useful, but authoritative only in the sense that they are the best figures available on the subject. Always retain a critical attitude.

SUMMARY

Although many people consider statistical tables the ultimate example of boring literature, they also commonly become deeply interested in certain tables. This suggests that whether or not a table is boring depends on the extent to which an individual is interested in the contents of the table.

A part of statistical technique is to know how to design a good table. Tabular design is relatively simple, though commonly ignored. Two primary concerns are involved in designing a good table. The first is to give the table a complete though concise heading. The second is to provide as much information as possible within the confines of a limited number of tabular cells, usually no more than a hundred or so.

Tables are designed to simplify large masses of information. As a result, statistical tables are confronted with the same problem that occurs in all communications: how do you simplify complex information without distorting it to the point where it becomes meaningless?

An example of a good table is Table 4.3 (page 76), which was drawn from the *Statistical Abstract of the United States*—an excellent source for models of good tabular design. A good table is defined as one that needs little additional commentary or interpretation to make its point. The tables in the *Statistical Abstract of the United States,* for example, are offered with no additional textual commentary. They stand by themselves.

There is a common misconception that numbers are cold or unfeeling. This misconception overlooks the fact that it is not the *numbers* that are cold but rather the *reader* of the numbers. A sensitive reader of tabular material will see within the numbers in a table as much cause for dismay or elation as he or she might find in any other source of information, including poetry, drama, or art.

Finally, this chapter offers suggestions, through an example, on how to read a table. The important thing is not to try to read a table as a whole but to approach it in terms of particular items of interest. If necessary, one can then move toward an examination of the larger aspects of the information contained in the table.

Thinking Things Through

1. Write a brief paragraph on the question of how a statistical table is a kind of "story." In what ways is a table similar to or different from what we generally think of as a story? (We want you to give a lot of thought to the question of the differences and similarities that exist between statistical and literary modes of communication, as well as to the question of how each is superior or inferior in describing human conditions.)
2. What constitutes a good table?
3. We know that stories (as a literary form) distort reality in order to get the story told. Discuss the ways in which both stories and statistical tables are forced to make compromises in order to deal with the problem of communicating complex events in a simplified form.
4. We've been told that 150,000 people have been killed in the Bosnian-Serbian conflict. Why is this horrific bit of information less emotion-provoking for many people than reading a narrative account such as *The Diary of Anne Frank*? After all, in the *Diary*, only a few people die. In the

statistical item we are told of the deaths of 150,000 people. How would you go about trying to get people to see that numbers can be as poignant in their meaning as the personal stories that people might have to tell?

5. Two men are given a fine of $50 for parking. One makes $100 a week; the other $5,000 a week. What is the difference between absolute and relative punitiveness in this instance? Show the difference in terms of specific numerical values. What are the implications for an ethics of justice? Does statistical reasoning have a place in evaluating ethical issues?
6. The Pacific Rim contains one billion potential consumers eagerly anticipating elevated standards of living and a lifestyle similar to that enjoyed by Americans. Economic growth in the Pacific Rim is extremely high at present, reaching levels of 10 percent per annum. You should have the compound interest formula memorized. Using it, let's say that the typical Pacific Rim consumer now has $500 per annum to spend. How long will it take before the average income of these one billion people is $50,000 per annum if the 10 percent per annum rate remains consistent?
7. Look up a relatively complex table dealing with material of interest to you that appears either in a book, journal, or newspaper article. Evaluate how well it conveys information. What are its strengths? Its weaknesses?

Notes

1. Although basic frequency distributions are perhaps the most informative source of statistical information for social scientists and those interested in social-statistical information, the trend in advanced journals has been to move away from simple frequency distribution tables, such as those shown in this chapter, toward complex statistical indexes. If so, this is a nice example of putting the cart before the horse. A clear presentation of frequency distributions can generally permit a seriously interested reader to generate complex indexes of his or her preference or check on the indexes used by the author of the article. However, if no frequency distributions are presented, only complex indexes, one can never be certain that the indexes are not distorted by eccentric aspects of frequency distributions. It is argued in this book that all good statistical reporting should include a clear statement of the fundamental data from which more complex indexes are derived.
2. The eighteenth-century philosopher-satirist Voltaire is supposed to have written a friend saying that he did not have time to write a short letter, so he was going to have to write a long one. Voltaire was aware of how difficult it is to be brief. (This anecdote has also been attributed to the French mathematician Blaise Pascal.)
3. The mock table shown in Table 4.1 is based on a table taken from a major journal. Because the table is being used as a poor example, we are disinclined to provide a specific reference. We would like to add that major journals have a tendency not to publish frequency distributions, preferring more intricate statistical procedures such as path analysis, complex multi-variate regression techniques, or LISREL. It is strictly a personal opinion, but we find basic frequency distributions more informative than more elaborated procedures. Certainly there is a place for sophisticated technique, but it should not be used merely for its own sake. It should be used only when it can provide more information, with greater certainty, than can be provided

by simpler devices. It is for this reason that we have devoted an entire chapter to the significance and the character of basic tabular presentations of data.

4. When doing computations, it does not hurt to use extended decimal values or figures that seem more precise than they are because such figures can minimize rounding errors across long summations or other arithmetic operations. When reporting results, however, such extended decimals can give a false impression of precision.
5. *The Statistical Abstract of the United States 1990: The National Data Book,* (Washington, D.C.: United States Government Printing Office) presents data compiled by the U.S. Department of Commerce, Bureau of the Census. The Abstract is updated each year. Any student in the social sciences who does not possess a relatively recent copy of this book as a basic reference source is displaying a serious lack of interest in his or her work. While there are numerous other statistical data sources, we like this one because it is inexpensive, comprehensive, and relevant to many modern issues. It also exemplifies how statistical tables, when well designed, are highly informative by themselves.
6. *The Statistical Abstract of the United States, 1990: The National Data Book* (Washington, D.C.: United States Government Printing Office), Table 392.
7. Thorstein Veblen, *The Theory of the Leisure Class* (New York: Macmillan, 1968). Veblen was born in 1857 and died in 1929; his writings provide an interesting contrast to modern economic writing. His works are also a kind of benchmark illustrating how extremely influential the statistical style has become. Despite his casual approach to quantitative data, Veblen is still worth reading. Few people in the social sciences today are capable of penetrating the follies of our time in the way that Veblen did.

5

ASSESSING THE FAMILIAR: AVERAGES

"I feel like a fugitive from th' law of averages."

—WILLIAM MAULDIN

"Ivan Ilych's life had been most simple and most ordinary and therefore most terrible."

—TOLSTOY

Day in, day out, we are caught up in the question of whether thousands of different things and happenings are "typical" or "average" or "ordinary" or whatever other term we might want to use to refer to what statisticians call measures of **central tendency.** We are equally interested in whether they are not typical. Averages are profoundly woven into our reactions to the world around us. Even the simplest, most ordinary statements are based on implicit averages:

"Gosh, you don't seem to be yourself lately." (You are not acting in a typical or average manner.)

"Siwash University dedicates itself to the pursuit of excellence." (Siwash will not settle for being average. In this respect colleges and universities are like the children in Garrison Keillor's Lake Wobegon—they are all above average.)

"Hey man, you're okay, I'm okay." (We're both typical.)

"Don't rock the boat." (Don't move things out of the usual or ordinary or typical rut.)

The assessment of what is or is not typical is fundamental to life. We cannot talk without relying constantly on a sense of what is average, typical, or ordinary. So profound is our reliance on typification that it might even have a genetic basis, insofar as our survival depends on such assessments. The average is not an idea cooked up by statisticians—it is grounded in a deeper intellectual matrix. We cannot overstate the significance of typification in human communications. As we shall see, it is also a primary concern in all statistical work.[1]

AVERAGES AND TYPICALITY

The average is useful because it is essential to our adaptations to the world. When we wrongly evaluate typicality, we experience problems in adjusting to the circumstances we have to deal with. The following passage is an example of a case in which a false sense of the typical might have created unwarranted anxiety and fear among young men. It is taken from the work of a British physician, William Acton, who published a work on the functions and disorders of the reproductive organs. The book appeared in 1857. In this passage Acton describes what supposedly happens to boys who habitually masturbate. (Later statistical studies revealed, of course, that masturbation is typical.[2])

> The frame is stunted and weak, the muscles undeveloped, the eye is sunken and heavy, the complexion is sallow, pasty, or covered with spots of acne, the hands are damp and cold, and the skin moist. The boy shuns the society of others, creeps about alone, joins with repugnance in the amusements of his school fellows. He cannot look anyone in the face, and becomes careless in dress and uncleanly in person. His intellect has become sluggish and enfeebled, and if his evil habits are persisted in, he may end in becoming a driveling idiot or a peevish valetudinarian. Such boys are to be seen in all stages of degeneration, but what we have described is but the result towards which they all are tending.[3]

This is a fine illustration of a false typification that surely caused a lot of people unnecessary grief. People are constantly concerned with the assessment of averages. Such concern is not something reserved for the statistical laboratory. If we make incorrect judgments, we can find ourselves in greater difficulty than when we make correct ones. There is no great surprise in any of this, but we need to be reminded of it occasionally.

Furthermore, the way we assess averages in day-to-day living is largely the same as the way the statistician does the job in the statistical laboratory.[4] Keep in mind that *the idea of the average is extremely basic to statistical work*. All statistical indexes, even the most complex, almost invariably rest on the determination of averages.

Averages are simple. Even so, they can be trickier than they seem. The fact that we have an array of numbers packed into a computer does not mean we can produce a meaningful average. The computer spits out answers when we ask it to make a calculation, but the things these calculations tell us are rarely obvious. This fact is demonstrated in the following passage, which argues that even though we have a huge mass of quantitative global climate data we cannot say for certain that there is such a thing as a "typical" global climate (despite the fact that people and scientific meteorologists commonly talk about global climates—and believe they know what they are talking about when they do so):

> *Does a climate exist?* That is, does the earth's weather have a long-term average? Most meteorologists, then as now, took the answer for granted. Surely any measurable behavior, no matter how it fluctuates, must have an average. *Yet, on reflection, it is far from obvious. . . .* [T]he average weather for the last 12,000 years has been notably different than the average for the previous 12,000, when most of North America was covered by ice. Was there one climate that changed to another for some physical reason? Or is there an even longer-term climate within which those periods were just fluctuations? *Or is it possible that a system like the weather may never converge to an average?* (italics added).[5]

It is also worth noting that the physical sciences struggle to establish various **constants** that are basic to science itself. One of the best-known constants in science is the speed of light. Constants are unique averages in that the individual expressions of the phenomena never vary from each other. The speed of light, in space, is *always* the same; moreover, it is always *precisely* the same.

As we shall see later in the chapters on relationship, discovering constant values provides a unique degree of control over events. The physical sciences are successful in part because they have established important constant values. To the best of our knowledge, however, the social sciences have yet to uncover a social constant.[6] **Variability** is the rule in social affairs. As we saw in the passage on planetary climate above, averages derived from variable conditions are not easy to assess. Much of what constitutes statistical technique and formal statistical analysis comes out of this observation. It is probably not an exaggeration to say that most statistical technique is, in one way or another, dedicated to establishing stable and reliable average values of some kind. The importance of averages in statistical work simply cannot be overemphasized.

Some questions worth pondering are (1) Why are people so interested in averages?, (2) Why do they put so much effort into evaluating averages?, and (3) Why are they so concerned about average values that if they do not know what such averages are, they go ahead and pretend to know what constitutes the typical case (as exemplified in the passage from the work of Dr. William Acton above)? It is one thing to know *how* to do statistics, such as computing averages. It is another thing to know *why* we want to do such things in the first place. But the why is as important as the how, because once you begin to see why, the question of how becomes easier to address.

It is also worth keeping in mind that what one average is doing, in terms of providing meaning, might be quite different from what another average is doing. Means, modes, and medians tell us different things. Also, a given arithmetic mean might be quite powerful as a statement of conditions among some set of events—if there is little variation within the set. It will be much weaker, however, if there is a great deal of variation. This aspect of typicality will be discussed when we come to the topic of variability or variation.

MEASURES OF CENTRAL TENDENCY

Three major forms of the typical are commonly used in social research (and in common discourse). Measures of what is typical or average are commonly referred to as measures of *central tendency* by statisticians. These measures are used in such diverse activities as assessing the average height of six-week-old corn plants in a Nebraska farm area, determining accurate measurements of mean distances to the stars, finding typical population growth rates for countries that rely on modern birth control methods, or establishing per capita income figures for developing nations.

Distinctions are often made among statistics for teachers, statistics for social scientists, statistics for business people, statistics for biologists, and so on. While the interests of these different groups vary, however, the basic forms of statistical reasoning are the same for any and all fields. There are only a few fundamental forms used in statistical reasoning in any field.[7] The content to which those forms are applied can be almost anything that comes to mind. Statistical imagination rests not on mastery of the forms but on mastery of the applications of those forms to whatever content is of interest to us.

The three basic types of central tendency are the *mode,* the *median,* and the *arithmetic mean.* These three measures of central tendency have different strengths and weaknesses. Moreover, in situations confronting the social scientist, they have the unsettling tendency to provide different values for the average. This raises the question of which measure is best in a given situation. What is typical when your measures of typicality disagree?

The Mode

The easiest way to establish typicality is by observing the most of whatever it is you are interested in, or the mode.

As elementary and straightforward as the mode is, and as commonly as it is used, it can still create confusion. Not all situations are easily typified. Here is a case that illustrates a circumstance in which it is impossible to draw on any particular measure of central tendency to describe what is typical.

Figure 5.1 is an idealized curve displaying the number of days throughout the year that are extremely sunny, moderately sunny, and very cloudy. For many regions of the United States, this diagram is a valid representation of what the weather is like over the year. Note that the curve has two distinct modes—one for very sunny days and one for very cloudy days. In other words, if we use the mode as our measure, a typical day is either a sunny day or a

Figure 5.1. An Idealized Curve Representing the Number of Cloudy and Sunny Days in the Midwestern United States

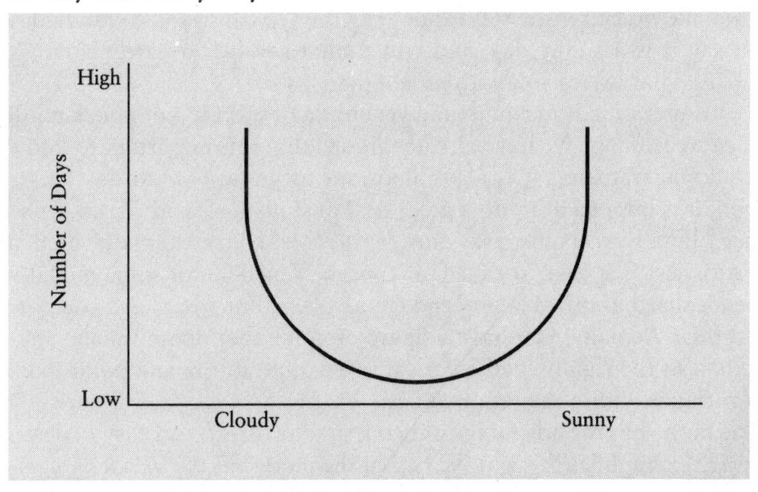

cloudy one: the typical case has two forms, each the opposite of the other. This would be like saying that the typical American is either poor or rich, tall or short, or employed or unemployed.

We should also notice that if we use one of the other measures of central tendency—the median or the mean—we would have to conclude that the typical day is in the center; that is, we would have to conclude that the typical day is one of partly cloudy weather. This does not make much sense because partly cloudy days are, overall, a distinct minority. It would be like saying that the typical American is a college professor, a jogger, or a vegetarian. You cannot typify an aggregate by referring to its minority cases. Figure 5.1 should be kept in mind when you are working with complex data and issues. A major point of this book is the fact that averages are trickier than they appear. Statistical indexes should not be calculated thoughtlessly. Even elemental measures of typicality can be deceiving.

Getting numerical answers to statistical problems is the simplest part of statistical work. The real difficulty lies in interpreting our answers after we calculate them. There is a tendency to create meanings for statistical indexes that go beyond what the indexes indicate. In this tendency, statistics is much like other communicative devices.[8] The lesson learned from Figure 5.1 should not be forgotten. It clearly demonstrates a situation in which we can calculate the mode, the mean, or the median—but that does not mean we have found the holy grail of typicality.

Distributions such as that between sunny days and cloudy days illustrated in Figure 5.1 are referred to as **bimodal distributions.** Whenever one has a distribution with several distinct modes, determining what is "typical" of the distribution becomes, as we have seen, problematic. Bimodal distributions warn you that you might have trouble settling on a single value as representative of the average for that distribution. The reason for this is that no matter where you move along the scale of sunny to cloudy days you get answers that do not make much sense. You cannot say the typical day is a cloudy day, you cannot say it is a sunny day, and you cannot say it is a partly cloudy/partly sunny day. That leaves you with no alternatives.

This illustration is meant to add yet another aspect to our understanding of statistical reasoning. We have already argued that it is important to know *why* we are doing statistics. It is also important to know *how* to do statistics. In addition, it is important to do a great deal of statistical work, or to get a lot of practice.[9] Finally, you must also consider *what* you have after you've calculated your statistics. It is easy to calculate a mean, a median, or some modal values for the idealized graph of sunny and cloudy days. But what have you got when you do this? Actually, you have a figure or valuc that doesn't make any sense other than in the rigidly mathematical sense of establishing a point that minimizes variance within the set of values.

The mode has the advantage of being simple, direct, and easy to determine. Figure 5.1 exemplifies this simplicity. But the mode has the *disadvantage,* when the number of events is small, of fluctuating wildly as a result of small changes in the data. For example, imagine that the fictional figures in Table 5.1 refer to salaries paid to the faculty of a social science division at a small college. It is apparent that the modal value for this group is $3,000. This is the amount most of the faculty receive. Say the teacher's union decides to use these data to embarrass the dean of the college. They publish a newspaper article with the headline, "Elmtown College Social Science Faculty Averages $3,000 per Year." The dean immediately retaliates by giving $1,000 raises to Professors Brown and Green. (What does this do to the mode?) Now the dean is able to say that although the average salary was once a mere $3,000, it is now a whopping $25,000. With an investment of only $2,000, the dean has brought about an apparent typical modal increase for the whole division of $22,000.

The mode is a deceptive measure when the number of cases is small. In such instances, it should be used only to gain a quick impression of what

Table 5.1. Annual Salaries, prior to Tax Deductions, for Eighteen Faculty Members in the Social Science Department at Elmtown College, 1983. (Data are fictional.)

Professor	*Salary*	*Professor*	*Salary*
Jones	$25,000	Diaz	10,000
Smith	25,000	Diehl	10,000
Doaks	25,000	Lear	10,000
Brown	24,000	Evans	5,000
Green	24,000	Keith	5,000
Adams	16,000	Lambert	3,000
Allen	16,000	Murphy	3,000
Berry	15,000	Patterson	3,000
Benvenuti	15,000	Schwartz	3,000

constitutes the typical case. When the number of cases is large and we do not need to worry about bimodality—that is to say, when there is a definite, single modal value—then the mode is a more effective way of gaining a sense of typical values.

The Median

The mode, as we have seen, answers the question of what is typical by suggesting it is whatever there is the most of in a set of specific observations. The median answers the same question by suggesting that the typical is whatever is *in the middle* of observations that have been ordered from the lowest to the highest value. It is the center value if there are an odd number of cases and the average of the two middle cases if there are an even number. Here is an example. In the array 1, 2, 3, 4, 5, 6, 7 the number 4 is in the middle and is the median. In the array 1, 2, 3, 4, 5, 6, 7, 8 the numbers 4 and 5 are in the middle, and the median is 4.5.

Because describing in detail how to find the median is surprisingly complicated for such a simple concept, and because we do not want to spend too much time on this topic, we shall rely on an elementary and artificial example. Consider the following distribution of college students by scores on a statistics test. The question is: What is the median score? (This illustration is based on fictive data.)

Score on Test	*No. of Students*
90–100	3
80–89	11
70–79	29
60–69	6
50–59	4
40–49	1
30–39	0
20–29	0
10–19	0
00–09	0

There are 54 students in the class. We want to know the score that would divide the class into two equal parts with 27 students getting that score or higher and 27 getting that score or lower.

If we count down from the top, we find that we have reached 27 students by the time we are in the 70–79 interval. We have 14 students when we reach the beginning of this interval. We must go into the 70–79 interval for 13 more students. There are a total of 29 students in this interval, so we want to move thirteen twenty-ninths of the way into the interval. That is 45 percent of the interval. The **true limits** of the interval (that is, the limits that determine the interval within which a specific observation will be placed) extend from 69.5 to 79.5, or a total of 10 units. We therefore move .45 times 10, or 4.5, units down from 79.5, to reach the middle point. The median proves to be equal to 75. This is the score that divides the class into two halves. One half of the students have scores equal to or higher than 75, and the other half have scores equal to or lower than 75.

The median is so basic that it appears in toddler's tales; consider, for example, the mother bear's chair and bed in the story of Goldilocks and the Three Bears. The median is typical because it is one way of expressing what stands in the middle of extremes—and extremes, by definition, cannot constitute what is typical.

The median is relatively uninfluenced by extremes and is, therefore, commonly used when distributions are extremely skewed. To see how the median is uninfluenced by extremes, take the illustrative case of statistics test scores above. Move the last 11 cases all the way down to the 0–9 interval and then recalculate the median. Why does its value remain the same as before, even though you have seriously altered the distribution?

The Arithmetic Mean

The mode is the most. The median is the middle. The arithmetic mean is not as simple as these two measures of the typical, but it is sufficiently simple to enable us to say that we all have had experience with it and have known how to determine it long before we learned the decimal system or basic arithmetic. Most of us have had the experience of getting together with other kids and sharing our marbles, chewing gum sticks, baseball cards, or whatever else we might have wanted to share. The sharing resulted in a mean value. The mean is what you get *if* things are evenly shared. Since, in the social sciences, things usually are not evenly shared, the mean is a pretend number, or a convenient fiction. (This is important to keep in mind, because people sometimes apply mean values to individuals in stereotypical ways—often with unpleasant consequences for those involved.)

Suppose the following group of children come together for a game of marbles:

Child	*No. of Marbles Brought by Each Child*
Carlos	36
Juanita	15
Edwin	18
Maria	41
Charles	25

Let us imagine that the children put all the marbles into a large can and start handing them around, one at a time, until they have used up all the marbles. After a while each child has twenty-seven marbles and the can is empty. The arithmetic mean is nothing more or less than parceling out some quality or thing so that each case being dealt with gets an equal share. The average number of marbles possessed by the children before sharing was twenty-seven, or the same number they have after sharing.

Note that, after sharing, each child possesses a number of marbles equal to the mean. Before the sharing, this was not the case. The mean tells you what the situation for each event under consideration would be, with regard to the distribution of a quality, if all events had the same value. This property of the mean will pop up again in discussions of probabilities (which are usually means), where we are frequently frustrated by the fact that even though the probability of a given event might be high, the event, nonetheless, might not occur. In our little example of the children and their marbles, the mean tells us that we should expect a particular kid to have twenty-seven marbles. However, not one of the kids has twenty-seven marbles. The mean (as well as probabilities and a lot of other statistical indexes) tells us what we might expect *if* things were sorted out according to some scheme. Statistical values are one thing—reality is something else.

As elementary and as important as the preceding discussion is, it is still something that a lot of people do not understand or really want to understand. It is often hard for students to accept the fact that the mean, at least for most social and psychological data, is not real; rather, it tells you what things would be like *if* everybody shared some quality. The point is that people do not share intelligence, or income, or height, or any of a number of other characteristics for which averages are determined on a daily basis. The average should not be seen as something real or tangible or existing in the real world. It is an abstraction. It resides in a place called "What if?". If all Americans equally shared the same amount of intelligence, we would all have IQ's of 100. If all Americans equally shared the annual income of the United States, we would each—every man, woman, and child—have had about $21,000 to spend in 1993.[10] The mean is simply the amount that every case would receive if things were shared equally.

Researchers recently reported that older single males are typically schizoid. Having read this comment you might, when you encounter an older single male, be inclined to wonder a bit about his schizoidal inclinations. He might have them; he might not. The average condition refers to *groups,* not to individuals. Seeing averages as something "real," and mistakenly applying them to individuals rather than groups, can create a great deal of confusion and, quite often, a lot of unnecessary misery.

It is important to see that averages are not, in fact, "real." Rather, they are intellectual constructions.[11] Although a measure of typicality such as the mean is not real, it is nonetheless useful. In fact, assessing averages is absolutely essential in our lives. However, almost without exception, they are not real.[12] Most social averages are not real in the sense that they simply do not exist. They are made-up values. People do not all equally share the same IQ. They do not all equally share the same height. They do not all equally share the same incomes. Yet the average tells us what people would have if they did share these things.

Why is it important to keep in mind the fact that averages are not real? The reason is that people often deal with the average as though it were, in fact, real. Suppose you happen to know that men make more money than women, on the average. This is true. If, however, you encounter two people, a man and a woman, you can only assume—not go beyond assuming—that the man makes more money than the woman. It just might be the case that she makes a great deal more money than he does. So powerful is the desire to see averages as realities, though, that we tend to "dump" them on individuals whether the individual fits the average or not.[13]

The fact that averages are not real allows us to say some things that at first do not seem to make much sense and yet *do,* finally, make a great deal of sense. For example, suppose twelve people own four airplanes. The average ownership is one-third of an airplane. One-third of an airplane is not going to fly very far and means nothing in terms of the real world. Yet it does make sense to say that if you were to divide the planes equally among the owners, each would get one-third of a plane. We mention this example because of the fact that people are occasionally either irritated or amused by seemingly peculiar statements from statistical observers of the human scene, such as the following:

- The average size of the American household in 1950 was 3.37, and in 1993 it was 2.84.
- The suicide rate in 1992 in America was 29.8 persons per 100,000.
- In 1992 there were 4.8 divorces per 1,000 population. (How, these amused or irritated people ask, can you have 0.37ths or 0.84ths of a person, or 0.8ths of a suicide or a divorce?)[14]

All such figures mean is that *if* you were to divide the total number of people in households by the total number of households, you would get 2.89 people. (The same is true for the numbers of suicides and divorces.) Such a figure is a fiction, and you should always keep in mind that it is a fiction; but it is also a *useful* fiction. Much is made of average values in economic, social, political, and other forms of policy analysis. Much is also made of averages in ordinary, day-to-day affairs. This basic treatment of the mean as a fiction or a construction can lead into considerations of even more elaborate mental constructions that have no real existence. *Probability models* (which will be discussed further in Chapters 8 and 9) are a wonderful case in point.

The arithmetic mean is the most commonly used measure of averages in statistical work. Unless there is a good reason to use some other measure, the arithmetic mean is generally preferred. One reason for this is that it minimizes probable error when we attempt to estimate what is typical of some set of observations. Probability and averages go together like ham and eggs. After all, we constantly assess the typicality of people and things in the world around us in order to minimize the errors we might make in our adjustments and responses to that world. Here is a case in point. The data in Table 5.2 (page 95) refer to the grades received by a fictional class in phrenology in the spring of 1994 at a state university in the Midwest.

These data are commonplace yet also interesting. The question is whether we should trust the mode, the median, or the mean. We can see immediately that B's prevail. A more careful consideration, however, requires that we take into account other ways of dealing with what is typical for the class with regard to grades. The mode tells us that the average grade is a B, which is worth 3 points. The median tells us that the two central grades are a B and a C:

1 1 2 2 3 3 3 4

The median point value, then, is 2.5.

Table 5.2. Grades Received by Students in Advanced Phrenology (Phrenology 407), Department of Phrenology, Midwestern University, 1993. (Data are fictional.)

Student	*Grade*	*Points*
Alice	B	3
Bill	D	1
Charlene	B	3
Don	C	2
Edward	B	3
Francine	C	2
Gary	A	4
Helen	D	1

(Where A = 4; B = 3; C = 2; D = 1, and F = 0.)

The arithmetic mean tells us the average is 2.375. Between the mode and the mean we have a difference of 0.625 points. One estimate (the mode) is a solid B; another estimate (the arithmetic mean) is more in the C range. Which measure of typicality should you trust?

The question is not so easy to answer as we might like to think. One way to deal with this problem is to ask ourselves how much we would be in error if we used the mode to guess what each student got as his or her grade. This degree of error can be easily determined by simply using the mode to estimate each student's grade and then seeing how right or wrong we are.

It is important to get the logic of this in mind right away. The average leads us to *expect* some value. In statistics, this value is, naturally enough, called an *expected value.* We have, as well, the actual value, referred to as the *observed value.* This does not deviate in the least from what we do all the time. Before we take a look at what happens when we compare what we would expect on the basis of the mode with what actually took place in the phrenology class grades in Table 5.2, let us consider a few common examples of the expected and the observed in everyday life:

- "Everybody told me that the movie at the Palace is great. I went to see it. Believe me, it's rotten."
- "Before we got married, Harold, you said you would find work and settle down. I really didn't believe you. But you have certainly lived up to your promise."
- "Your scholastic aptitude scores show quite limited abilities. The fact that you have attained Phi Beta Kappa membership is probably because you are an overachiever."

In each of these comments there is an expectation and then an observed, or actual, performance that either deviates from or conforms to the expectation. Expectations that are violated usually cause problems. Calling the Phi Beta Kappa scholar an overachiever is an attempt to retain the integrity of the expectation by inserting a kind of "correction" into the observation. Thus the expectation or belief can be sustained while the evidence or experience is, if not ignored, at least reduced in worth.

Now, using Table 5.3, we can come back to considering an average as an expectation and how that expectation compares with reality. First we will look at the mode as an expected value. Then we will see what happens when we use the median to predict grades. Finally, we will consider the arithmetic mean. The mean, median, and mode are expected values, and the actual grade received is the observed value. We are interested in how much difference, *on the average,* we get using each measure of typicality. (Already we can begin to see that statistics is a matter of averages within averages within averages.)

Computations for Average Error, Using the Mode as a Predictor, Based on Data in Table 5.2.

Actual Grade	*Predicted Grade: Based on the Mode*	*Amount of Error in Prediction*
(Observed)	***(Expected)***	***(Observed minus Expected)***
1	3	−2
1	3	−2
2	3	−1
2	3	−1
3	3	0
3	3	0
3	3	0
4	3	+1
	SUM OF ERRORS =	−5

If we do the same thing for the median, we get the following results:

Computations for Average Error, Using the Median as a Predictor, Based on Data in Table 5.2.

Actual Grade	*Predicted Grade: Based on the Median*	*Amount of Error in Prediction*
(Observed)	***(Expected)***	***(Observed minus Expected)***
1	2.5	−1.5
1	2.5	−1.5
2	2.5	−0.5
2	2.5	−0.5
3	2.5	+0.5
3	2.5	+0.5
3	2.5	+0.5
4	2.5	+1.5
	SUM OF ERRORS =	−1.0

This exercise informs us that the median would give us a sum of errors equal to −1.0, and the mode −5.0. If we are interested in using some kind of typical value to reduce our average error, then we see that we get different sums of error with different estimates of typicality. The mode generates more error than the median.

Now we shall try the mean and see how well it predicts grades.

Computations for Average Error, Using the Mean as a Predictor, Based on Data in Table 5.2.

Actual Grade	*Predicted Grade: Based on the Mean*	*Amount of Error in Prediction*
(Observed)	***(Expected)***	***(Observed minus Expected)***
1	2.375	−1.375
1	2.375	−1.375
2	2.375	−0.375
2	2.375	−0.375
3	2.375	+0.625
3	2.375	+0.625
3	2.375	+0.625
4	2.375	+1.625
	SUM OF ERRORS =	0.000

What's this? The sum of errors for the mean is equal to zero? This suggests that the mean has correctly predicted each student's grade. But obviously this is not so—the mean did not predict even one student's grade with perfect accuracy. Yet the sum of errors is equal to zero. This peculiar result arises from the fact that, with the mean, positive errors are canceled out by negative errors. We need to get rid of the pluses and minuses if we are going to evaluate the predictive power of the several measures of the typical or average case. One way of doing this is to simply ignore the signs—to pretend they do not exist, as it were, and sum only positive values. This is the sum of absolute errors. When we do this, the three measures come out, comparatively, as follows:

Sum of Absolute Error Values		
Mode	***Median***	***Mean***
7	7	7

The three measures of typicality produce the same sum of absolute errors. It makes little difference, in terms of absolute error, which measure of typicality you use, even though the three measures give you different values for the average. There is, however, another measure of error that is commonly used in statistics: the sum of the *squared* differences between observed and expected values in the distribution. This is called the **sum of squares** and it is a heavily used statistic, especially in research contexts.

While the absolute deviation values simply ignore signs, there is another way to get rid of the signs: they can be eliminated by dealing with the square of the error. All we have to do, then, is add these squared figures. (Remember that a negative number, when multiplied by itself, results in a positive number. For example, −2 times −2 equals +4.) When we obtain the the sum of squared error values for the data in Table 5.2, we get the following results:

Sum of Squared Error Values		
Mode	*Median*	*Mean*
11	8	7.875

The mean is the value that minimizes the total degree of error, when error is dealt with as the sum of the squared differences between the expected and observed values. If this sounds a bit involved, go back to the data and see for yourself that when you square the errors and add these figures, you get a smaller value for the mean than for the other two measures of central tendency. In this instance the difference between the mean and the median is slight in terms of reducing error. But it is the mean that has the mathematical property of always producing the smallest amount of error when the sum of squares is used as the criterion for error.

In working with statistics, it is important to understand the idea of observed and expected values, the notion of error, the idea of summing up and averaging errors, and the practice of summing up the squared values of the differences between observed and expected values. The sum of squared errors, known as the *sum of squares,* is particularly important in advanced statistical analysis.

AVERAGES, GROUPS, AND INDIVIDUALS

Once again, keep in mind that averages do not refer to individuals. *They refer only to groups.* Averages inform us of how things would be equally shared within the group (*if* such things were, in fact, equally shared). The average grade of 2.375 points for the phrenology course in Table 5.2 represents what every student would receive if all the points given out to everybody in the class had been shared equally. The truth of the matter is that the grades were not shared equally. Some people got high grades, some low.

In fact, if you use the mean to predict any individual grade in the class, you will be wrong eight out of eight times—one hundred percent of the time! So, it is obvious that the mean can be misleading when applied to the individual. Yet averages are used by people every day to ascertain the character of individuals. For example, say you meet someone and are told she is a professor. You carry in your head a typification that goes with the category "professor." You assume that this woman is well-read and intellectual, because these are supposed to be typical traits of professors. However, averages are for groups. In the individual case, this professor might be poorly read and not the least bit intellectual; or, she might be a walking encyclopedia with an IQ in the genius range.

The average, then, refers to values obtained from groups of observations. It is valid only when it is applied to groups of individuals or observations. This is true of any kind of average—the mode, the median, the arithmetic mean, or anything else. The application of the average to the individual is an abuse of a perfectly reasonable statistical procedure. Such potential abuse should not be seen as the fault of the procedure itself, however.

USING EXTREME VALUES AS TYPICALITIES

So strong is the inclination to think in terms of some kind of typification that people are willing to use extreme values as proper estimates of the average. Advertising does this regularly. For example, when a car is advertised as getting up to forty-five miles per gallon of gas, we might think that the car *always* gets forty-five miles per gallon of gas. In this instance we are allowing an extreme case to represent the typical case. Note that when we use extremes to represent the typical, we make considerably greater errors than when we use the mean. In fact, the whole point of the mean, or any other average value, is to minimize the sum of errors or average error made in estimating individual values. Consider the fictive data in Table 5.3.

Table 5.3. Mileage Ratings for Ten Randomly Selected Belch-Fire 8 Automobiles, with Error Values Based on the Most Extreme Case and on the Arithmetic Mean. (Data are fictional.)

Car	*Mileage*	*Mileage minus Extreme*	*Mileage minus Mean*
1	15	15 − 35 = −20	15 − 18 = −3
2	17	17 − 35 = −18	17 − 18 = −1
3	12	12 − 35 = −23	12 − 18 = −6
4	14	14 − 35 = −21	14 − 18 = −4
5	21	21 − 35 = −14	21 − 18 = +3
6	18	18 − 35 = −17	18 − 18 = 0
7	15	15 − 35 = −20	15 − 18 = −3
8	16	16 − 35 = −19	16 − 18 = −2
9	17	17 − 35 = −18	17 − 18 = −1
10	35	35 − 35 = 0	35 − 18 = +17

When we sum the errors based on deviations from the extreme case (the car that got thirty-five miles to the gallon), squaring first to get rid of the negative values, our sum of squares equals 3,264. When we use the mean we get 374 as the sum of squares—roughly one-tenth the error we got when we used the extreme. As obvious as the difference in error is for the two estimates (the mean of 18 and the extreme case of the tenth car that got thirty-five miles per gallon), we still tend to fall for the idea that the extreme is typical and that what is typical characterizes the individual case.

Some trouble has been taken here to make a worthwhile point. When we foolishly buy some product in the belief that it will conform to the extreme case, the damage is essentially economic—the waste in billions of dollars that comes from ill-informed purchasing in this country. And, when such thinking involves referring to an extreme person (such as Saddam Hussein) to characterize a group of people (the people of Iraq), this kind of error can lead to intensely emotional and vicious misconceptions with consequent massive suffering.

A common form of this kind of thinking occurs when we have had a bad experience with a person from a particular group, causing us to then typify the entire group on the basis of the experience. In other words, we establish a presumed typical value on the basis of a single, possibly extreme case. Say you have been sexually harassed by Professor X of the Phrenology Department. You might then conclude that this is typical behavior for all professors of phrenology. Later, we will look at this same fallacy in another context—that of sampling.

When we meet a stranger, it is wise to assume that he or she is ordinary if we have no knowledge to the contrary. In a state of ignorance, the average is your safest bet. That qualification—*in a state of ignorance*—is an important one. As you become more informed, you generally begin changing your estimate of the individual case. It is a fallacious use of the mean to apply it as a fact for each individual case when it is merely a good working assumption (in fact, it is the best *assumption* you can draw on). You should always remember that it is nothing more than an assumption when it is applied to individual cases.

There is one situation in which the mean is perfectly reliable as a predictor of individual values: in cases where there is no variability. *Constants* are means with no variability. There are, however, no constants in the social world. What are the implications of this? (See note 12, page 111.)

THE WEIGHTED AVERAGE

One kind of average that causes a bit of confusion among those first encountering it is the **weighted mean.** A straightforward example of the weighted mean is found in the suggestion of an American industrialist a few years ago. He proposed that Americans should be allowed to vote in proportion to the wealth they control. A person with ten million dollars might get ten votes, and anyone with less than a million dollars would get only one vote. In other words, the vote should be weighted by wealth. Instead of having one vote per person, we would have the average vote value weighted by the wealth of the voter.

The weighted mean refers to "weighting" observations, values, or events according to a consistent procedure (in the above instance, for example, according to some measure of individual wealth). The weighted mean can be illustrated by going back to the example of the marbles used earlier to discuss the nature of the mean. Suppose the marbles that Carlos collected were valuable "aggies"; those collected by Juanita and Edwin were special "shooters"; and those collected by Maria and Charles were just plain old ordinary marbles. Further, let us say the aggies are three times as valuable, and the shooters are twice as valuable, as the ordinary marbles. Then, in sharing, we might want to weight the marbles by their value. This is easy enough to do.

Child	*Marbles*	*Weight*	*Weighted Marbles*
Carlos	36	3	108
Juanita	15	2	30
Edwin	18	2	36
Maria	41	1	41
Charles	25	1	25

We noted before that the unweighted mean was 27 marbles. Now, with the marbles weighted in this way, the sum of the weighted marbles is 240, and the weighted mean is 48. The average marble value, then, is 48 per individual.

Table 5.4 provides a more complicated example using real data. The principle involved is exactly the same. The simple arithmetic mean for the data in Table 5.4 is equal to the sum of the salaries divided by the number of states, or

$121,244 divided by 8 = $15,156

The average teacher income for these states, as a whole, was about $15,156 in 1979. However, we also know that some of the states are a lot larger than others. Some have many more classroom teachers. It might be interesting to see how the average is influenced if we weight the average in terms of how many teachers a state has, as shown in Table 5.5 (page 102).

Table 5.4. Average Annual Salary of Classroom Teachers for Selected States, Elementary and Secondary Public Schools, United States, 1979. (Source: *Statistical Abstract of the United States,* 1979.)

State	*Annual Salary of Classroom Teachers*
Arizona	$16,860
California	17,890
Colorado	14,616
Montana	13,293
Nevada	14,970
New Mexico	15,525
Utah	13,588
Wyoming	14,502
TOTAL	121,244

Table 5.5. Average Annual Salary of Classroom Teachers and Number of Classroom Teachers for Selected States, Elementary and Secondary Public Schools, United States, 1979. (Source: *Statistical Abstract of the United States*, 1979.)

State		*Annual Salary of Classroom Teachers (A)*	*Number of Classroom Teachers (B)*
Arizona		$16,860	24,000
California		17,890	205,700
Colorado		14,616	28,700
Montana		13,293	9,600
Nevada		14,970	6,300
New Mexico		15,525	13,700
Utah		13,588	12,800
Wyoming		14,502	5,000
	TOTAL	121,244	305,800

To obtain the weighted mean, we simply multiply each figure in Column A (salary) by the figure in Column B (Number of Teachers). Column B provides the weights to be used for modifying the unweighted original mean.

State	*Products of A × B*
Arizona	404,640,000
California	3,679,973,000
Colorado	419,479,200
Montana	127,612,800
Nevada	94,311,000
New Mexico	212,692,500
Utah	173,926,400
Wyoming	72,510,000
TOTAL	5,185,144,900

If we sum these products we have the sum of Column A times Column B. When this value is divided by the sum of Column B, or all the teachers in the region, we obtain the weighted mean:

$5,185,144,900 divided by 305,800 = $16,956

This is an average almost $2,000 higher than the one reported for the unweighted data. If you were a faculty union leader for the Western States Federation of Teachers, which mean would you want to publicize? On the other hand, if you were a state legislator representing the western region at a conference and wanted to ease teacher salary discontent, which one would you use? Why is the weighted mean larger?

These questions will gradually involve us in a complex question—one that, to our knowledge, no statistician has ever answered especially well. The question is this: when averages—means, modes, medians, weighted and unweighted values, or other measures—differ, which statement of the average is correct? Or true? Or valid? One answer, of course, is that they are all true, correct, and valid. However, we see that they have different consequences or truth values in different contexts.

Does this suggest, then, that statistical information is granted validity by its context? The idea is repugnant to most of us. The truth, we like to think, is the truth, regardless of context. We do not want contextual relativity generating ambiguity and uncertainty for us. We expect more of our statistics, but statistical reasoning, properly understood, warns us that the resolution of such a question calls for a better understanding of statistical procedures, not a rejection of them. Or, to cite an old cliché, here is an example of a situation in which people are too often willing to toss the baby out with the bath water.

Much of the power of physics as a system of laws derives from the fact that the basic statistical regularities it has uncovered are valid regardless of context. Unfortunately, this is not the case for social research. This problem of contextual relativity deserves a little more attention before we conclude this discussion. The question of contextual relativity has to be raised because how we answer it determines how we think about statistical problems and how we go about solving them. In considering this question, we also begin to approach fundamental questions about how we think in general and how we come to understand the world around us.

Table 5.6 presents typical income values for the American people in 1987 using the mean, the mode, and the median, along with several different definitions of the income recipient. The first problem we encounter in considering the data in this table is that the three measures of typicality give us three different answers. How we define what we mean by "people" also gives different answers. We get different values, depending on whether we are referring to families, households, or individuals.

There is more to trying to get a precise statement of the American income than this little table can suggest, but this should be enough to illustrate the

Table 5.6. Mean, Median, and Modal Estimates of Income for Americans in 1987 by Different Income-Receiving Categories.

	Income of the American People (in 1987 dollars)		
Income Recipient	***Mode***	***Mean***	***Median***
All consumer units	—	27,326	20,943
One-person consumer unit	—	15,006	—
Six-person consumer unit	—	29,458	—
All households	50,000+	—	26,061*
All households	20,000	33,526	25,986**
Household discretionary income	—	12,232	—

*Calculated on the basis of an older technique.
**Calculated on the basis of a different technique now used by the U.S. Census Bureau.

The data are official and were obtained from the Statistical Abstract of the United States, *1990.*

point we're trying to make: we can calculate an average, but that does not imply a solid value. It is this aspect of statistics that sometimes leads people to despair. It is easy to come to the conclusion that we cannot rely on what social statistics have to tell us, leaving us haunted by the satiric expression that there are lies, damned lies, and statistics.

But let's think more about this. If we cannot rely on statistics, then what can we rely on? We are forced by our natures to rely on some kind of statistical or folk statistical conception of things. We should always be aware, however, that even the simplest statistical manipulation is trickier than we generally suspect. Great care must be taken to qualify what we mean when we say that average income is a particular value. Even measures as simple as means, modes, and medians keep giving us different values. In science, the fact that we keep generating different values for an observation that is supposed to have a stable value suggests that we should be careful about what we are doing and what we are claiming.

The data in Table 5.6 give us wide-ranging measures of typicality; the largest of these measures is over four times greater than the smallest value. On the surface, this is like suggesting that the average height for American people is something like, say, three feet—or maybe twelve feet, depending. But depending on what? That is the question. If people are standing in holes, then the average might be three feet. If they are on ladders, it might be twelve. In connection with measurements of height, the problem sounds silly. But we often accept reported income figures without bothering to determine whether the data were taken with people standing in holes (after taxes, for example) or on ladders (before taxation and other essential expenditures).

In the "all households" category where we can compare three valid forms of measures of typicality, we get three different values, and the mean is the largest of the three. It is the largest because extreme incomes are generally *quite* extreme; their extremity tends to lie in the upper range, pulling the rest of the distribution along with them. For example, in 1990 the leading executive in the United States had an income of $188 million. If we assume that this individual paid no taxes (as is typical of the extremely wealthy in the United States), then the executive's discretionary income was extremely high—perhaps close to $180 million. On this basis, one lone person was the equivalent of about 16,000 American households. A single income was able to pull the average up 16,000 times higher than a typical household's effect. That is why the mean is higher than the median.

If you are not impressed by what a factor of 16,000 suggests, look at it this way. Imagine a pathway with zero income as the starting point of the path, the typical householder's income represented by a distance of one foot away from zero, the income of two households represented by a distance of two feet, and so on. To get on down the path from the householder to the executive, you would have to take a three-mile hike. That's a lot of leverage.

Standard statistics textbooks go into great detail on how skewing affects measures of typicality. In the case of the executive's income, the distribution is pulled out toward extremely high, or positive, values, pulling the arithmetic mean along. Though textbooks warn about the problem of this kind of skewing, it is often ignored in social science research. Instead, the assumption is made that everything is "normal." If a distribution is normal and not skewed, then there is no problem in at least one sense: the mean, the mode, and the median all fall on a common value. That is one reason why one hopes for data that are symmetrically distributed in a normal form. But what do you use as a measure of the typical in those cases in which data are *not* symmetrically distributed? This is something to think about. It is not an easy question to answer, however.

The necessity of pinning down averages by carefully defining their character, as well as all of the things that might disrupt a good estimate of the average, gets us into what statisticians call problems of control. Control is a complex, highly developed aspect of statistical work. We encountered an ordinary control device when we talked earlier about "standardizing" incomes. When we compare 1970 income values with later values and use a "standard dollar" for both periods, we are, in effect, controlling the influence of inflation. One of the most commonly used standardizing devices is the percentage, which gives values based on the assumption that the sample being examined always consists of precisely 100 cases. While complex control and standardization techniques constitute entire studies in their own right, it is good to recognize at the outset that even the most ordinary statistical devices, such as the percentage, represent one of the common concerns of statistical thinking: reporting values in standardized forms.

The primary purpose of this discussion of averages has been to move away from purely mechanical treatments of the average into more thoughtful questioning. If nothing else, such philosophical musing enables us to see how difficult it is to come to grips, in any definitive and precise manner, with the stuff that we call economic or psychological or social reality. It is the beginning of intellectual maturity to accept one's limits rather than try to hide them in the cloak of illusion. In a sense, that is what the careful and thoughtful study of statistical thinking is all about.

SUMMARY

Typification is central to all human communication and thought. Typification is a generic term for the assessment of what we think of as average, or commonplace, or usual, or ordinary. Typification is used in literary works and in journalistic accounts, just as it is in statistics. In statistical work, typification is based on the analysis of numerical information. This permits the use of three primary forms of assessment of what statisticians call *measures of central tendency:* (1) the mode, (2) the median, and (3) the arithmetic mean.

Any measure of typicality refers to a group rather than to individuals within the group. If a typical value is drawn from a set of values that are all the same, then the typicality is constant and can be used to refer to individual as well as aggregate situations. If a typical value is drawn from a set within which values vary from one another, however, then typicality must be understood as a group phenomenon and should only be used as a best guess when approaching an individual case without other information.

In most statistical work in the social or behavioral sciences, the arithmetic mean is preferred because it has the property of minimizing the sum of squared errors.

In skewed distributions it is possible to get three different values for what constitutes the typical case. This poses a problem with respect to establishing a correct or single value as being representative or typical of the whole.

THINKING THINGS THROUGH

1. If you look at a single page of a telephone directory you will see a listing of telephone numbers. If you sum the telephone numbers and then divide by the total, you will have the arithmetic mean telephone number for the people listed on that page. Is this average number meaningful? Why or why not?
2. Baseball players with a batting average of .250 presumably get one-fourth of a hit each time they bat. If they bat four times, they should make it on base. How are batting averages obtained? Why do they vary tremendously (for example, a single player can have a high average at the beginning of the season and a low one at the end of the season)? What is the relationship between the idea of typical batting activity and probability in this case? What, more broadly, is the relationship between typicality and probability in any situation? When typical values vary, as in the case of the baseball players' averages, can you have much faith in what you are calling typical activity?
3. First indicate what you believe to be typical conditions for each of the cases outlined below. Then turn to the Notes at the end of this chapter for data that are currently considered statistically reliable averages for these cases.[15] What sorts of confusion, errors, or misrepresentation might result from any discrepancies between what you believed to be typical and what the data suggest as actually being typical for these cases?

   Case 1. Median school years completed for people 25 years of age and over in the United States in 1988.

	Estimated	Actual
For Whites:	______	______
For Blacks:	______	______
For Hispanics:	______	______

   Case 2. Average amount of money spent by Americans in one year (1987) on the following:

	Estimated	Actual
Veterinary care of dogs:	______	______
Tuition costs at public institutions of higher learning:	______	______
Tuition costs at private institutions of higher learning:	______	______

Case 3.

a. Average weekly earnings in 1970 and 1988 for the following groups:

	Estimate		Actual	
	1970	1988	1970	1988
Manufacturing workers	______	______	______	______
Service workers	______	______	______	______
Retail trade workers	______	______	______	______

b. Average weekly earnings in 1970 and 1988 in terms of 1988 dollars.*

	Estimate		Actual	
	1970	1988	1970	1988
Manufacturing workers	______	______	______	______
Service workers	______	______	______	______
Retail trade workers	______	______	______	______

*Here you will need to find a way of standardizing for 1988.

What effect does standardizing for inflation have on these measures of typical weekly incomes? (This exercise is intended, incidentally, to demonstrate that we usually have some notion of what is average, but it rarely corresponds to the true average. We therefore tend to carry out actions with knowledge that does not minimize error.)

4. If you obtained the percentage of the population that was nonwhite in 1990 for each region of the United States and calculated the mean from these data, it would probably differ from the percentage recorded for the United States overall. The percentage of nonwhite people in the United States in 1990 was 12.4, determined by total population figures for the entire country. Calculate the average from the values given below for the different regions of the United States. Why is this figure different from the value of 12.4 reported by the U.S. Census Bureau? What might you do to the regional data to produce a mean closer to 12.4?

Region	*Percent Nonwhite*
Northeast	4.6
Middle Atlantic	13.6
East North Central	12.2
West North Central	5.0
South Atlantic	21.0
East South Central	20.0
West South Central	14.6
Mountain	2.5
Pacific	6.8

5. Why is typification a major statistical and, more broadly, a major communications concern?

6. What do we mean when we say that for virtually all social data, the mean or any other measure of central tendency should be treated as a fiction?

7. What is the sum of squares and how is it determined?

8. Provide two examples of a set of numerical values that are not capable of generating a meaningful average.
9. If you sum the absolute deviations from the mean and divide by the number of cases, you have the mean absolute deviation. Means imply typification. What is being typified in this instance?
10. What are the consequences of using an extreme value within a set of values as a measure of central tendency, instead of a measure such as the arithmetic mean?
11. Statistical problems, in a technical sense, usually begin with a *data base,* or a set of numbers or values that call for summarization. Determine the median, mode, and mean for the following data set. The data refer to the number of murders per 100,000 population for American states and the District of Columbia in 1988 (source: *Statistical Abstract of the United States,* 1990, Table 285). To simplify the data set, the names of the states have been omitted, and only the rates are presented. (Note that you will find it easier to work with *grouped data* to determine the mode or the median.)

Data Set

3.1	5.5	1.7	59.5	9.4	3.6	5.1
2.3	5.4	8.0	7.8	9.9	2.5	10.4
2.0	1.8	4.9	8.6	5.7	5.7	4.0
3.5	6.4	3.1	7.8	8.7	11.5	
4.1	8.6	3.6	9.3	11.6	8.5	
5.4	10.8	3.4	11.7	7.4	2.8	
12.5	3.0	5.2	11.4	12.1	10.5	
5.3	2.9	9.7	6.2	2.6	5.7	

Notice the one extremely high rate of 59.5. If you leave it in, how will it affect the mean?

The next table, a grouping of the homicide data, is used here as a working table—that is, a table to be used for working out the calculation of the assignment.

Homicide Rate	*Number of States*
16+	1
12–15	4
8–11	16
4–7	17
0–3	13

Notice that the open-ended interval at the top of the table permits you to assign values as high as or higher than 59.5. If we were to include this specific value in the grouping, we would have to create ten more intervals, each with a width equal to four units. This gives you some idea of how deviant the 59.5 homicide rate is, and how much influence it is going to have on a measure such as the arithmetic mean.

To make things more puzzling and push your thinking about solving problems of typicality a bit further, let us mention that the U.S. Census Bureau reports the mean homicide rate as 8.4 for the United States as a whole, including the District of Columbia. You will obtain a smaller mean for the state data. Why?

12. The next set of data is similar to that used in exercise 11. Calculate the mean, mode, and median for these data. Are there any ***outliers*** (unusually deviant values)? If so, how do they affect the computation of central tendency values? The data pertain to the number of recorded abortions, in 1985, for every 1,000 live births. Again, the states have been omitted to simplify the data base. The original data can be found in Table 102 of *The Statistical Abstract of the United States,* 1990.

308	746	486	140	185	315	320	373	283
419	672	246	268	379	333	288	116	611
448	348	257	264	228	142	155	641	412
533	357	248	451	397	159	125	458	
572	202	261	480	465	240	438	374	
550	372	230	1,186	189	269	219	640	

Before you make your calculations, roughly estimate what the typical value should be. Look up Table 102 in the 1990 *Statistical Abstract* and see what the table states as the average for all states. Does it differ from your calculation? If so, how might you account for the difference?

If you have a statistical computer program such as SPSS or STATPAC (a simple system), you might want to enter the above values into a data base of some kind and have the computer check your calculations. Where do you and the computer differ, if at all?

Remember, though, that it is important to do these problems first by hand, the hard way, to gain a sense of what the computer is doing when you use one. It is also important to notice that sometimes a computer will give you different results than you get when you do a problem using pencil and paper. You should be aware of why such differences exist (if they do). A computer generally calculates statistical values in a special way. This can create small differences between your obtained values and those of the computer, at least in cases when you work with data that are arrayed in terms of a frequency distribution.

While trying to be perfectly precise is largely an academic exercise when it comes to social data, it is good to be as precise as one can be. For most social data, however, small differences in values are not particularly disturbing, especially when the data themselves are relatively imprecise or imperfectly recorded.

13. Briefly discuss why observations that generate constants are so useful in statistical descriptions of reality. Can you think of any constants in human social affairs? If sociologists and psychologists cannot establish constants, what are the implications for the development of these fields as sciences?

Notes

1. Typification is also a primary concern of literature. It is the task of the literary artist to typify, through literary devices, some human quality or character. Richard Wright attempts to typify the black experience in *Native Son.* Mark Twain

typifies one form of American life at the end of the nineteenth century in works such as *Life on the Mississippi* and *Huckleberry Finn.* Typification is central to both statistical and literary forms. However, the literary form is extremely complex and any discussion of the manner in which it arrives at its typifications would require much more space than that given to the simpler discussion of statistical typifications.

2. The introduction of statistical studies into the realm of sexual behavior is an interesting example of what we mean by statistics having potential moral implications—in this instance, extremely profound moral implications. The precise counting of various aspects of human sexual conduct is relatively new, and it has had an effect on how we now deal with questions of sexual morality. The literature on the statistics of sexual activity is voluminous. Pioneer work was done in the studies of Kinsey, recorded in *Sexual Behavior in the Human Male* (Philadelphia: W. B. Saunders, 1948), and followed later by the more elaborate statistical and clinical studies of Masters and Johnson recorded in *Human Sexuality* (Boston: Little, Brown, 1985). Here is an example of a topic that was once the exclusive domain of moralists being given over to statisticians. It is interesting to note that this transformation has been referred to as the "modernization" of sex by one writer; see Paul A. Robinson, *The Modernization of Sex: Havelock Ellis, Alfred Kinsey, William Masters, and Virginia Johnson* (New York: Harper and Row, 1976).
3. Quoted in Steven Marcus, *The Other Victorians: A Study of Sexuality and Pornography in Mid-Nineteenth Century England* (New York: Basic Books, 1966), 19.
4. Folk forms of determining typicalities are much more complicated than formal statistical approaches that rely on simple numerical data. The ways in which people reach judgments about what constitutes a typical Japanese person, movie star, homeless person, victim of domestic abuse case, or other similarly complex person are extremely idiosyncratic and cannot be discussed in an elemental work of this nature. At the same time, anyone interested in statistics should not ignore this problem simply because it is not precisely resolvable. The question remains: how do people reach such certain conclusions about the various typicalities in which they believe, even when there is no formal procedure for determining those typicalities? It is not a light or easy question.
5. James Gleick, *Chaos: Making a New Science* (New York: Penguin, 1987), 168.
6. We are indebted to Professor Rolf Kjolseth for pointing out that what seem to be social constants—such as conflict, poverty, crime, and so on—are really ubiquitous qualities. They are found in nearly all societies, but are not *constant* throughout all societies. Ubiquity and constant typicality are quite different things.
7. While the statistical method is identical across all applications, this does not mean the same results will be obtained no matter what subject we probe. Some applications of statistical methods (for example, quantum mechanics and quality control techniques in specific manufacturing situations) produce astonishingly precise results. Other applications result in less precise answers. Clinical psychology and the social sciences, though reliant on statistics for the past century, have only been able to achieve crude precision. This does not mean that statistical studies in the social sciences should be abandoned. It does mean, however, that serious attention should be paid to the constraints on statistical methods that are operative any time a person undertakes the study of some social issue.

8. The tendency to "load" meanings into any communication, whether literary or statistical in form, is extremely common. For an examination of how people read statistical information according to their own special interests, or how statistics can be used to get people to see what the statistician wants them to see, several books are worth considering: Grahm Burchell, Colin Gordon, and Peter Miller, eds., *The Foucault Effect, Studies in Governmentality* (Chicago: University of Chicago Press, 1991); Abram J. Jaffe, *Misused Statistics: Straight Talk for Twisted Numbers* (New York: M. Dekker, 1987); Leland Garson Neuberg, *Conceptual Anomalies in Economics and Statistics* (Cambridge: Cambridge University Press, 1989); and Mark Maier, *The Data Game: Controversies in Social Science Statistics* (Armonk, NY: M. E. Sharpe, 1991). It should be mentioned here that economists, perhaps more than any other group of social scientists, have become increasingly disenchanted with the extent to which they have overrelied on statistical information. Curiously enough, measurement error seems to be the central problem in economic prediction. If economists have problems with measuring such simple concepts as demand or gross national product, then more serious consideration must be given to the even more complex measurement problems encountered in psychology and sociology.
9. An excellent book that approaches statistics by having its readers do hundreds of statistical exercises using several standard statistical computer programs is James W. Grimm and Paul R. Wozniak's *Basic Social Statistics and Quantitative Research Methods* (Belmont, CA: Wadsworth, 1990). This book relies almost entirely on the concept of learning through doing. There is much to be said for this. On the other hand, sometimes we become quite adept at "going through the motions" without really knowing *why* we are going through them. Our point here is that statistical knowledge requires practice, experience, and a constantly questioning attitude. It is often too easy to become lulled into carelessness by the numerical forms that modern computers are so readily capable of generating.
10. Source: *Statistical Abstract of the United States,* 1994, Table 691.
11. Because typifications are central to all communications and because typifications apply to collectivities or aggregates rather than individuals, the implication is that almost any application of a typification to an individual case will create some degree of semantic or communicative confusion. This is clearly seen in the case of applying what are called *stereotypes* to individual members of ethnic or racial groups. (It is often not as easy to see the problem in a broader statistical sense, however.) In sum, the application of averages to individuals is warranted only as a working assumption, never as a valid scientific attribution.
12. When a set of values is constant—that is, when no variation exists within the set—the mean determined from those values always identifies the individual case. In this instance the mean is extremely effective as a predictor and can be said to be "real." In social studies, however, constants do not exist in the same sense in which they are found in physical reality. This is a severe limitation for the development of a precise science of human affairs.
13. Notice that the mean, in a case such as this, is being used probabilistically. We see the man and woman and we think that the man probably makes more money than the woman. It might be a correct general probabilistic assessment in principle, but in the individual application it could very easily be completely incorrect.
14. Source: *Statistical Abstract of the United States,* 1994.

15. Data for Exercise #3:

Case 1: Whites, 12.7 Blacks, 12.4 Hispanics, 12.0

Case 2: Dogs, $82.66 Public School, $1,160 Private School, $6,316

Case 3:

Manufacturing Workers		**Service Workers**		**Retail Trade Workers**	
1970	*1988*	*1970*	*1988*	*1970*	*1988*
$133	$418	$97	$290	$82	$194
$370*	$418	$270*	$290	$230*	$294

*In terms of 1988 dollars; 1988 dollars are roughly 2.8 times 1970 dollars.

6

ASSESSING THE UNFAMILIAR: DEVIATION

"Let us not be blind to our differences . . . ."

—John F. Kennedy

"Our slow, unreckoning hearts will be . . . / Unable to fear what is too strange."

—Richard Wilbur

In human affairs even a slight discrepancy from the usual, typical, or average can have consequences that seem out of proportion to the actual deviation. For example, you go to a party and stain your stylish new jacket. The stain is a little one, but it spoils the evening. The stain sets you apart. You have trouble with your presentation of self. It is a small difference that has big consequences. There are hundreds of moments, like this one, when we are concerned with assessing whatever is different, peculiar, out of the ordinary, odd, or strange—in other words, that which deviates from the typical or normal.

Political slogans make much of presumed differences. These differences usually imply that "we" are superior to "them." From this attitude come slogans such as "*Deutschland uber alles*" or statements such as "Russia is the locus of evil." Whatever the slogan, the fact of difference is seen to be important in social life. It is also an important feature of statistical analysis. In the world of common sense, the determination of differences calls for two elemental judgments. The first requires that we have some idea of what is average or typical. The second requires that we determine the extent to which some individual, event, or happening is different from the typical case. Assessing the unfamiliar is nothing more than noting a difference—in this instance, the difference between a specific observation and a mean or typical value that we believe applies to that observation.

THE RANGE

People rely heavily on averages, such as the mode or the median, to tell them what is characteristic of a group. They also often mistakenly rely on averages to judge individuals.[1] They might also notice that in some groups, all the individual members conform closely to some average, while in other groups this does not happen—individual members may vary considerably from the mean and from each other. If we want a clearer picture of what an average value is telling us, it is necessary to establish the extent to which there is *variation* around the mean.

Table 6.1 presents two sets of course grades. The material is fictional.

Table 6.1. Course Grades for Two Classes with the Same Mean but Different Variability. (The data are fictional.)

	No. of Students	
Grade	*Professor Skeen's Class*	*Professor Sterling's Class*
A	25	0
B	0	0
C	0	50
D	0	0
F	25	0

Both classes have exactly the same mean grade of C. In one class the probability of getting a C is zero. In the other class the probability of getting a C is perfect; a student cannot get anything *but* a C. Notice, immediately, that as variation around the mean approaches zero, the likelihood of anything other than the mean being true also approaches zero. Conversely, as variation around the mean increases, the likelihood of an individual event having a value other than the mean also increases. The fact that both classes have the same mean value for grades does not mean that the two classes are statistically the same. One class is homogeneous, the other heterogeneous. In one, variability is zero, while in the other it has a value greater than zero.

Notice, also, that the word *probability* is used above. *Variability* and *probability* are closely related terms. This relation will be taken up in greater detail in later discussions of probability. If there is no variability, then things are certain and we do not need to talk about uncertainty or error. *Uncertainty* is another term implying the presence of variability and the need to think in terms of possible or probable happenings rather than certain ones.

Variability, then, is an important feature of any set of observations. But how do we assess variability? The simplest and perhaps most common way is to use the **range.** The range is the difference between the highest and lowest values. In Professor Skeen's class (in Table 6.1), the highest grade is an A and the lowest is an F. The range is from A to F with the mean equal to C. In Professor Sterling's class, the range is zero (from C to C) with the mean also equal to C. The class grades differ in terms of variability, not central tendency; this is a most important difference.

The range is good for a quick estimate of things, but it relies on only two values. This makes it unreliable. For example, if only one of Professor Sterling's students received an A while another got an F, the range for the two classes would be the same. The two classes, however, would remain radically different in character.

Because it relies on only two values, the range, when obtained from a sample, is likely to *underestimate* the amount of variability in a given distribution. You can see this in an ordinary deck of cards. The range for the deck is from one (the Ace) to thirteen (the King). The chance of drawing a one and then a thirteen (putting the cards back after you look at them) is 1/13 times 1/13, or 1/169. Or, you could draw a thirteen and then a one, which would cut the probability of 1/169 in half to 1/85—still a small probability of drawing the true range. Since this is the only true estimate of the range, all other draws will be in error. They will all underestimate the true range. (If, for example, you drew the six of hearts and the six of spades, the range would be from six to six, or zero—a completely false estimate of variability in the deck.) In general, the range is not to be trusted. It is a whimsical estimate of variability—especially when a sample is small. In fact, variability, no matter how you measure it, will tend to be underestimated by samples, though with samples containing more than a hundred cases, the degree of underestimation tends to become negligible.

Despite the fact that the range is a capricious measure of variability, it is better to have a sense of the range than to have no sense of variability at all. We are inclined to avoid taking variability into consideration. Instead we tend to lock on to some average value for a group of people and then apply that average to all members of the group, without giving much thought to the fact that in virtually all instances, human social groups are variable.

Each of us is a sampling of the world. However, no matter how astonishingly varied our experiences might be, no human being has ever experienced the full range of events that go on in the world. Nothing brings this home like *The Guinness Book of Records,*[2] which is little more than a record of the range for different events drawing on all that is known about those events as they occur throughout the world over a long span of time. Few of us ever personally encountered the lightest living adult human being or the heaviest. Probably no one has encountered both. The lightest living adult human being on record was Lucia Darate who, at the age of 17, weighed 4.7 pounds. The heaviest human being was Jon Minnoch, who was estimated to weigh over 1,400 pounds at one point in his life. If we rely on this particular range to provide an estimate of variability in the weight of human beings, we would get a value somewhere around 1,396 pounds.

In sum, the range is too unstable for most serious statistical work. It relies on only two observations. Perhaps we could get more reliable and solid estimates of variability if we relied on *all* observations (implying, then, that we would be dealing with average variability). The most acceptable measures of variation do just that.

THE MEAN ABSOLUTE DEVIATION

Rather than draw on a few observations to establish the extent to which variety is a part of what we are observing, it might be a good idea to draw on as many observations as possible. There are three fundamental statistical devices that do this: the **mean absolute deviation,** the **standard deviation,** and the **semi-interquartile range.** The most important of these measures is the standard deviation. To gain some idea of what a standard deviation is all about, it is good to start with the mean absolute deviation. The semi-interquartile range will be addressed briefly at the end of this discussion.

The mean absolute deviation can be compared with the range in the following array of numbers:

100 50 50 50 50 50 50 50 50 50 50 50 50 50 50 50 50 50 50 0

Here the range is from 0 to 100. The mean for the array is 50. The absolute deviations from the mean are

50 0 0 0 0 0 0 0 0 0 0 0 0 0 0 0 0 0 0 50,

and the sum of absolute deviations is 100. There are 20 cases, so the mean (or average) absolute deviation is 100/20, or 5. The average absolute deviation from the mean is equal to 5 units; obviously, the variability is being caused by the two extreme cases at each end of the array. The range is equal to 100, but the absolute mean or average deviation is only 5. You will rarely see a distribution that looks like this, but it works well as an illustration of the effect of outliers on statistical indexes. It is used here to make clear the difference between the range, a value that depends only on two extreme values, and an average variation that takes all deviations into account and provides an estimate for the typical or shared deviation for each case in the distribution.

With the above array we have a mean, for the array, of 50, with an average absolute deviation of 5 units. We do not expect, typically, to encounter great deviations from the mean for the distribution in this case. The smaller the deviations from the mean, on the average, the more reliable the mean becomes for estimating individual cases. Reliable indicators of variability are essential for assessing the utility of various statistical indexes.

THE STANDARD DEVIATION

The standard deviation is a slightly different version of the absolute mean deviation. If you understand the absolute mean deviation, you should find the standard deviation familiar stuff. Its job is the same—it tells us something about how much variation, *on the average,* is characteristic of a set of observations in which we are interested. The standard deviation is a workhorse statistic for measuring variation. A good grasp of its character, how it is obtained, and how it is interpreted, is necessary for even the most elementary understanding of statistical procedures. The following are the major advantages of the standard deviation:

1. It makes use of all observations.
2. It does not arbitrarily dismiss signs. It handles the problem of signs by squaring all deviations from the mean.
3. It is mathematically related to various features of the normal distribution.

The standard deviation is a central idea in statistical analysis. A large part of its value comes from the close relationship it has with the normal curve (a relationship that will be discussed in the next chapter). For the moment, the important thing is not to let the standard deviation appear to be more than it is. All the standard deviation amounts to is a measure of the extent to which observations differ, *on the average,* from the mean. *It is a measure of the average amount of deviation from the average.* Is there a lot of difference, on the average? Is there very little difference, on the average? This is the question answered by the standard deviation. To calculate the standard deviation use the following steps:

1. Find the mean for the distribution. (What's typical?)
2. Find the difference between the mean and each observation. (How is each case different from the typical?)
3. Square the difference. (Get rid of those negative signs.)
4. Sum the squares. (You need the sum to get an average.)
5. Divide the sum of squares by the total number of observations. (You now have the average squared deviation.)
6. Obtain the square root of the average to come back to the original character of the measure. (That's it!)

The above steps should *not* be memorized and followed mechanically. The nature of the standard deviation should be understood so well that you can calculate its value as readily as you would that of the mean.

Let's calculate the standard deviation for the two sets of numbers, set A and set B, listed below. First we will get the mean for both sets, then the deviations from the mean in each set, then the squared deviations, then the average of the squared deviations, and then the square root of that average. It sounds complicated, but it is just a fundamental way to get an average deviation that has the additional advantage of being extremely useful when events are normally distributed.

	Set A	Difference from Mean	Difference Squared	Set B	Difference from Mean	Difference Squared
	1	−2	4	80	+40	1600
	2	−1	1	60	+20	400
	3	0	0	40	0	0
	4	+1	1	20	−20	400
	5	+2	4	0	−40	1600
Sum	15	0	10	200	0	4000
Mean	3		2	40		800

Square Root of Mean Squared Difference = 1.41 for Set A, 28.28 for Set B.

We now have a standard procedure for comparing variability in the two sets of values. Before calculating the standard deviation, you should look at the distribution and get a rough estimate of whether there is a little or a lot of variability. In this instance it is obvious that Set A's numbers do not, on the average, deviate greatly from the mean. Set B's numbers, however, deviate a total of 120 absolute units from the mean. That's an average of 24 units. The standard deviation provides a slightly different value, but basically the same value we would get with the absolute deviations. Why not use the absolute deviations instead of going to the trouble to obtain the standard deviation then? The answer to that question has to do with the fact that the standard deviation, commonly abbreviated SD, is closely related to the normal distribution, a subject we will address in the next chapter.

THE SEMI-INTERQUARTILE RANGE

An occasionally used measure of variability is the semi-interquartile range. To obtain this value we simply find the score or value that defines the interval between the upper 75 percent of a distribution and the lower 25 percent. The point of this procedure is to have a range that does not depend on just the two extremes of a distribution, its lower and upper values.

In the following instance of grades for a fictional class of 240 students, the semi-interquartile range is between the middle of the 80–89 interval and the 60–69 interval. A grade of 85 defines the upper 75th percentile. A grade of 65 defines the lower 25th percentile. The semi-interquartile range is, therefore, 20 points. This figure at least warns us that there is variability in the scores. Its use is relatively limited to this purpose. The standard deviation is a preferred measure of variability.

Score		*Cumulative N*	*Cumulative Percent*
90–99	60	240	100.0
80–89	40	180	75.0
70–79	80	140	33.3
60–69	30	60	25.0
50–59	20	30	12.5
0–49	10	10	4.2

Before leaving this discussion of differences, variation, variety, variance, deviation, and the like, it is necessary to point out, again, that the notion of variation (or lack of it) is a part of ordinary life. Here are five quotations from various sources that reveal an interest in variation. Note in these quotations the extent to which the author implies either no variation whatsoever or tremendous degrees of variability.

- "*Semper eadem.*" ("Always the same"—the motto of Elizabeth I.)
- "That the king can do no wrong is a necessary and fundamental principle of the English constitution." (Sir William Blackstone)
- "When everyone is somebodee [sic], / Then no one's anybody." (W. S. Gilbert)
- "Now, my suspicion is that the universe is not only queerer than we suppose, but queerer than we *can* suppose." (J. B. S. Haldane)
- "Bang! Now the animal / Is dead and dumb and done. / Nevermore to peep again, creep again, leap again, / Eat or sleep or drink again, oh, what fun!" (Walter de la Mare)

What is especially interesting in ordinary conversation is to begin to see the extent to which variability is discounted or ignored. Variability introduces uncertainty, and people do not generally feel comfortable with uncertainty. It is, for example, one thing to say that hard work will produce success. It is something else to say that hard work might or might not produce success (a more realistic appraisal of how things actually take place in the real world). Notice, though, that the more realistic statement has less persuasive or rhetorical appeal.

SUMMARY

We cannot establish the degree of reliability of a measure of typicality or central tendency by simply knowing what its value is. We must also know how much variability is associated with our measure of typicality. If there is a lot of variability, the measure is unreliable as a predictor or indicator of individual cases. If there is no variability, then any measure of typicality we might want to use will be perfectly reliable as an indication of the value for any individual within the set from which the measure was obtained. In the natural sciences, certain averages acquire a special utility by virtue of the fact that they show no variability around the average. These are called constants. While the physical sciences have dozens of valued constants that are central to much of the most precise work in science, the social and behavioral sciences have not, as yet, established any typicalities involving zero variances.

Three measures of variability are commonly used in statistics. The range is a quick but unreliable measure of variability; however, using the range is a lot better than thinking there is no variability at all. We should immediately gain some notion of the range of variability within any set of data we are working with.

A second measure is the mean (or average) absolute deviation—a measure of the extent to which each individual in a set of individuals differs from the mean for the set. It is an *average* deviation. The mean absolute deviation would be the most popular measure of deviation were it not for the fact that its dismissal of signs is arbitrary and, more important, it does not relate in a simple mathematical fashion to the normal curve.

The most commonly used measure of variability is the standard deviation, which is the square root of the average of squared deviation from the mean. The standard deviation, like the average absolute deviation, is an average. But it has the particular advantage of being mathematically related to the normal distribution—a fact that gives it special utility in statistical analysis.

Thinking Things Through

1. What are the relative advantages of the range, the mean absolute deviation, and the standard deviation as measures of variability?
2. What are the consequences of variability for human social relations? What institutional forces try to promote variability? What social forces try to diminish it? Do you think the modern adulation and promotion of social diversity is a consistently good thing? Or does it contain hidden problems?
3. Do you think people tend to ignore variability when they talk about averages? Provide some examples of cases in which people recognize variability and others in which they do not. Is there any kind of pattern in whether variability is acknowledged or not?
4. Suppose you get a job as an instructor at a training facility. You can teach either of two classes. One class is essentially a group of mediocre, average, or typical students who are pretty much alike in interests, ability, and motivation. The second class is divided between two groups: one is dull and untalented, the other brilliant and highly motivated. Because your work schedule is demanding, you want to pick the class that will be easier to handle. Which one would you choose? Why? Are they different in terms of average ability for the class as a whole?
5. You obtain a sample and find that the mean for the sample is a large number; let us say it is 8,762,987. Conversely, in another sample the value for the mean is zero. Do these values have an effect on variability within

the samples? Why or why not? (Thinking this question through, even though it is simple, can enhance your awareness of the extent to which typicality and variability tend to be independent of each other. Be careful, however; there are exceptions.)

6. Two auto repair shops have been repairing cars for the past ten years. Each has worked on at least three thousand cars. The following are the means and *SD*'s for repair billings for both shops:

	Shop 1	*Shop 2*
Mean	$424.33	$455.89
SD	200.67	32.31

Both shops offer basically the same quality of repair. On the basis of just the above information, which shop would you take your car to? Why? (You gain or lose certain advantages with either choice. This exercise tries to show that sometimes a statistical decision rests on the character of the person making it as much as it does on the actual data.)

7. Without looking up any formulas or steps for solving the problem, find the standard deviation for each of the following arrays of values. Do the calculation in your head—no pencil or paper allowed (approximate the value for square roots). Until you can do this exercise, you probably do not have a working grasp of the standard deviation as an average of squared deviations.
   a. 1, 2, 3
   b. 68, 70, 72
   c. 527, 527, 527
   d. 40, 50, 60

8. Calculate the *SD*'s for the two sets of values presented below. Why does the second set have a high standard deviation? How is this related to the earlier discussion of the mean in Chapter 5 (pages 85–112)?

Set 1: 33, 34, 35, 36, 37, 38, 39, 40

Set 2: 33, 34, 35, 36, 37, 38, 39, 160

Notes

1. It is extremely important to avoid the error of using the mean to establish the character of the individuals from which the mean was drawn. This error was discussed at length in Chapter 5, and it will be emphasized over and over throughout this book. Using the mean to determine what is true of individuals in a set works *only* when there is no variability around the mean or other measure of central tendency. However, with social data, there is always a great deal of variability to deal with.
2. *The Guinness Book of World Records* (New York: Bantam, 1985). This book is updated each year. It is more than a mere source of amusement or entertainment. For someone who is seeking a deeper awareness of statistical reasoning, it provides example after example of how limited our own sampling of the range of things

going on in the world around us is. It brings home the elementary statistical argument that smaller samples tend to underestimate variability. As a consequence of this tendency to underestimate, much of classical inductive statistics has made various adjustments in the estimates of variability obtained from small samples. As samples approach one hundred ($N = 100$) cases in size, however, these adjustments become minor for estimates of variability such as the standard deviation.

7

THE NORMAL CURVE

"If the Greeks had known about the normal curve, they would have worshipped it."

—Source unknown

"I shall never believe that God plays dice with the world."

—Albert Einstein

"Common-looking people are the best in the world; that is the reason the Lord makes so many of them."

—Abraham Lincoln

When we talk today in schoolyards and on campuses about "the curve" and grading "on the curve," the reference is to one of the best known of all statistical models: the **normal curve** or **Gaussian distribution.** It is sometimes called the *bell curve* for its bell shape. What makes the normal curve of special interest is that it appears in so many places and that it is generated by a simple random procedure.[1] It is also of interest because in situations where it is possible to assume normal distributions, it is often possible to evaluate the strengths or limits of statistical generalizations. In addition, the standard deviation is mathematically related to the normal curve in ways that make the assumption of normalcy especially desirable and useful. Unfortunately, people all too often assume matters are normally distributed when, in fact, they might not be.

THE NORMAL CURVE ILLUSTRATED

The normal curve can be appreciated at two levels. First, it is extremely common (it appears in an astonishing variety of places), and it is aesthetically interesting. The second level is a practical one: the normal curve enables us to estimate probabilities quickly. Calculating exact probabilities, especially with large samples, is time-consuming (though modern computers can now solve such problems relatively easily).

To get some idea of what a curious matter this business of the normal distribution is, we need first to look at a few of the many places where it lurks. We need to develop a "sense" for where it is likely to be found.

The Birds

Here is an example from a bird lover's backyard. This bird lover put up a bird feeder one spring at the top of and near the center of a fence in her backyard. After a few months the birds had left a tell-tale pattern of their visits in the form of splatters on the fence. The splatters were far down on the fence at the point of the feeder and then gracefully tapered off on either side, forming a nearly perfect, upside-down normal distribution.

The Bird Feeder

Our bird lover also had a feeder hanging from an eave that the birds visited. They dropped the shells of seeds on the concrete below. The scattered seeds formed a nearly perfect circle with a heavy concentration of seeds directly below the feeder and a lighter and lighter concentration of seed shells as distance from the center increased. If you were to take a "slice" of this distribution, cutting across any diameter of the circle, you would get an approximation of a normal curve.

The Antique

If you walk up an old staircase you will commonly notice that the front edges of the stairs have a familiar worn pattern in them. They are heavily worn in the center with less wear appearing as you move away on both sides from the center—forming a bell-shaped curve of wear.

This is known as a normal curve of wear and is typical of random wear patterns in old furniture and other items. So characteristic is this curve that it is used by antique dealers to test whether a particular antique is a forgery or not. A forger might, for instance, sand in an arc rather than a bell curve. By superimposing what an actual curve of wear ought to look like, then seeing if there is a difference, a dealer can determine whether or not an antique is a fake.

The Horse Race

At the race track, it is common to see horses distributed as illustrated in the line below. (While this is a common distribution during a race there are, of course, many exceptions.)

S S S B B BBBB BB BB L L L

In this line, *S* stands for stragglers, *B* for the bunch, and *L* for the leaders. What is interesting is that there are a lot of horses bunched up in the middle with a few stretching out in front and back. This is what a normal curve is like—a bunch of things in the middle, with the ends tapering off into the traditional bell-shaped form.

The Hand Rail

This is a variant of the antique example. If you go into a frequently visited place like the stacks in the library, you might notice that the banister posts at the bottom of the stairs have the paint worn off of them at points where people have grabbed hold and swung themselves around. Because people are normally distributed with respect to height, the curve of wear on the upright post looks like a mathematician could have created it.

The Popcorn Party

When you pop popcorn, listen for a moment. First a kernel or two pop alone. Then a few more. Then some more. Then a burst of kernels. Wait. A few more pop. Then, after a long while, one last kernel pops. If you were to diagram the noise of the popping corn, you would begin with a few kernels for the first few intervals of time. The level would increase until it reached a peak and then there would be a tapering off. Your diagram would take the form of a normal distribution.

The Sound of Laughter

A teaching assistant in a large class at a state university told us this tale. She proctored an examination for a class of more than five hundred students. The instructor had inserted a funny question in the middle of the examination. After fifteen or so minutes of silent mental effort, she heard a student snicker. Then a few others snickered. Then the volume of laughter rose throughout the room. Then the noise subsided and there was, again, silence. After a few more minutes there was a chuckle. Finally, after yet a few more minutes there was a deep, guttural "Har, har, har" from a dense fellow who was just catching up with the rest of the class. If the volume of laughter were diagrammed, it would look like a well-defined normal distribution, slightly skewed to the lower end.

The Maze

Imagine a maze such that when you walk through the first door you are confronted by a barrier and must go to the right or left of the barrier. After you pass through this barrier you find yourself facing another. You must determine whether to go to the right or left of the barrier by the toss of a coin. Suppose that you must contend with ten of these barriers, one after the other. Each person going through this maze, then, must make ten selections. If one hundred people went through the maze, they would distribute in a normal pattern. Only a few would go to the far left or the far right. Most would fall in the center. (The likelihood of someone *always* going to the right or to the left is $1/2^{10}$, or about one in a thousand. We would, therefore, not expect many of our group to fall at the far extremes.)

Every once in a while you should take a look around you and see if you can find an example of the normal curve. For example, when you are driving down the highway you might notice that an oil slick covers the center of the road where cars have randomly leaked oil. The slick is heavy in the center and grows dim at its edges. It is an approximation to the normal curve. If you are target shooting, you will notice that most of your shots are in a cluster with a few near the center of the cluster and a very few missing the cluster by a large degree. Your target displays a normal distribution in a circular form. If you diagrammed a slice of the target, taken through the center of the cluster, you would commonly get a close approximation to the normal distribution.

We could go on with more tales about the normal distribution. If you were able to measure it, you would probably find that your doctor has cases in which he or she exceeds normal skill level and others in which he or she falls below it. Most of the time, though, your doctor is probably between these extremes. The same is probably true of your garage mechanic, your friends and, most interestingly, yourself. One is tempted to suggest that life itself is just a great big normal curve. However, that is the trouble with the normal curve. We can get carried away with it. Actually, it must be thought about most carefully. Something we are interested in *might* have the property of being normally distributed—but then again it might not.

This discussion is designed to help you see that the normal curve is not just a device used by educators, psychologists, or a few other professional groups to do their statistical work. It appears in thousands and thousands of settings, and you should develop an imagination for suspecting where it might appear or where it should not appear. For example, are the heights of the crests of ocean waves ridden by surfers normally distributed? Is the extent to which you have ordinary days, really super days, and really down days more or less normally distributed? Thinking about such things can give you some intuitive ideas regarding why the normal curve is as common as it is. It is a powerful descriptor because it describes such a variety of events.

USING THE NORMAL CURVE

When we can assume that some quality is normally distributed, we can also do other things, such as determine the probability of a specific observation, make relatively reliable estimates of how an individual performed on different tests, or establish the reliability of different samples. Let's look at a normal distribution that involves one hundred individuals ranked according to their score on an index of success (see Figure 7.1). (Actually, success is probably not normally distributed. If we equate monetary worth with success, for example, then success is positively skewed; that is, there are a few extreme cases that extend way out on the positive side of the curve.) In Figure 7.1, the index of success ranges from zero to one hundred. The typically successful person has fifty points according to this measure.

Figure 7.1. A Crude Approximation of 100 Individuals Normally Distributed by Success Scores (Diagram is illustrative only.)

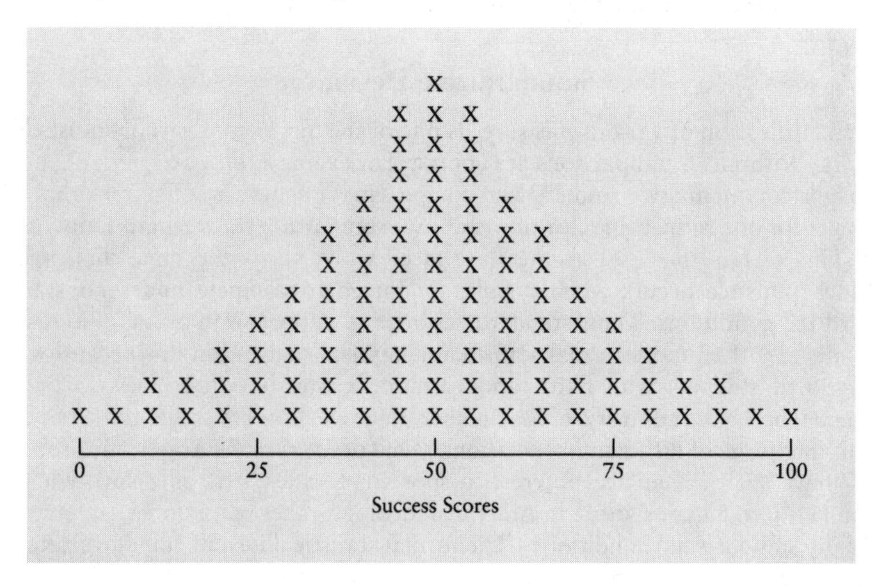

Frequency refers simply to how frequently something occurs. Let's look at the normal curve in Figure 7.1 and solve some simple problems. First of all, how frequently do we find a score of 50 in this distribution? All we have to do is count the X's above 50 and we find there are 12. How frequently does a score of 100 occur? We count the X's above 100 and find only one. A score of 50 is 12 times more likely than a score of 100. How many X's have a score of 50 or above? We can find this by adding all of the X's with a score of 50 or better; we find there are 50 X's in this category. Exactly half the distribution falls within a score of 50 to 100. How many have a score of 75 or better? We count those with a score of 75 or better and find that there are 13 such cases.

This illustration shows us an interesting property of the normal distribution. We can translate scores into percentages. We can also use these percentages to tell us something about probabilities. For example, go back to the problems in the paragraph above and ask the question: What is the probability of a score of 50? There are 12 such cases, so the probability is 12 out of 100 total cases, or a probability of 0.12. If you selected a case at random from the total of 100 cases, what is the probability that you would get a score anywhere between 50 and 100? There are 50 cases of such scores, so the probability would be 50 out of 100, or 50 percent (0.50). Now, what would be the probability, if you drew one of the X's out of the distribution by chance, of getting a zero score? There is only one of these, so the probability would be 1 out of 100, or 1 percent (0.01).

This is where the normal distribution shows its muscle: it can be used to make quick and easy estimates of probability. For the moment, set this feature on the back burner to simmer slowly while we investigate another aspect of the normal curve—how it relates to the standard deviation.

Standardizing Deviance

Standardization of various measures is one of the major concerns of statistical work. Without it, comparisons are not fair. For example, in sports it would not be fair to compare two runners who compete in the same race on a track that is longer for one runner than for the other. We standardize track length simply by making certain that both runners are faced with exactly the same challenge. Much injustice occurs when people are forced to compete under nonstandardized conditions. Thus, standardization is as important in ordinary human affairs as it is in more formal statistical analysis. We mention this because we have had students who find standardization repressive, or bad. We would argue that, to the contrary, it can be liberating when we are trying to be fair in our appraisals of different observations. (One of the reasons Americans probably have such a maniacal interest in sports is because it is one place where competition, a major virtue in American ideology, takes place under extremely rigidly standardized conditions.) The formal statistical idea of standardization and the folk notion of fairness are not far removed from each other. The same students who complain that the statistician's obsession with standardization is "uptight" are often quick to complain if they think an exam is not graded "fairly"—that is, in a standard manner.

Suppose you score 80 on a math exam and 70 on a sociology exam. On which exam did you get the better score? If you think before you answer, you will say, "It depends." On what does it depend? It depends on what your scores mean relative to all the other scores that were made on each exam. We need to know two basic things: (1) What was the average for each exam? and (2) How far above or below the average was your score? If the examinations were normally distributed, we can compare your performance on both tests.

Suppose the mathematics test mean score was 85, and the sociology test mean score was 75. You were five points below average in both tests. Does this mean you came up with the same performance on both? Once again, it depends. On what does it depend? It depends on how variable the scores were on the two tests. Suppose, for example, the range on the mathematics test was from 80 to 90. In this case, you got the lowest grade possible on the test. On the sociology test, suppose the range was from 0 to 150. In this case, you got a grade fairly close to the middle of the distribution—if it was a symmetrical distribution—and performed much better on the sociology test than on the mathematics test.

The standard deviation solves this last problem; it tells us how far a particular case is from the mean in terms of standard deviation units. In the math exam you were not far from the mean in absolute terms, but the exam displayed very little variability. Because there was little variability, even though you were not far from the mean in absolute terms, you were quite far from it relative to the limited variability in the distribution. You were at the extreme bottom of the variability in the exam. In the sociology exam, however, variability was high and you were near the middle of it. So, you did far better, *everything considered,* on the sociology exam.

Let's make the same point, this time using the concept of standard deviation. The standard deviation allows us to locate a score or value along a continuum, so that we always have a sense of how far that score is from the mean for the total distribution in terms of standard deviation units. If the score is three standard deviations from the mean, it is a long way from the mean. If it is one standard deviation from the mean, it is not too far away. If it is zero standard deviation units from the mean, it does not differ from the mean at all. This is important. To understand the utility of standard deviations, you have to gain a sense of how they relate to averages and provide a sense of variability as a standardized quality. If your distribution of cases is normal, the standard deviation relates to the average in precise and specific ways.

Look at the standard deviations, in Figure 7.2, for a normally distributed set of events and see how they relate to the mean. What does it mean to be zero, one, two, three, or even four standard deviations above or below the mean? It is important to gain a sense of what the standard deviation is telling you about deviation before moving on to the next stage of the discussion.

Figure 7.2. Areas Under a Normal Curve Determined by the Distance from the Mean in Increments of Standard Deviation

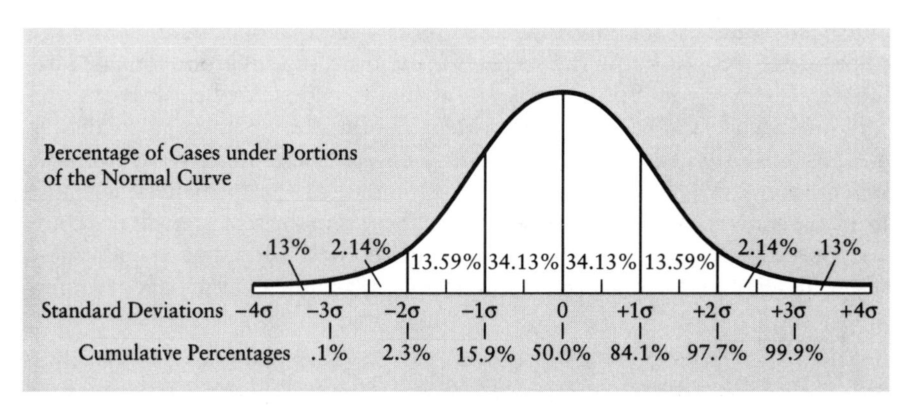

What does it mean to be three standard deviations above the mean? It means that a person or an event is exceptional—whatever the scale might be. If you are three standard deviations above the mean with respect to intelligence, you are extremely intelligent. Not many people are above you and a lot are below you. The nice thing about the normal curve is that when you and others are distributed along such a curve on the basis of, say, IQ scores, we can tell precisely how many people are above you and how many are below you if you have an IQ that is three standard deviations above the mean.

This kind of thinking calls for patient practice and carefully looking the scene over. You must, if you are to have a better understanding of what the normal curve is about, see clearly the relationship between the standard deviation and the general character of the curve. In brief, this relationship is one in which any particular value for a standard deviation divides the curve into two parts: one part is above the standard deviation and the other part is below it.

If the distance of the standard deviation unit from the mean is zero, how is the curve divided? By just guessing, estimate what percentage of the curve rests on either side of a line drawn above one standard deviation beyond the mean. What percentage rests on either side of a line drawn above two standard deviations? Three standard deviations? Four standard deviations?

Suppose an IQ test has a mean of 100 and a standard deviation of 10, and Marianna has a test score of 130. Then Marianna is three standard deviations above the mean. Let's presume she has also taken a general education test that has a mean of 130 and gotten a score of 140. The general education test has a standard deviation of 20. Marianna is not especially far from the mean. She is only half a standard deviation above the mean in terms of education. Marianna is, then, exceptionally intelligent, but only moderately above average in terms of her tested educational skills. So, we can say she is more intelligent than she is educated.

This is, in part, the power of the standard deviation. It permits us to compare, in *standardized* ways, the extent to which things differ from each other. Now we have an idea of why it is called a *standard* deviation. (Actually, when you think about it, the idea of a standard deviation is something only a statistician could dream up.) If we relied on raw scores alone, we could not say much. If we did not know the means for the distributions from which Marianna comes, we would not even know if she was above or below average in both intelligence and education. Knowledge of the mean permits us to answer the question of whether she is above or below average. If we want to know how far above or below average she is, or compare her performances, we must also have some idea of how much variance there is in the scales involved. Once we know the standard deviations, we can see that Marianna is extremely unique in terms of intelligence and not especially unique in terms of education.

THE NORMAL CURVE AND PROBABILITY

To say something is unique or common is also to suggest that it is improbable or probable. If something is typical, it is likely, or usual, or what is most possible. If something is extremely uncommon, then it is also improbable, or unlikely, or what is not usually the case. There is nothing complicated about this kind of reasoning. We think this way all the time. We make thousands of assessments daily of the ordinary, the typical, the unusual, the unlikely, the "weird," and so on.

This is what makes the newspaper interesting reading as you sip your morning cup of coffee—the newspaper, by definition, is concerned with what is news, not what is ordinary. So, one reads about a murder, a theft, a person who wins the lottery, an honors student in a local high school, a speech by a powerful politician, a battle on some distant shore, or a house that burns down. All of these events are atypical. Our most common source of information about the world is heavily biased in the way that it describes that world. If we rely on newspapers to evaluate our ordinary lives, we are likely to make some serious errors. For example, it is easy to get the impression from the news that crime is lurking in every doorway. While crime is a part of what is going on in the world, it is also true that during any particular year we probably will not be personally involved in a criminal assault.[2]

Let's get back to the normal distribution. In the previous section you were asked to estimate how the SD (standard deviation) can be used to divide the normal curve into two sections. You were also asked to estimate how much of the curve each section takes up. Now it is time to see more precisely the relationship between the standard deviation and the percentage of the curve that lies on either side of the standard deviation.

Table 7.1 is an abbreviated version of a standard table that appears in nearly all statistical texts. It shows the areas on either side of the standard deviation from the mean for selected values of the standard deviation. Notice that when the standard deviation unit from the mean is equal to zero—that is, when it lies right on the mean—the normal curve is divided into two equal halves. When we are one standard deviation above the mean, about 84 percent of the curve lies below the standard deviation and about 16 percent above it. When we are two standard deviations above the mean, about 98 percent lies below the standard deviation and only 2 percent above it. By the time we are three standard deviations above (or below) the mean, the curve is divided into two parts; the larger part takes up over 99.9 percent of the curve and only one-tenth of one percent remains in the smaller part.

Now we can use the standard deviation for obtaining some notions of how probable or likely a particular event might be. We can ask, for example, how likely a particular score is to fall within the area that extends above a point that is on or over 1.50 standard deviations from the mean (refer to Table 7.1 to find the answer). You will see that 7 percent of the cases fall on or above this point. Therefore, the probability of an event of this kind is about 7 out of 100, or 0.07. The likelihood of anyone's getting a score 1.50 standard deviations above the mean on any kind of normally distributed event is relatively low.

More than anything else, the normal curve is used in the assessment of probabilities of various kinds. Later we shall see its applicability to the vexing problem of how good a particular sample is. Only when we gain a sense of the relationship between the normal curve, standard deviations, and probabilities

Table 7.1. Areas in the Normal Curve above and below Specified Values of the Standard Deviation

Standard Deviation Units from the Mean	*Area in Larger Part (%)*	*Area in Smaller Part (%)*	*Probability of Falling in Smaller Part*
0.00	50	50	0.50
0.10	54	46	0.46
0.20	58	42	0.42
0.30	62	38	0.38
0.40	66	34	0.34
0.50	69	31	0.31
0.60	73	27	0.27
0.70	76	24	0.24
0.80	79	21	0.21
0.90	82	18	0.18
1.00	84	16	0.16
1.10	86	14	0.14
1.20	88	12	0.12
1.30	90	10	0.10
1.40	92	8	0.08
1.50	93	7	0.07
1.60	95	5	0.05
1.70	96	4	0.04
1.80	96	4	0.04
1.90	97	3	0.03
2.00	98	2	0.02
2.10	98	2	0.02
2.20	99	1	0.01
2.30	99	1	0.01
2.40	99	1	0.01
2.50	99	1	0.01
2.60	99.5	0.5	0.005
2.70	99.7	0.3	0.003
2.80	99.7	0.3	0.003
2.90	99.8	0.2	0.002
3.00	99.9	0.1	0.001
3.10	99.9	0.1	0.001
3.20	99.9	0.1	0.001
3.30	99.95	0.05	0.0005
3.40	99.97	0.03	0.0003
3.50	99.98	0.02	0.0002

can we begin to see how we might use the normal curve to tell us whether we probably have a good sample or probably have a bad one.

At the moment, you should try to gain a good intuitive sense of how standard deviations are related to probabilities. You should be able to make crude estimates. Later you can check your intuition against more specific figures. But first it is important to be able to make rough guesses that show you have a general idea of how the standard deviation relates to probability values—always, of course, *assuming that events are distributed normally.*

If we know a teacher with evaluations that place her 2.9 standard deviations above the mean, how typical is she? This is pretty far up the scale. She sounds like about one out of a hundred. We can look at Table 7.1 and see that she is actually closer to being one out of a thousand. If we know a doctor whose income places her 0.8 standard deviations below the mean, how typical is she? She is about one standard deviation below the mean. She is close to being in the lower third of all doctors ranked by income. (Be careful of the normal curve assumption, however. Income usually is not normally distributed. Therefore, this assessment must be looked on as a rough estimate.)

The normal curve is a great time saver for estimating probabilities for normally distributed events. As we will discover in the following chapters, determining precise probabilities can take a great deal of time and calculation, though modern computers *have* made these calculations simpler. Given the many problems that confront social researchers, estimates of probability are, perhaps, as good as precise determinations in much social research.

THE Z–SCORE

The **z-score** is a figure that tells us how far an individual or event is from the mean in terms of standard deviations. It is appropriately used only when a distribution is normal (though not all researchers are completely reliable when it comes to presenting z-score values for normally distributed data only). To obtain the z-score you need to find out how far an individual measure is from the mean and then divide that amount by the standard deviation.

Suppose, for example, that you have taken an ethnic tolerance test and scored 17 on it. The mean for the test is 12 and the standard deviation for the distribution is equal to 3. You were 5 points above the mean. If you divide this by 3, you find that you were 1.66 standard deviations from the mean. This information can be used to see how typical or unusual your ethnic tolerance score is by turning to Table 7.1. You are more than 1.60 standard deviations above the mean, so your score is higher than that obtained by 95 percent of those taking the test.

The following are the steps for establishing the z-score:

a. Subtract the mean from the score (be careful to note whether you get a negative or positive value).
b. Divide the difference by the standard deviation.

In algebraic expression, the formula is:

$$z\text{-score} = \frac{\text{score} - \mu x}{SD}$$

For example, suppose you swallow a pill that makes you 7′3″ tall. The mean height of people, let us assume, is 5′10″. The standard deviation is 3 inches. Height is normally distributed. Your z-score for height is as follows:

a. 7′3″ − 5′10″ = 17 inches.
b. 17 inches divided by 3 = 5.67 = z-score.

After swallowing the pill you find yourself 5.67 standard deviations above the mean in height. This makes you an extremely deviant individual indeed.

The nature of the z-score is relatively simple. It will be seen, in a different form, in the discussion of sampling, Chapter 10. This second form, derived from the **standard error,** is used to determine the reliability of generalizations based on samples. A sample that is two or so standard errors from a certain value is extremely "deviant" or unusual. In such an instance, either the presumed value or the sample becomes suspect. We mention this because a lot of people have trouble with standard errors. But the standard error is just a dressed-up version of our old friend the standard deviation. Anything that is two standard deviations from some value is relatively unique. Any sample that is two standard errors from some value is unique in the same way.

SUMMARY

As we saw at the beginning of this chapter, the normal curve appears in a great variety of settings. More important for statistical analysis, if an event is distributed normally, we can use the standard deviation to tell us a great deal about the probability of individual cases. Especially significant is the fact that the standard deviation can be used to tell us about the probable occurrence of events. Something that has a high standard deviation value is extremely unlikely or improbable. Something that has a standard deviation close to zero, which implies little deviation from the mean, is more ordinary, typical, or probable. The normal curve is especially useful in evaluating the reliability of samples. (This is really just a formalization of our commonsense understanding of the world around us. That is, we know that what is odd or weird is also usually rare or improbable.)

THINKING THINGS THROUGH

1. Why is the assumption of normalcy necessary if you are using the standard deviation to determine probabilities or percentages for a given distribution?
2. Can you think of some places where the normal curve appears that were not mentioned in this chapter?
3. What happens when a distribution has a value for the SD equal to zero? Is such a distribution normal? Is this kind of distribution superior to other

distributions with respect to predicting the character or value for individual cases? Using the z-score, show why or why not.

4. Suppose you discover that the check-out clerk at the store where you buy groceries has cheated you out of a small sum each time you get change. You realize that this has occurred on twelve separate occasions, with one exception. What do you conclude? Do you need the assumption of normalcy to come to a conclusion?

5. The discussion of the normal curve in this chapter involves two fundamental concepts: averages and variability around averages. The discussion deals only with the case of variability that is "normally" arranged around the mean. What happens when you have a lot of normal variance? What happens when you have very little? Why is normally distributed variability easier to deal with than variability that is not normally distributed? (This question is central to seeing some of the problems that arise in other statistical areas such as sampling, the examination of correlations, and the analysis of variance.)

6. Using Table 7.1, solve the following problems:
   a. Duane, Alice, and Kim take three different examinations. The test scores for all three are normally distributed. Duane's test provides him with a score of 33, where the mean and SD of the test scores for the group as a whole are 55 and 11. Alice's score is 137 on a test where the mean for the group is 125 and the SD is 13. Kim's score is 1,486, with a mean for the group of 1,000 and an SD of 500. Which person did best? What is the probability value associated with each person's score (the probability of getting a score that high or higher)?
   b. Kamal tells Priscilla that she is "one out of a thousand" when it comes to charm and beauty. What would Priscilla's z-score be if this were true? What do you have to assume to answer this question?
   c. Suppose you are applying to medical school, and you receive a letter from the dean telling you that you scored 1.20 SD's above the mean on the medical school application test. The medical school only takes people in the upper ninety percent of scorers. Should you think about reconsidering your future? Why?

7. Lincoln said that God must have especially liked ordinary people because he made so many of them. If we define as "ordinary" anyone who is not above the seventy-fifth or below the twenty-fifth percentile, (the semi-interquartile range) what are the z-score limits that define God's "ordinary" people?

8. Many social groups achieve a sense of being special by putting people through tests that make them feel distinctive. The U.S. Marines, for example, test their recruits rather severely. If they test too severely, however, they will lose all their recruits. On the other hand, if they test too easily, they will not provide the recruits with a sense of accomplishment. Suppose it is decided that in order to stay in your unit in the Marines, you must do as well as or better than the lowest ten percent. You take three tests. The following are the means and SD's for these three tests:

   Test 1: Mean = 45; SD = 6.

   Test 2: Mean = 1,245; SD = 185.

   Test 3: Mean = 1.45; SD = 0.04.

What would be the minimum grade on all three tests that would permit you to stay with the unit?

Notes

1. The concept of randomness seems simple enough, and it is generally assumed that we know what we are talking about when we speak of random numbers. The random number tables used by social scientists are sufficient for the tasks they are assigned. However, at deeper levels, the concept of randomness proves extremely difficult to deal with. For example, recent tests reveal that the most sophisticated random number generators in use today have biases built into them, which means that the numbers they generate are not truly random. For delicate problems in physics, requiring truly random numbers in huge amounts, this bias has created headaches.
2. The situation is more complicated than this simple assertion implies. While it is true that most people are not assaulted or criminalized in ways that commonly make up newspaper material, it is also true that some forms of assault are so common that they do not appear in the newspapers as often as they should. It was not until the past four or five years that domestic abuse and child abuse, both extremely common forms of assault, began to appear regularly as issues of concern in newspaper articles and columns.

8

PROBABILITY, CHANCE, AND UNCERTAINTY: PLAYING THE PERCENTAGES (PART I)

"The theory of probabilities is at bottom nothing but common sense reduced to calculus."

—Pierre Simon De Laplace

"We do not what we ought; / What we ought not, we do; / And lean upon the thought / That chance will bring us through."

—Matthew Arnold

"Chance is perhaps the pseudonym of God when He did not want to sign."

—Anatole France

For the individual who seeks total control over his or her affairs, the idea that life is largely a matter of chance is disturbing; the fact is, though, we cannot control all aspects of our lives. In fact, it is an engaging philosophical amusement to wonder just how much of it we *do* control. Some argue we can control whatever we wish—that what we want to be is up to us. Others argue that we control nothing—that life is a matter of luck, chance, fate, or happenstance. This is a profound problem, and we cannot deal with it here. Instead, we shall take a limited look at chance and probability.

Let's begin with a notion most of us are familiar with: "playing the percentages." This is a good place to begin because we know what percentages are, and it does not take much thought to see that percentages and probabilities are closely related. For example, suppose a lottery has 99.99 percent losing tickets and 0.01 percent winning tickets. You buy a ticket. What is the likelihood that you bought the winning ticket? Obviously playing the percentages is an uncertain way of doing things, though there is always the possibility of beating the odds. Adventure movies draw on our awareness of probabilities, creating tension by placing the star of the film in situations where the likelihood of survival has a low percentage value (and yet the star always survives). As ordinary as these examples are, they lead immediately into an awareness that reliance on probabilities, which are a form of average, is tricky.

We talked earlier about how important it is not to use averages to describe individual cases. This is especially true with probabilities, and for the same reason. Probabilities are values that pertain to entire distributions of events rather than to individuals. A really good estimate of probability is useful as a device that converges toward a specific figure over an infinitely *repeated* series of happenings. For example, it is easy to see that the probability of getting heads in the toss of a coin is 0.50 (heads constitutes fifty percent of the coin, tails constitutes the other fifty percent). If you toss a coin over and over and over and over and over, you will be right fifty percent of the time if you call heads each time, call tails each time, randomly call heads or tails, or call heads and then tails in a consistent fashion each time. In fact, *no matter what you do,* you are not going to be able to create a system that will enable you to do better than converge toward fifty percent accuracy at guessing individual outcomes over the long run. This is a hard lesson for some people to learn. Truly probabilistic events cannot be transformed into nonprobabilistic events by any kind of system.

PROBABILITY, BEATING THE ODDS, AND AVERAGES

Probabilities do not refer to specific cases, any more than the arithmetic mean necessarily describes any specific individual. Probabilities are averages. Moreover, the average event they refer to happens in a place characterized by an infinite repetition of events. In order to think about probability, you have to think about repetitive acts. The coin is tossed over and over and over and over and over. After a while, it begins to show a typical pattern of heads and tails. This pattern might, for a while, favor heads, or it might favor tails. In the long run, however, the numbers of heads and tails will balance out. (Keep in mind that no one has ever tossed a coin an infinite number of times.)

Sports provide the most highly standardized repetitive forms of activity for modern mass audiences. It is in the realm of sports that we seem to hear the most talk about odds, beating the average, probabilities, chances, and so on.

We have asserted that a probable event such as a coin toss cannot be predicted with better than fifty percent certainty. It should be obvious that if you can predict the outcome accurately, you are no longer faced with the problem of uncertainty. Truly probabilistic events are immune to being "doped out," or controlled by some kind of system. We mention this because some people think that studying statistics will enable them to beat the odds. But this can happen only if something that people believe is a matter of chance is, in fact, not really a matter of chance. It is one of the most difficult of statistical realities to accept, but pure chance events always force you to live with uncertainty.

Probability in Everyday Life

The fact that no system exists for transforming truly probable events into nonprobable events becomes especially profound when we realize that a great variety of human social experiences are probable rather than certain. For example, communication is a probable rather than a mechanical process. We try to convey a message to someone else. The other person might or might not comprehend the content of our communication. Teachers, for example, are commonly astonished by the distortions of their lectures that appear in final examinations, especially in essay exams.

The phrase "You can't please 'em all" refers to the fact that human interaction is uncertain (that is, variable) and therefore loaded with probability implications. For example, no matter how good a movie might be, a fair percentage of the viewing audience will give it a bad rating. This is a good place to introduce an aspect of probability that we shall talk about later: the problem of independence. Suppose you see a movie with a friend and you give the movie a four-star rating. What is the probability your friend will give it a one-star rating? It's likely to be low because you probably attend movies with friends who share similar tastes. If you liked the movie, your friend probably liked the movie. In dealing with probability, you must always remain alert to the problem of whether the events under discussion are truly independent of each other or are related. If the events are not completely independent, this makes mathematical solutions to probability problems considerably more complicated.[1]

Let's come back to considering how common probability is in day-to-day life. Here are seven illustrations of ordinary moments that involve the casual assessment of probabilities. Think of several additional examples of your own before moving along with your thoughts about the nature of probability and chance.

1. You marry. Your assessment of the probability of a happy, life-long wedded experience is high. (Empirical statistics suggest something different.)
2. You are a child in the ghetto. You devote all your time to playing basketball instead of studying. You think you have a chance to make $100,000 a year by becoming a great sports hero. (Your chances of making such a salary in the world of professional sports are actually very low.)
3. You refuse to go aerobatic flying with a friend because you think it is dangerous. That is to say, you are convinced that the probability of getting hurt is high. Instead you go out on a crowded highway on your bicycle for a fifty-mile tour. (While probabilities here are difficult to determine, bicycles on highways are certainly quite dangerous.)
4. You have a chance to go out with a Phi Beta Kappa and turn it down because you know that brainy students are bad dates. (Here you have set a probability level at certainty by reducing variability to zero and then acted accordingly, when the evidence suggests that intelligent people are probably no more or less fun than anyone else.)

5. You are working as a clinical psychologist with a patient who is certain he is going to be killed by a meteorite that is going to fall on his head. The case is interesting because you know the probability of such an event is so low as to be absurd. (Six years after you leave the case you read in the papers that your patient was struck and killed by space debris from an old American satellite that fell out of orbit.)
6. You have a hamburger at a fast food outlet. There is a small but definite probability your meal will seriously poison you. (Why does this probability not deter you from ordering your sandwich?)
7. You are on a safari in the heart of Africa, looking for a rare—that is, improbable—type of orchid. You come across the camp of some other travelers. To your amazement, a member of the other camp is your old high school English teacher. Your amazement derives from the fact that both of you are aware that this is a seemingly improbable event. (There are so many possibilities for improbable things to happen, however, that nearly everybody experiences, at some time or other, an incredibly weird or improbable event.)

The point of these examples is to get you to relax before entering the thorny thickets of probability. Most of what is to come is, as Laplace points out in the quotation at the beginning of this chapter, simply good common sense. What we have tried to do in this discussion is to make you more consciously aware of thought processes that you have relied on since before you could walk.

What Is Probability?

What is probability? Mathematicians and philosophers are still puzzling over this question, and we are not about to resolve it here. Instead, we shall grab this philosophical bull by its horns and define probability as simply as it is possible to define it. Probability is an assessment of the extent to which something falls, *on the average,* between the two extremes of being perfectly certain of occurring or perfectly certain of not occurring in a specified time period.

The probability that we who are alive today shall all die within the next two hundred years is virtually certain. The probability that we shall not show any signs of aging as we grow older is zero. In between such perfect certainty of occurrence or certainty of nonoccurrence, however, are various other probabilities. The probability that we shall have an automobile accident in a given year is roughly one in four or five. The probability that our first child will be a boy is slightly greater than the probability it will be a girl. The probability that nuclear war will occur this year is greater than zero and less than certainty, but we do not know how to determine this probability with any accuracy.[2]

Percentages and Lotteries

The first thing, then, about probabilities is that they can be expressed in terms of a number. This number is 1.00 for events that are certain to occur. It is 0.00 for events that are certain not to occur. It is something in between for events that might or might not occur.

The toss of a coin is equally likely to produce a head or a tail. Therefore, the probability of a head is one out of two, or 0.50. The probability of tossing a six on the toss of a single die is one out of six (since the die is six-sided), or 0.17. The probability of drawing a heart from a deck of regular playing cards is one out of four, or 0.25. Notice that these numbers are also the percentage values for these events.

The numerical statement of probability is a ratio—as is a percentage. It is a matter of drawing a line, then putting one number above the line and another below it. If we want to determine the probability of winning the lottery we have to determine two numbers. What might these two numbers be? What would you guess goes on top of the line and what below? What kind of ratio do you get? What is the percentage of losers (people who do not hit the big winning ticket)?

The lottery problem is straightforward. The number on top of the line is the lone winner. This is the *probable event,* or the event with which we are concerned—also commonly called the *favored event.* We want to know the probability of being a winner. There is only one big winner, so the first number is 1. The second number is the total number of people playing in the lottery. This will be a large number. A typical lottery involves several million players. Suppose we assume 4,236,785 players. Then the probability (P) of winning is computed as follows:

$$P = 1 \text{ divided by } 4{,}236{,}785 = 0.0000002$$

We can easily convert this figure into a percentage. The percentage of losers in this case is 99.99998 (Actually, it is larger because the lottery corporation also takes its "cut").

A person's chances of winning the lottery are no better than one in five million. (When you buy a ticket, what, for a fleeting moment, do you think your chances are? To what extent do lotteries depend on people incorrectly assessing their probabilities of winning?)

Making Mistakes

At this point there is nothing to probability other than putting one number above another and doing simple division. The difficulty with probability, however, is that nearly everyone, including probability experts, gets fooled into putting the wrong figures above and below the line. They wind up concluding that a particular probability is either too low or too high, and they make bad judgments accordingly.

Here are some common ways in which people put the wrong number above and below the line:

1. Chuck wants to date Tamiko. He does not ask her, however, because, as he likes to put it, "I don't stand a chance." He assumes Tamiko gets asked out a lot, so he has put a large number under the line. He also assumes Tamiko only responds favorably (the probable or favored event) to one or two requests. He puts a small number above the line. He is not aware, though, that everyone else thinks the way he does and, as a result, no one asks

Tamiko out. The number below the line is actually small. Tamiko has, in fact, had only two requests, and she responded favorably to both. If we solved for *P* in this case, what value would we get? How did Chuck's assessment of probabilities generate bad judgment?

2. Suppose you came from a high school where you never had to study for tests and you always managed to get good grades. You took a total of 76 tests over four years in school, and you got good grades on 75 of them—a ratio of 0.99. How do you assess your likelihood of getting a good grade? Now, however, you find yourself in a university setting where the ratio of good grades to tests given for the student population overall is 0.32. What is likely to happen to you in your freshman year? Why?
3. Terrorists kill sixteen tourists in a country you were planning to visit. You decide to visit another country where thirty tourists were killed in a train wreck. You are trying to decide which place has the higher probability of being safe. What else do you need to know to determine which country is best? (As it turns out, five million tourists visited the place where the terrorists carried out their vile deed. The country where the train wreck happened had sixteen thousand tourists. How does this change the probabilities?)[3]

Probability is tricky because it demands that we know *exactly* what goes above and below the dividing line before we make our final calculations. Once we know that, and know it well, then we can assign a specific, numerical value to a particular probability. It is important to understand that if we cannot put specific values above and below the line, then we cannot know the probability of occurrence of any event with accuracy.

Even the most mathematically sophisticated people in the world get drawn into schemes to beat probabilistic games, and they often make serious blunders. In the 1970s, a group of physicists and mathematicians at the University of California at Santa Cruz were certain they could beat the roulette tables in Las Vegas. They tried extremely clever tricks, including hiding a small computer in a shoe, but they were unable to beat the odds. Perhaps that is yet another way of defining probability: it is a sequence of events for which the predictability of individual occurrences is, *in principle,* unmanageable, while the collective character of the events, viewed as a whole, is stable. We know, for example, that car accidents are more probable on Memorial Day and Labor Day. However, we do not know which specific individuals are going to have the accidents—and there is no system that can solve this problem. If we could answer the question of who would have the accidents, then having the accident would no longer be a matter of chance.

MONTE CARLO PROBABILITIES

Early studies in probability were based on gambling problems. Gambling situations offer the simplest introduction to probability. Let's begin with the simplest of gambling games. You toss a penny and bet whether it will come up heads or tails. We assume the coin is fair—that is, that it is not rigged in some fashion. We also assume that each toss is independent of what happened in the previous toss.[4] Suppose we guess heads. There are only two possibilities, of which one will be the favored event (a head). Before you read the next paragraph, use your intuition to answer this question: What goes above and below the probability line?

We know, already, that we want to put the favored event in the numerator, and that consists of one condition, "heads." Therefore, we shall put this possibility (1) above the line. But what goes under the line? Only two things can happen, a "head" or a "tail," so we put them (2) under the line. *What goes below the line is a careful summation of all possible things that can happen along with the favored event:*

$$\frac{\text{HEAD}}{\text{HEAD or TAIL}} = \tfrac{1}{2} = 0.50$$

There are two possibilities below the line and one above. The ratio is, then, $\frac{1}{2}$, or 0.50; thus, $P = 0.50$.

What is the probability of getting two heads, one after the other, in two tosses? We are looking at a single event—just two heads. There are four things that can happen in two tosses (H = heads and T = tails):

HH *HT* *TH* *TT*

We are interested in only one favored event, *HH*. There is one favored event and four things that can happen. The probability is easy enough. It is equal to 1 divided by 4, or 0.25. So, $P = \frac{1}{4}$ or 0.25. The trick, then, is to have a sense of what goes above the line and what goes below.

What if you have fifty coins? How many different ways could they fall? Two coins can show four different ways of coming up or four different **combinations.** Obviously fifty coins can be combined a huge number of ways. To find out how many ways a number of events (N) can be combined (where the events are divided into two classes such as heads or tails, greens or blues, winners or losers, and the like), simply raise 2 to the number of events. Where $N = 50$ coins, the number of ways such coins can be combined is quite large. Here are three possible ways in which fifty coins could be combined (H = heads and T = tails):

HHTHHTTTTHHHHTTTTHHHTHTHTTTTHHTTTTHHTHTTHHTHTHHTHH
TTTTTHHHHHTTTTHHHHHTTTTTTHHHTTHHTTTHHHHHHTTTHTHHHH
TTHHHTTTHHHTHHTHHTHHTHHTHHTHHTHHTTTTTTTTHHHHHTTTTT

This is already enough to give anyone a headache—and these are only three of the many possible combinations. Obviously this approach to probability problems, though it has the advantage of being clear, has the disadvantage of driving a person crazy. It would be nice to have a simple way to find out how many possibilities exist for fifty coins, and there does happen to be an easy way out of the problem. Here it is:

> Where there are two possible outcomes, in this instance a head or a tail, the total number of outcomes[5] for N number of trials is equal to 2^N.

Recall that for the case where we tossed the coin two times, we had two trials. Therefore, all possible outcomes would equal

$$2^2 = 2 \times 2 = 4$$

Now let's answer the question of how many possible outcomes we would have with fifty tosses of the coin. We would have

$$2^{50} = 1{,}125{,}900{,}000{,}000{,}000 \text{ possibilities}$$

Who would want to sit and write down all of these possibilities? It is sufficient simply to know the number of possibilities, because now we have a specific value to put in the denominator of the probability ratio. Of this large number of possibilities there is only one that consists of all fifty tosses being heads. Therefore, the possibility of tossing fifty straight heads with an unbiased coin is about one out of one quadrillion.

Now, something that is important to understand is this: although the probability is extremely small, it is still possible. If you toss those coins long enough, sooner or later all fifty are going to come up heads. But it would most likely take a lot of coin tossing (of course, it could happen on the first toss—you never quite know how things are going to go in the world of probable events). Probabilities are a statement of what you can expect, *on the average* (there's that concept again). If you tossed fifty coins once a second, *on the average* it would take you a long time to hit fifty heads in a row.

There are 60 × 60 × 24 × 365 (seconds times minutes times hours times days), or 31,516,000 seconds in a year. We would have to toss fifty coins 1.1 quadrillion times, on the average, to get all heads. Since 31,516,000 goes into 1.1 quadrillion about 36 million times, it would take over 36 million years, on the average, to hit fifty coins all in a row, on a free toss with all coins independent of each other. The only point to this little exercise is to gain a deeper impression of what a low probability we get for an event of this kind. Again, think of a student trying to guess his or her way through a fifty item true-false test. If each item is a pure chance event, it will take the student about 500,000 lifetimes, on the average, to make a perfect score on the test. How many students are aware of how low the probability of such a venture is? What other factors interfere with the assumption of independence and pure chance in this illustration?

Remember, we are talking about what we expect *on the average*. It might take longer or it might happen sooner; this is an imaginary average. That does not make it useless, but it is good to keep in mind that statistical reason plays with some interesting images, and that a lot of it relies on imagination as well as on observation.

What you should focus on at this point, however, is that the formula 2^N makes things easier as we move into the realm of determining how things can combine in different ways. In this discussion of probability we shall not move beyond simple two-way problems—heads or tails, men or women, black or white, true or false, and so on. Even so, the number of ways two things can combine in different-sized groups or over a large number of trials is not readily intuited or visualized, which means we must depend on a few formulas. At the same time, however, we should not abandon our intuitive sense of things. Mathematics and intuition should work back and forth, each acting as a check on the errors that can come from a one-sided approach to statistical work.

People often feel uncomfortable with large numbers, so it might be a good idea to relax for a moment. Besides, most of us do not intend to make a living tossing coins, so it may seem that the relevance of this discussion is slipping away. So let's bring things back to focus with an ordinary application of what we have just learned.

The logic of combinations applies nicely, for example, to true-false tests that are now so so popular in American schools. Such tests are popular for a number of reasons. Some students like them because they feel they have a chance of getting an A on a true-false test even if they did not study. From the above discussion of coins, however, we can see that on a fifty-item test, the likelihood of getting every answer right purely by chance (the equivalent of tossing fifty straight heads) is so low as to be ridiculous.

What would be the probability of getting every item right, strictly by chance, on a true-false test that consisted of twenty items? Before you move on to the answer in the next paragraph, stop for a moment and try to solve the problem on your own. There is only one way all twenty items could be answered correctly (C = a correct answer):

CCCCCCCCCCCCCCCCCCCC.

Now, how many different ways might the test be answered? That is solved by raising two to the twentieth power. What goes in the numerator? What goes in the denominator? If you have it right, all you have to do is make the division with a calculator or estimate it in your head.

Here is the solution: The answer would be one out of two raised to the twentieth power, or

$$P = \left(\tfrac{1}{2}\right)^{20} = \frac{1}{1{,}048{,}576} \text{ (or about one in a million)}$$

Let's pause now to consolidate what we have just discussed. Probability is not easy and is best taken in small doses. To determine if you have a good sense of what has gone before, see how you fare with the following questions. If you have trouble with any of them, go back and review the above material before going on.

1. There are thirteen spades in a regular deck of fifty-two playing cards. If you were determining the probability of getting a spade on a random draw of a card, what would you put in the numerator and what would you put in the denominator of the probability ratio? What does the ratio prove to be? Is this in keeping with your intuition?
2. There are two red queens, two black jacks, and one three of hearts in a regular deck of cards. What would you put in the numerator of the proba-

bility ratio to determine the probability of getting any one of these cards on a single draw? What is the probability?

3. You decide to boost your chances of winning the lottery by buying 40,000 tickets. A total of 7,387,543 lottery tickets are sold. What goes in the numerator and what goes in the denominator to enable you to figure out your chances of winning? What are your chances? If your chances of winning are a particular value, then what would your chance of not winning be?
4. A radio station is sponsoring a promotional gimmick in which listeners each week receive a gold star or a blue star if they send in a self-addressed card. The stars are given out randomly—that is, each person's chance of getting a gold star or a blue star is 0.50. At the end of six months the station has said it will award a prize of $50,000 to any person who has collected twenty-four gold stars. What is the probability of winning this prize?
5. It is a season in which no competing team in the Big 8 dominates any other team. Every game will be pretty much up to luck. The teams have to play a schedule of twelve games. What is the probability of any given team having an undefeated season?

If these problems seemed pretty easy (as they should have), then we can move along to something that might challenge you a bit more. Keep in mind, as we move through the thickets of probability, that our primary concern is to determine what goes above and below the line that establishes the probability ratio.

One of the above questions was about something we have not discussed, but which your intuition should allow you to answer. The question asks for the probability of something *not* happening, if you know the probability that it will happen. If, for example, the probability of drawing a spade from a deck of cards is 0.25, what is the probability of not drawing a spade? It would have to be 1.00 minus 0.25, or 0.75. If your probability of living to the age of 65 is 0.87, then your probability of not making it to the age of 65 is 0.13. This simple rule is useful in solving numerous probability problems.

PLAYING WITH COMBINATIONS

We have seen how to determine how many different combinations of heads and tails you can get for any given number (N) of coin tosses (the number) is 2 raised to the Nth power). We now need to complicate the problem a bit further. Consider the case of a true-false test of three items. Any given item can be answered right or wrong. We already know, from our knowledge of the formula of 2 to the Nth power, that we can only get eight different possibilities, that is $2^3 = 8$. We can plot all eight possible happenings by making a little chart, as shown below. One possibility is that all three are correct. Another is that all three are wrong. In between these two possibilities we have various mixes of right (R) and wrong (W) answers.

RRR	WRW
RRW	RWR
RWW	WWR
WRR	WWW

For this test, let us assume that having two or more right answers means you pass the test. We can look at the chart and see right away that there are four ways (or four favored possibilities) in which you can get two or more right answers. These favored possibilities are

RRR	RRW
WRR	RWR

You should not have much trouble figuring out what the probability of passing just by chance might be. That is, what goes in the numerator and what goes in the denominator of the probability ratio? The number of favored events is four and the number of all possible happenings is eight. The probability of passing, then, is

$$P = \tfrac{4}{8} = 0.50$$

The logic is clear enough. However, we are stuck with the fact that as soon as we start moving into big numbers, the problem of charting the combinations becomes impossibly tedious. It would be nice to have an easier way to do things. Once again a simple formula leads the way. Before we take a look at the formula, however, we should rely on our intuition. Without a good intuitive grasp, we are likely to rely too mechanically and unthinkingly on formulas.

Think about the following situation for a moment. Suppose a particular school, wanting to be absolutely fair in its treatment of students, selects students for admittance purely by chance. Within the community from which the students are selected there are the same number of girls and boys. The school selects twenty new students every year. Now, what we want you to think about intuitively is how you would react to the events listed below. Would a particular event be highly probable, improbable, or somewhere in between? Just using your intuition, indicate which choice you would make.

Event	*Probability*		
1. All girl students selected.	High	Medium	Low
2. No girl students selected.	High	Medium	Low
3. Between 10 and 13 boy students selected.	High	Medium	Low
4. Between 9 and 14 girl students selected.	High	Medium	Low
5. Between 0 and 20 boy students selected.	High	Medium	Low
6. Exactly 10 girl students selected.	High	Medium	Low

If your intuition is humming the way it should be, you are aware that the probability that all twenty students would be girls or boys is 1 over 2 raised to the 20th power—a low probability. The answer to both #1 and #2 above, then, is low. Your intuition should tell you that getting between 10 and 13 or between 9 and 14 of one group or the other should be fairly high; this should be a common kind of happening, in part because it can happen in so many different ways. The answer to #3 and #4, then, should be high. Obviously, the range of 0 to 20 boy students covers everything that can happen out of all possibilities and is therefore dead certain, so the answer to #5 should be high. Question #6 is tough. Our intuition should have trouble with this one. We know that getting a mix is common. But getting an exact fifty-fifty mix does not seem all that likely. Here we might want to hedge a bit with an answer of medium, or perhaps low.

A simple formula gives us a precise way of dealing with the possibility listed as item 6. above. If we have 20 individuals and they can consist of any arrangement of boys and girls, going from all 20 being boys to all 20 being girls, then let *r* be a number representing how many boys are in the total set of 20 individuals. We first want to find out how many ways this can happen. Let's first consider four situations: all 20 are boy students; all 20 are girl students; 19 are boy students; and 19 are girl students. If *r* is the number of boy students, then we are talking about $r = 20$, $r = 0$, $r = 19$, and $r = 1$. The solution to these four cases happens to be very easy.

There can be only one way in which all the students are boys (B) or all are girls (G):

B B B B B B B B B B B B B B B B B B B B
GGGGGGGGGGGGGGGGGGGG

There can be only twenty ways in which there is just one boy or one girl in the group. Here is the case for one girl:

GBBBBBBBBBBBBBBBBBBB	BBBBBBBBBBGBBBBBBBBB
BGBBBBBBBBBBBBBBBBBB	BBBBBBBBBBBGBBBBBBBB
BBGBBBBBBBBBBBBBBBBB	BBBBBBBBBBBBGBBBBBBB
BBBGBBBBBBBBBBBBBBBB	BBBBBBBBBBBBBGBBBBBB
BBBBGBBBBBBBBBBBBBBB	BBBBBBBBBBBBBBGBBBBB
BBBBBGBBBBBBBBBBBBBB	BBBBBBBBBBBBBBBGBBBB
BBBBBBGBBBBBBBBBBBBB	BBBBBBBBBBBBBBBBGBBB
BBBBBBBGBBBBBBBBBBBB	BBBBBBBBBBBBBBBBBGBB
BBBBBBBBGBBBBBBBBBBB	BBBBBBBBBBBBBBBBBBGB
BBBBBBBBBGBBBBBBBBBB	BBBBBBBBBBBBBBBBBBBG

Now, if you try to produce a diagram like this for the case of ten boys and ten girls, you are going to go crazy trying to list all the combinations that are possible. Let's use a formula, then, to solve this problem and see what kind of an answer we get. This formula looks complicated when working out a specific problem, but it is simple enough if you work through it slowly:

$$\frac{N!}{r!\,(N-r)!}$$

The exclamation points here are not for emphasis. In this case the exclamation point is a mathematical symbol for the *factorial;* it simply asks you to take a number and multiply it by its next lower integer, then by the next lower, and so on. The whole thing is easily illustrated by showing that where $N = 6$, $N!$ would equal

$$6 \times 5 \times 4 \times 3 \times 2 \times 1 \text{ or } 720$$

(We should also mention, at this point, that it is conventional to let 0! equal 1.)

Now, N in our school problem is equal to 20. We can set the number of girl students at 10 and solve for the problem we encountered in event #6. The solution is

$$\frac{20!}{10!(20-10)!} \text{ or } \frac{20!}{10!10!}$$

Solving this problem is not as difficult as it looks, and it certainly is nowhere near as difficult as trying to diagram all of the combinations possible. We need to be patient here.

The value in the numerator is 20!. This is a matter of multiplying 20 × 19 × 18 × 17 . . . . . until we reach 5 × 4 × 3 × 2 × 1. We have 10! times 10! in the numerator. Here is what we have to calculate:

$$\frac{20 \times 19 \times 18 \times 17 \times 16 \times 15 \times 14 \times 13 \times 12 \times 11 \times 10 \times 9 \times 8 \times 7 \times 6 \times 5 \times 4 \times 3 \times 2 \times 1}{10! \times 10!}$$

If nothing else, this gives you some idea of what your computer has to go through to make this calculation. Before leaping to your calculator, however, you should do at least one or two problems of this sort by hand to begin to see that they are not as intimidating as they seem to be. In the above division problem there are a lot of values in the denominator that can be divided into values in the numerator, thereby simplifying things considerably. Cancel out as much as you can. For example, the numerator has 10 × 9 × 8 × 7 × 6 × . . . and in the denominator we have 10!. The two values are the same and can therefore cancel each other out. So, we can simplify the problem to the following:

$$\frac{20 \times 19 \times 18 \times 17 \times 16 \times 15 \times 14 \times 13 \times 12 \times 11}{10 \times 9 \times 8 \times 7 \times 6 \times 5 \times 4 \times 3 \times 2 \times 1}$$

This value can be reduced, through further cancellations, to

$$\frac{19 \times 2 \times 17 \times 2 \times 13 \times 11}{1} = 184{,}756$$

There are 184,756 ways you could get a mix of 10 girls and 10 boys in the class of 20 students. Who would want to diagram that many combinations? The formula is a lot easier to deal with. Remember, too, that there are 2^N, or 1,048,576 total possibilities in selecting the class of 20 students.

So, we now know all we need to know to determine exactly what the probability of getting a perfect mix of ten girls and ten boys selected by the school. It would be

$$184{,}756/2^N = 184{,}756/1{,}048{,}576 = 0.18$$

Our intuition was right in telling us that an exact mix of ten boy students and ten girls would be somewhat unlikely on a pure chance basis.

Let's stop at this point and catch up on all that we have discussed up to now—it's a great deal of information. To hit the high points:

1. We have seen that probability is just a matter of what is to be put above and below a line of division.
2. The event whose probability we are interested in goes on top of the line, and all of the possibilities from which it might spring are placed below the line.
3. When we are interested in the probability of various mixes of events, we need to rely on basic formulas that can tell us how many mixes are possible.
4. The formula 2^N gives us the total number of ways in which two possibilities can show up in N trials. For example, if you were faced with the possibility of either meeting or not meeting a friend on each day of the month of April, there would be 2^{31} different ways this could happen, starting with not meeting her on any of the 31 days, meeting her only one of the 31 days, meeting her on any two days of the month, any three days of the month, and so on up to meeting her on all 31 days.
5. If you want to know how many ways a specific mix might occur, you need the formula

$$\frac{N!}{r!(N-r)!}$$

to tell you the answer. For example, how many ways could you possibly meet your friend 5 times out of the month and miss her the other 26 times? The answer is 169,911 different ways. Obviously this is a point where a computer comes in very handy and intuition is not readily able to deal with the large numbers involved.

Now let's return to the problem of the school that picks twenty students each year at random from a community of girls and boys. Twenty students are selected for each new class. How many ways can the selection go? In Table 8.1 the numbers of possible mixes are presented for each possible case. Notice in this table that the total number of ways things can take place is 1,048,576—the same number we get by taking 2 to the 20th power. There is something else about this table that you should have noticed: the distribution looks normal.

This example allows us to gain a new perspective on some of the issues that arise in the area of human relations and sex discrimination. Suppose the school board, on a given year, selects a new class that has seven girls and thirteen boys. Is this evidence of discrimination? The answer is not easy to determine. The probability of getting this many girls, or fewer, is roughly 0.13, or about 13 out of 100 times. (Simply add up the percentages that involve getting 7, 6, 5, 4, 3, 2, 1, or 0 girls in a class of 20. It turns out to be 13.2 percent.) This is a fairly

Table 8.1. Mathematical Probability Distribution for Number of Boys or Girls Selected at Random from a Sample with $N = 20$, from a Population with an Equal Number of Either Sex

Number of Boys Selected	*Number of Girls Selected*	*Number of Different Ways This Can Happen*	*Percent of Total*
0	20	1	0.00+
1	19	20	0.00+
2	18	190	0.02
3	17	1,140	0.11
4	16	4,845	0.46
5	15	15,504	1.48
6	14	38,760	3.70
7	13	77,520	7.39
8	12	125,970	12.01
9	11	167,960	16.02
10	10	184,756	17.62
11	9	167,960	16.02
12	8	125,970	12.01
13	7	77,520	7.39
14	6	38,760	3.70
15	5	15,504	1.48
16	4	4,845	0.46
17	3	1,140	0.11
18	2	190	0.02
19	1	20	0.00+
20	0	1	0.00+
	TOTAL	1,048,576	100.00

likely occurrence. It could readily happen by chance. However, notice that the determination of whether a probability of 0.13 is a likely or unlikely event is, finally, an arbitrary decision. A spokesperson for sexual equality might see this as an improbable event and suspect discrimination. Much as we might like to have our human problems dealt with by machines and mechanical forms of logic, there is no way in which statistics of this nature can be made to solve human problems in any sort of mechanical fashion.

This discussion has been fairly lengthy. But the topic of probability is worth some further comments. So, the discussion of probability will continue for a while longer in Chapter 9.

SUMMARY

Probability is a fundamental concept in statistics as well as a problem encountered in our day-to-day lives. Estimating probabilities requires a good knowledge of how events are distributed within the populations that concern us. For example, if we want to estimate the probability of having a girl or a boy baby when we plan a family, we need to know the general distribution of infants by sex before we can begin. Once we know this distribution, the problem is solvable. But what if we plan to have a family of five children? How can we determine the likelihood that our family will be all girls, four girls, three, two, one, or no girls? To solve this problem we need to know three basic things:

a. The likelihood of having a girl baby or a boy baby. We establish this by seeing the ratio of girls to all infants in the greater population. Let's say it is 0.50.
b. The total number of kinds of mixes of girls and boys we could have with a family of five children. This is 2 raised to the 5th power, or 32.
c. The number of mixes or combinations we could have for any partticular type of family, say a family with three girls. This is equal to $N!/r!(N - r)!$, where $r = 3$ and $N = 5$, or

$$\frac{5 \times 4 \times 3 \times 2 \times 1}{(3 \times 2 \times 1)(2 \times 1)} = \frac{120}{12} = 10$$

Once these three values are established, the probability of having three girls is equal to the number of ways we can have three girls (10), divided by all the different kinds of family we might have (32). The probability is, therefore, $\frac{10}{32}$, or 0.3125.

The probabilities for all possible combinations of girls and boys in a five-child household are shown in the following table. You should know the logic behind this table before going on to the next chapter.

Number of Girls	*Number of Boys*	*Possible Ways of Occurrence*	*Probability*
5	0	1	.03125
4	1	5	.15625
3	2	10	.31250
2	3	10	.31250
1	4	5	.15625
0	5	1	.03125
Total		32	1.00000

There are two basic reasons for studying probability. First of all, it gives us at least a sense of what might happen under given circumstances. Remember, however, that probabilistic knowledge is uncertain knowledge (with the exception of cases where $P = 1.00$ or 0.00). If you have trouble remembering this, just keep thinking of how you would feel about your car if your brakes worked in a probabilistic fashion; let's say they could be counted on ninety percent of the time ($P = 0.90$) with regard to their effectiveness. They would be significantly effective, but that ten percent uncertainty, though very low, would be enough to make your car impossibly dangerous to drive.

The second reason for studying probability is more important than it might seem to be at first. To know that something happened by accident rather than by some causal agent can often have a powerful emotional influence on one's thinking. For example, it is one thing if you are accidentally injured. It is quite a different matter to be intentionally injured by someone else. Probability plays a strong role in how we respond to what happens to us. Therefore, the assessment of probabilities is not only a technical problem in statistical analysis, but also a profound problem in day-to-day human affairs.

THINKING THINGS THROUGH

1. Sometimes an animal will advance into a fight with another animal and at other times it will retreat. Do you think the animal makes an assessment of risks (another word for probability)? Are probability problems deeper than the sense we obtain of them from formal exercises in statistics? How might an animal's assessment of risks be distorted? (For example, might hunger make an animal more willing to take on a situation in which it is likely to lose?)
2. Einstein was convinced that God does not "play dice" with the world. To what extent do you think people's lives are a matter of chance and to what extent do you believe people have control over their lives? (We asked a statistician colleague of ours this question and his response was instant: People have twenty percent control. How do you think he arrived at this figure?) Obviously there is no way to come to a precise answer for this problem, yet people get seriously entangled in it. It is not a light matter. To what extent do you think your life is a matter of chance and forces outside your control? How did you reach this conclusion?
3. What minimal conditions are required for a formal solution of probability problems as statisticians generally deal with them?
4. Suppose there are two gangs that roam a neighborhood, the reds and the blues. A particular corner is randomly patrolled by one or the other gang every night. You have to walk past the corner three nights in a row. What is the probability that it will be patrolled by reds all three nights, two nights, one of the nights, or none of the nights?
5. Say you decide to have four children. The probability of having a girl, let us say for simplicity, is 0.50. What is the probability of having a family of four boys, three boys, two boys, one boy, or no boys?
6. The World Series is being played between two teams that are evenly matched. How is "evenly matched" defined in probability terms? The winning team must win four out of seven games. What is the probability that the series will end after just four games? What is the probability that the series will go a full seven games?
7. You believe in full equality for men and women and whenever you go into a particular situation, you expect to find an equal number of men and women on the job. Suppose you enter a class in the field of social work, and the class has seventeen women and five men in it. If this were simply a random happening, what would the probability be of having this number of women, or more, in the class? Would you be inclined to accept this sex

distribution as a random event, or would you presume that something other than chance was behind the distribution?

8. You find yourself playing cards with a man named Doc. The game is simple. If you pull a black card from the deck, he gives you a dollar. If you pull a red card from the deck, you give him a dollar. You make ten draws and pull one black card. Do you suspect Doc of cheating? What are the precise probabilities that establish your suspicions? Suppose you found he was perfectly honest and that what happened was simply the "luck of the draw." How would this change your feelings?

Notes

1. Identical twins, reared apart, are not as "free" of each other as nonidentical siblings or unrelated children living apart. This means that the probable life experiences of identical twins are different from those of other individuals. This circumstance is mentioned again in the discussion of correlation in Chapter 11. The assumption of independence is important in probabilities and also important in correlation analysis, suggesting that the two are logically related.
2. As long as a given probability of a favored event is greater than zero, no matter how small the probability, given sufficient time the favored event will occur. This simple fact about probability made intellectuals extremely nervous during the Cold War. Now that the probability of nuclear conflict appears to be lower, we have lost our concern. However, the fact remains that if the probability is greater than zero, no matter how small the probability, given enough time nuclear war will take place.
3. In assessing probabilities, people are commonly influenced by bad counting. They tend to count what they see, and they tend to see what frightens or irritates them more than they see ordinary events. So, for example, people tend to count victims of terrorism or airplane crashes where they often ignore victims of train wrecks. They see red lights when driving and do not notice the green lights they drive through. This latter error results in people believing they are more likely to encounter red lights than green ones.
4. Ignoring this assumption leads to what is commonly called the gambler's fallacy. This fallacy consists of thinking that if five heads come up in a row, then the probability of a tail appearing is now greater than 0.50. Actually, it is still the same—unless someone has "loaded" the coin so that it always comes up heads.
5. Some people want to know what all the various combinations look like. But this is not important. A few cases, like those presented for $N = 50$, are sufficient for an idea of how varied the set can be. What is important is how many cases there are in the set.

9

PROBABILITY, CHANCE AND UNCERTAINTY: PLAYING THE PERCENTAGES (PART II)

"Good luck lies in odd numbers."

—Shakespeare

"Something must be left to chance. . . ."

—Horatio Nelson

"Although men flatter themselves with their great actions, they are not so often the result of great design as of chance."

—La Rochefoucauld

Our consideration of probability in these two chapters is limited to events having one or the other of two qualities—heads or tails, male or female, black or white, young or old, and so on. So far we have only dealt with situations in which either of the two qualities has an equal chance of occurring. However, qualities are commonly unequally distributed. This chapter addresses what must be done when we are faced with probabilities obtained from unequally distributed qualities. For example, what is the likelihood of drawing five police officers or five nonpolice–officers in a random sample, when $N = 5$ and the proportion of police officers in the general population is 0.001 (instead of 0.50)? Your intuition should tell you right away that the probability of getting a randomly drawn sample of five police officers under these conditions would be extremely small, while the probability of getting five nonofficers would be extremely high.

THE ADDITION AND MULTIPLICATION RULES

We shall now examine some more examples of probabilities that involve combinations. Before we do, however, we need to consider two extremely fundamental rules in probability: the addition rule and the multiplication rule.

The Addition Rule

Suppose you are selecting people at random from a telephone directory for a survey you want to conduct. What is the probability that you will select either a woman or a man? You might be inclined to say about fifty percent. But remember that the question is not asking for the probability of drawing a man. Nor is it asking for the probability of drawing a woman. It is asking what the probability would be of drawing *either* a man *or* a woman. How can you miss, once you see what the question is asking? What is the probability your selection will be either a woman or a man? It has to be one or the other. Even if there is only one woman in a directory with a million names in it, you must select either a man or a woman.

When we talk about selecting either this *or* that, we imply addition. If we reach our hand into a candy jar that has chocolates (C) and bonbons (B) and pick two candies at random, four things can happen:

CC CB BC BB

There are three events that include chocolates and three that include bonbons. Getting a chocolate *or* a bonbon on two random draws from the jar includes everything that can happen. You must get CC, BB, BC, or CB. In each instance you wind up with a chocolate *or* a bonbon. Therefore, the probability of getting one or the other is 1.00, or the probability of a chocolate *plus* the probability of a bonbon.

Suppose the jar holds five chocolates, two bonbons, and three taffy pieces. Then the probability of getting a chocolate, or a bonbon, or a taffy when you reach into the jar is $0.5 + 0.2 + 0.3 = 1.00$. One of these events must take place because all possibilities are covered. The probability of getting a chocolate or a bonbon would be $0.5 + 0.2 = 0.7$. This is the same as asking for the probability of not selecting a taffy, or $1.0 - 0.3 = 0.7$.

If a teacher allocates grades as in Table 9.1, what is the probability of getting an A *or* a C? Of getting an A *or* a B *or* a C? What is the probability of getting a D *or* an F? Of getting an A *or* an F?

Your chance of getting an A or a C when randomly selecting individuals from this class is $(10 + 40)$ divided by $100 = 0.50$. It is simply a matter of adding the A's and the C's and then dividing by the total.

Using the **addition rule** requires interpreting questions and making certain what those questions are asking for. Consider the following question: What is the probability of failing in college or getting through? If you rush, you might think this question is asking you for the probability of failing. Or, you might

Table 9.1. Distribution of Grades in a Hypothetical Class (illustration)

Grade	*Number*
A	10
B	20
C	40
D	20
F	10

think it is asking for the probability of getting through. Or, you might even think it is asking you for the probability of both failing and getting through. Once you are clear on what questions are asking, you can figure out probabilities with confidence. Unfortunately, it is easy for nearly anyone to get confused when it comes to interpreting what a probability problem is asking—no matter how carefully the question is stated.

The Multiplication Rule

Now we are going to change the question: what is the probability of getting a male on the first draw from the telephone directory *and* then getting a female on the second draw? This probability requires multiplying the first chance event by the second—an application of the **multiplication rule.** If the probability of getting a male on the first draw is 0.50 and the probability of getting a female on the second draw is also 0.50, then the probability of getting a male *and* then a female in two consecutive draws is 0.50 times 0.50, or 0.25.

In two tosses of a coin, what is the probability of getting a head (H) *or* a tail (T)? (It has to equal one—all possibilities are exhausted.) What is the probability of getting heads on the first toss *and* then repeating with another head? The following four things can happen:

HH HT TH TT

As you can see, HH is one of the possibilities. Once again, the probability would be 0.50 times 0.50, or 0.25.

If you were sampling the class grade list in Table 9.1, what would be the probability of drawing an A and then drawing a C, assuming replacement (that is, assuming that after you drew an A, it was put back into the list)? It would be equal to 0.10 times 0.40, or 0.04.

The logic of problems like these has applications for sampling. At the present time there are over 5.5 billion people on the planet. The tallest person is a single individual. So is the shortest. Therefore, the probability of meeting the tallest individual is 1 divided by at least 5,500,000,000, and the same is true for the shortest person—assuming a purely random encounter. (This assumption, of course, is not warranted because people this unique are usually sheltered from the public in some manner.) The probability of meeting one *or the other* of these people would be small: 1 in 2,750,000,000. To encounter *both*—the tallest *and* the shortest—by accident would be amazing; in this case the probability is roughly 0.00000000000000000003. This is a small probability, indeed. You are not likely to sample, randomly, the true range of height among human adults. When we talk about sampling we will see that it is important to be aware that samples tend to underestimate variability. We now see why: *small samples are not likely to pick up improbable events found in the extreme tails of a distribution.*

Mastering the Addition and Multiplication Rules

Although they are relatively straightforward, people often have problems with the addition and multiplication rules. Pause now for a few minutes and go through each of the following six questions. Be certain to answer each one before you go on to the next. If you have trouble with any question, get some help with it before you move on to the final considerations of probability that bring this discussion to a close. (The answers to the six questions are given at the end of the chapter, on page 165.)

1. Several years ago, a woman in Colorado claimed she purchased ten $50 winners and two larger winners in the state lottery because she had a "feeling in her tummy" when she bought the tickets. The probability of getting a $50 winner is 0.001. Let's say the probability for the other two winners was 0.0001. Obviously this woman bought some other tickets as well, but to simplify the board let's say she won the ten $50 tickets and the two other winners in twelve straight purchases. What is the probability of such an event? Lottery officials and a local academic statistician who were quoted in the Denver papers claimed that this was a "reasonable" possibility. After you work out the probability, are you inclined to agree with them?
2. You are doing research in which you want to draw some legal documents randomly from a library of 10,000 documents. Of these documents, 40 percent deal with murders, 30 percent deal with robberies, and another 30 percent deal with assault. What is the probability of drawing a murder and then a robbery in just two draws? What is the probability of drawing a murder or a robbery in just two draws?
3. Let us say that the probability of having an accident in a certain factory in a given year is 0.10. The factory has 10,000 workers. If you have an accident three years in a row, you are considered accident-prone and are fired as an insurance risk. By this criterion, how many people might be called accident-prone who were simply the victims of chance? If the probability of having an accident in a given year is 0.10, then the probability of not having an accident is 0.90. What would be the chances of not having an accident five years in a row?
4. Suppose the probability of nuclear war in any given year were 0.01. Then the probability of not having war would be 0.99. What would be the probability of not having war for a period of fifty years? (A pocket calculator can handle this problem easily by simply squaring 0.99 fifty times.)

Give the probabilities of either having, or not having, nuclear war in the fifty-year interval.

5. The probability of a deprived child making it out of the inner city sometime during his or her life is, let us say, 0.01. If only chance were involved, what would be the probability of a boy and his sister making it out? (Notice that if a boy makes it out, we intuitively sense that pure randomness might not be at work if his sister also makes it out. The thing that helped the boy might also be helping the girl. This is a violation of the rule of independence in the calculation of pure random probability problems.)
6. Suppose you rely on chance to get you through a true-false test with fifteen questions. The probability of getting any single question correct is 0.50. What is the probability of getting the first five questions right? What is the probability of getting all fifteen questions correct?

If you have gone through these problems, have seen how the questions are asked, and now have a good sense of what the answers are, you are ready to move on to the following discussion of probability as one of the elements of statistical reason.

WHEN THINGS ARE NOT FIFTY–FIFTY

Up to this point we have primarily considered situations in which things are fifty-fifty. We talked, for example, about tossing coins and drawing men and women from populations that were equally divided. Now we want to consider situations in which things are not equally divided; indeed, these kinds of unequal distributions are far more common in the real world. Most social categories (such as professors, Catholics, police officers, and so on) do not comprise fifty percent of the total population.

Let's consider the following kind of problem, which can be a serious one in the arena of social and political disputes. Say there are eighteen teams in a football league—let's call it the Pro League. There are no Jewish quarterbacks. Accusations of discrimination are leveled against the managers and owners of the Pro League, who insist there has been no discrimination. While this dispute cannot be definitively resolved, we can at least ask whether this situation—the absence of Jewish quarterbacks—might have happened by chance.

The question we are interested in is the probability of having no Jewish quarterbacks in a league of this kind. To make calculations simple, let us say that twenty percent of the population from which professional athletes are drawn is Jewish. The probability of not getting a Jewish quarterback in any single instance, purely by chance, would be 0.80. The probability of not getting a Jewish quarterback 18 times straight would be 0.80 raised to the 18th power. This proves to be about 2 out of 100, or about 1 in 50 (on the average). It could happen by chance, but it is not a commonplace sort of thing. If the Pro League picked all new quarterbacks each year, and if no discrimination were taking place, there would be, on the average, two years each century when no Jewish quarterbacks would make the line-up.

We determined the probability of getting all eighteen quarterbacks from the non-Jewish population of athletes and none from the Jewish population. We can now go ahead and find out what the probability would be of getting any number of Jewish quarterbacks, up to all eighteen teams having a Jewish quarterback.

We need to know two things. First, we need to know what the probability might be for any particular combination. Then we need to know how many ways that combination can take place. For example, what would the probability be for a combination that involved six Jewish players (J) and twelve non-Jewish players (N)? What would the probability be of getting a J, *and* a J, *and* a J, and so on, and an N, *and* an N, *and* an N, and so on? It would be

$$0.20^6 \times 0.80^{12} = 0.000064 \times 0.068719 = 0.0000044$$

However, this is only one way in which we can get a combination of six Jewish and twelve non-Jewish players. How many other ways can this happen? It can happen

$$\frac{18!}{6!\,(18-6)!} \quad \text{ways.}$$

Using the computer to solve this factorial problem (though the calculation is easy enough with a hand calculator), we find that there are 18,564 possible combinations. We want to know what the probability would be of getting one of this particular mix, *or* another, *or* another, for a total of 18,564 times. That suggests adding the above probability (0.0000044) 18,564 times:

$$18{,}564 \times 0.0000044 = 0.08$$

The probability of getting six Jewish quarterbacks in the eighteen-team league would be less than 1 in 10 on a purely chance basis. The complete distribution of possibilities is presented in Table 9.2.

The probability of having a Jewish quarterback on each of the eighteen teams, just by chance, would be 0.20 raised to the 18th power, or 0.0000000000002621439—about three chances in ten trillion trials.

Notice in Table 9.2 that events are no longer distributed normally as they were when we were drawing cases from fifty-fifty distributions. The curve is skewed. Notice that the maximum probability is for three or four Jewish quarterbacks. Why isn't it for, say, eight?

CHANCE AND BAD FEELINGS

We have worked with some imposing numbers to come to several conclusions about probabilities. Is all this effort worth it? To answer this question we have to come back to at least a little philosophical musing. On the one hand we can

Table 9.2. A Hypothetical Distribution of Probabilities of Obtaining 0, 1, 2 . . . 18 Jewish Quarterbacks Playing for Any of the Eighteen Teams in the Pro League (where the population from which players are drawn is 20 percent Jewish and selection is purely random)

No. Jewish Quarterbacks	*Probability of Specific Combination (A)*	*No. Combinations of This Kind (B)*	*Probability of this Event (A × B)*
0	.01801	1	.02
1	.00450	18	.08
2	.00113	153	.17
3	.00028	816	.23
4	.00007	3,060	.21
5	.00002	8,568	.17
6	.00000+	18,564	.08
7	.00000+	31,824	.03
8	.00000+	43,758	.01
9	.00000+	48,620	.00+
.	....	....	.00+
.	....	....	.00+
18	.00000+		.00+
		Total	1.00

simply say no—it is not worth the effort. On the other hand, though, we might consider the fact that a great deal of heartbreak, anger, and even madness can come out of the tricks played on us by chance and the fact that nature is always tossing various kinds of dice in front of us.

In Table 9.2 we saw that it could happen by chance every one out of fifty times that no Jewish quarterbacks would get chosen for the eighteen teams of the Pro League. Thus, this could happen without discrimination and with the owners of the Pro League being perfectly fair. So, to assume that discrimination was occurring in this case when, in fact, chance was at work, would be an error. Statisticians call this a **Type 1 error.** Such an error occurs when you dismiss chance as causing a condition and assume that something is "behind" whatever it is that is going on—in this case discrimination against Jewish players. In this case, such an assumption would probably be wrong. Chance appears to be the only thing involved here.

There is, of course, another error you could make: you could assume there was no discrimination when, in fact, there was. Statisticians call this a **Type 2 error.** In this case, you would be masking an injustice under the name of chance, something you certainly would not want to do.

The important thing to note here is that we are concerned with whether (a) a random, impersonal force is creating some pattern before us, or (b) that pattern comes from something other than chance, such as discrimination, cheating, human meanness, weakness, or some other intrusive factor. This is no small matter. The judgments we make here determine whether we should become embroiled in battle or remain calm. They can also determine whether we promote justice or further aggravate the problems of injustice.

Of course, there are always those who use a different logic, arguing, for example, that if there are no Jewish quarterbacks in the league it is obvious evidence of discrimination. To such people, chance is simply an intellectualizing of political corruption. From a statistical point of view, by rejecting chance on all occasions these people are constantly making Type 1 errors. The other side of the coin, however, is to accept chance as the force behind everything. Some biologists and naturalists have moved in this direction in modern times. They argue, for example, that the twists and turns of evolution were chance happenings. They reject the idea that a principle or grand determinant might have been behind it all. If there is a God, God is merely a dice player and we are the result of the roll of the dice.[1]

The question of God as a creator or merely a dice player has gone on in science for several hundred years. One modern scientist, more baffled by the problems of probability than earlier thinkers, waffled on the issue by suggesting that God is a dice player who tosses loaded dice.

We only mention this question (of the existence of a God as creator) to point out that while probability problems are commonly dealt with in a purely technical sense, probability itself is a broad concept that moves throughout the entire domain of human intellectual activities. In the literary novel probability is called coincidence. In religion the improbable is called a miracle. In business, trends are used to make forecasts that are probabilities (of a quite uncertain character, incidentally). In medicine, probability or improbability is called a prognosis. Regardless of the term being used, the logic always remains the same. A probability is an expectation of a favored event occurring among all possible events, *if you were to observe an infinite number of happenings.* But an infinite number of observations is impossible to make. Like the mean, probability is a fiction, but it is an extremely useful and logically interesting fiction.

Chance plays a huge role in human social relations. This is, possibly, the most profound finding to come out of modern quantitative social science research. For example, people from different professional groups develop quite different perceptions of the world as a result of how their encounters with people are influenced by the chance events that led to their professional affiliations. Police officers, for example, may tend to become cynical about the world, seeing it as made up of crooks and other corrupt individuals. College professors, on the other hand, may be overly inclined to think that people are studious, reasonable, and considerate.[2]

More important than the probability illustrations in these chapters is that you begin to think about the implications of chance, and how it influences our relations with each other and how we come to know the world. A case in point is the pride many men have as a result of siring large families of boys. Assuming that boys constitute fifty percent of live births (actually, the ratio is slightly higher), and that sex of the child is a randomly distributed event, what would the probability be of having a family of six, five, four, three, two, one, or no boys? Out of a population of a million fathers who have sired six children, how many will have at least three boy babies? How many will have none? Doing this little exercise can offer further insight into how pure chance even has relevance to our self-esteem and our sense of worth.

SUMMARY

When we draw from one or the other of two groups, A or B, that make up a total population, and the two groups are not equal in number, we get probability distributions that are not normal. Using the addition and multiplication rules, in conjunction with the formula for combinations, we can obtain probability distributions that enable us to determine the likelihood of drawing any number of A's or B's from a given number of drawings (N).

Type 1 errors (which involve an underestimation of the effects of chance) and Type 2 errors (which involve an overestimation of the effects of chance) are common in considerations of probability in ordinary life as well as in statistical research. Whether we attribute happenings to chance or to an outside influence of some kind can have serious consequences for how we react to what we observe.

THINKING THINGS THROUGH

1. We often think some things are probable when they are not, and that other things are improbable when they are quite probable. For example, the probability of a winning lottery ticket being made up of the numbers 1 2 3 4 5 6 7 8 is the same as that for a ticket with the number sequence 3 7 2 6 7 8 9 2. However, most lottery players (at least among those we've questioned) think the first ticket has a lower probability of winning than the second. What is wrong with their thinking?
2. The probability of being a top movie star is, let us say, one in a million. This would produce roughly 250 movie stars for the population of the United States. Suppose you do a study and find that the newest generation of 250 movie stars contains fifteen people whose parents were movie stars. Does this provide you with the right to suggest that nepotism operates in a land where opportunity is supposed to be free and equal (that is, subject to the free play of talent unimpeded by anything other than random events)? How might the concept of probability be used to deal with ethical issues such as equality of opportunity, or to provide a way of measuring racism or other forms of intolerance?
3. Let's assume that both the Republican and the Democratic parties in the United States are extremely conscientious about trying to create policies that will enhance the welfare of the American people. In other words, the probability that Republicans will do something "bad" are no greater than

that the Democrats will do so. Suppose you keep clippings from the local paper reporting negative and positive comments about both parties. You find that out of eighteen clippings, twelve present negative remarks about the Democrats. Is this sufficient evidence of journalistic bias by the paper? (Remember, this question is filled with assumptions.)

4. What kind of statistical error is the paranoid individual inclined to make? What kind of error is the saintly or naive person inclined to make?
5. Hubert Lumpkopf proudly boasts that he made a million dollars by the time he was forty years old and he never got past the sixth grade. "Education is a waste of time," he loves to boast. A great deal of anecdotal material like this influences people as they read about such instances in the media. In statistical terms, Lumpkopf is telling us that the probability of education leading to success is zero. This is a silly argument, of course. What, in technical terms, makes Lumpkopf's assessment of probability such a bad one? If you wanted to determine precisely the probability of being successful as a result of getting a Ph.D., what would you have to do? By the way, is the value you get from the precise calculation the same thing as a percentage? (We ask this to remind you that probability is not far removed from that everyday workhorse statistic we call the percentage.)
6. The data below represent the distribution of given population characteristics in the United States for 1990. Assuming independence in the values we are working with, determine the probability of selecting, at random, each of the following combinations of characteristics:
   a. White, 65 or more years old, divorced, non-college education.
   b. Female, younger than 65 years old, nonwhite.
   c. Nonwhite, female, 65 years old or over, divorced, non-college education.

Category	Percent
White	84
Female	52
65 years old or over	12
4 years of college or more	20
Divorced	8

7. The likelihood that you will be involved in an automobile accident of a relatively serious nature in any year is roughly one in four. Draw up a probability distribution that would show the probabilities for having no accidents, one, two, three . . . nine, ten, eleven, or twelve accidents for a person driving for a period of twelve years. Assuming that any accident costs the insurance company an average of three thousand dollars, how much money would an insurance company have to pay out over a twelve-year period for one million drivers with coverage?
8. This problem is elaborate, but it is designed to get you to see that probability is something you can play with. In fact, that is what a lot of mathematical and statistical problems are—intellectual toys to be played with. In this problem we want to consider a possible paradox that comes out of the ratio scale. Remember, ratio scales are supposed to tell you if something is, let us say, twice as good as something else.

In this problem consider two ball players, Tom and Hank. Tom has a batting average of .150 and Hank has one of .300. Hank is, therefore, twice as good as Tom. Presume each player hits three times during any game. What is the likelihood that Tom will get a hit during a given game? What is the probability that Hank will get a hit? Is Hank twice as likely to get a hit as Tom? Shouldn't he be, if he's twice as good?

9. This is a continuation of the situation set up in problem 8. Let's say that hitting at least once in each of ten consecutive games constitutes a "grand achievement" for a player. What is the probability that Hank will accomplish this? What is the probability that Tom will accomplish it? Is Hank twice as likely to accomplish a grand achievement as Tom? Does this suggest a problem with respect to the use of simple ratio scales as basic measures of relative performance?

ANSWERS TO QUESTIONS BEGINNING ON PAGE 158

1. The outrageous probability of this event is a decimal consisting of a one preceded by 38 zeroes, or
   0.000000000000000000000000000000000000001.
   The universe is believed to be only 15 to 20 billion years old, and if you bought only ten $50 tickets every second, you might have to wait roughly 32,000,000,000,000,000,000,000 years, on the average, to accomplish what the lady did with her winners—or about 150 billion universe eons. This is a nice example, we think, of how monetary interests can influence how one interprets a given statistical finding. Social scientists generally reject something as being a chance event if it has a one in twenty or one in a hundred chance of occurrence. Here, however, an event that has a considerably lower probability is being accepted as something one might reasonably expect as a result of chance.
2. You would expect to draw one and then the other 0.4 × 0.3, or twelve-hundredths of the time. To get one or the other would occur 0.4 + 0.3, or seven-tenths of the time.
3. Ten people were fired because of chance. The probability of not having an accident five years in a row would be 0.59.
4. The probability of not having a nuclear war fifty years in a row would be 0.99 raised to the 50th power, or 0.61. Therefore, the probability of having a nuclear war in this interval would be equal to 1.0 − 0.61, or 0.39. However, keep in mind that no one knows what the real probabilities are for such an event. We can only hope that it is much lower than this. At the same time, such figures can instruct us on how careful we need to be to prevent the possibility of nuclcar disaster.
5. The answer is 0.01 × 0.01, or 0.0001.
6. Five correct questions = 0.5 raised to the 5th power, or 0.03. All fifteen correct would be 0.5 raised to the 15th power. You would have about one chance out of 32,768 of getting a perfect grade just by chance.

Notes

1. We introduce this speculation merely to point out that statistics and theology are not as distinct from each other as some members of either camp sometimes like to think.
2. The work of Professors Eric Poole and R. Regoli shows that police workers are especially prone to take a cynical view of human motives. References to their work can be found in John D. Hewitt, *Criminal Justice in America* (New York: Garland Publishing, 1985).

10

SAMPLES

"By a small sample we may judge of the whole piece."

—CERVANTES

"We know in part, and we prophesy in part. . . ."

—ST. PAUL

"To see a world in a grain of sand. . . ."

—WILLIAM BLAKE

When people tell us that they based a conclusion on a sample, we are on guard. Our intuition tells us that samples are tricky—and our intuition is right. However, until we think about the matter more seriously, we have little awareness of just how tricky samples are. For example, our intuition might tell us that a large sample is better than a small one. This may or may not be true. Our intuition might also tell us that it is better to know the whole story than one based on samples. Once again, this may or may not be true.

Oddly enough, our intuition also commonly tells us that there is nothing wrong with generalizing from a single experience (A statistician would call this a sample with $N = 1$). Certainly in human affairs this tendency to generalize from singular experiences nearly always leads to wrong conclusions. Nonetheless, many people refuse to believe that such conclusions are wrong. Even statisticians succumb to the temptation to generalize when the data are extremely limited.

You should get firmly in mind at the outset of this discussion of sampling that *virtually everything we know is based on some kind of sample.* Statisticians are not the only people who rely on samples. We all rely on samples, and we do so nearly all the time. (The generalization set forth in the previous sentence, for example, is based on a sample.) Our individual lives are unique samplings of the world; each of us obtains a different vision because we sample the world in different ways. Wealthy, white Americans, for example, get a different sampling of the world than, say, Asian Americans; women get a different sampling of life than men; the police get a different sampling of life than college professors. Each group then promotes different generalizations about life, based on how that group samples the world. Such groups occupy different "realities" primarily because they are sampling reality in different ways.[1]

We make generalizations based on our personal experiences. These personal experiences are *limited* and necessarily *biased.* Our generalizations will inevitably deviate from those of other people. We are likely, in some instances, to think that people who see reality in radically different ways from us are crazy. They might be; they might not.

In the arguments that follow, the relationship between probability and evaluating the merit of samples is broadly developed. The realm of sampling is a place where nothing is certain, everything is estimated, and we must rely on probability to guide us in determining how much confidence we can place in our estimates.

SAMPLING AND THE PROBLEM OF PRECISION

The basic question we want to consider is the following:

> What are the limits within which the findings of a particular sample can probably be relied on as a basis for generalizing to a larger, even infinite, population?

Suppose you meet a stranger at a party and, after five minutes of conversation, you conclude that this is a person of decent character. You made a generalization on the basis of a small and undoubtedly highly biased sample. After all, people usually try to make good impressions at parties.

Notice, in this ordinary case, that you do not specify lower or upper limits to your estimate of the person's character. You presume the stranger is "pretty decent." Such vague language allows a lot of leeway when it comes to figuring out exactly what was observed. If we had a scale of decency ranging from 0 to 100, where would you set the limits? How *likely* do you think you are to be correct in your assessment of the person's true decency? Would you say you pegged the person exactly at 93? Or would you place this person somewhere between 85 and 95, or between 70 and 100? Notice that as you make the limits broader for your guess, you gain more confidence in your estimate. That is to say, the person's real character *more likely* falls between broad limits than between narrow ones. If you said the limits were between 0 and 100 you could be perfectly confident, but you would have said nothing. Thus, you acquire perfect confidence at the cost of making an empty assertion. This problem is central in sampling: how much precision do you want, and how much confidence do you want to place in your degree of precision? (You can learn a great deal about safe but empty generalizations by listening to public political oratory, for example.)

Normal ambiguities in social language are an interesting response to this problem. If we give latitude to our meanings, we can be more confident of ourselves. If we try to be more precise, we lose credibility when we are found to be off the mark. Politicians are sensitive to this problem when making public statements; that is why they tend to keep their assertions extremely broad.

We might say someone's intelligence is "keen." We would not, however, suggest that the person had an IQ of 148, unless we had this precise information. The term "keen" is general, as are most social terms. There are many reasons for this generality. But in relation to the ways in which statistical forms are used in ordinary discourse, it is interesting to see how people give the appearance of a precise use of language without really becoming involved with rigorously defined limits. As a statistician might put it, in ordinary discourse, people tend to keep their **confidence intervals** broad.

The natural hedging that is a part of all ordinary conversation is an example of broadening confidence intervals. Grace might say that Robert is "dumb." In saying this she is applying a general quality to Robert's character (based on her sampling of Robert's actions). Margaret replies that Robert can't be too dumb since he got an A+ in microorganic neurophysiology. Grace then says it's obvious that Robert is dumb because he watches football games on television. Thus, Grace has made the limits for the quality "dumb" so broad that they now include Robert and most other Americans.

The point here is that confidence intervals appear in ordinary conversations, just as they do in statistical analysis. The logical forms are the same in common discourse and in the elaborate methods of academic statistics. Where Grace selects a single observation to sustain her generalization, the statistician relies on random observations of the population about which generalizations will be made. The term *random* appears again and again in sampling; its importance becomes apparent as you begin to see how often, in ordinary conversation, the ideal of randomly selected observations is ignored. In ordinary argument and debate, the biased sample is the norm (our own samplings of what we read in the paper and of the arguments we engage in with other people lead us to this generalization). Biased samples are especially common in ordinary conversations.

PROBABILITY AND SAMPLING

The connection between sampling and probability is quickly seen in a common practical joke. Now and then a bridge foursome finds itself dealt four perfect hands: thirteen spades, thirteen hearts, thirteen diamonds, and thirteen clubs. Everyone laughs and wants to know who screwed up the deal. In other words, the sample dealt to the players is recognized as phony or unreliable or *nonrandom*—even though such a distribution of the cards could, in fact, happen by chance. The odds for this event are so astoundingly low that people tend to conclude that a practical joker was at work; in making this conclusion they are virtually (though not perfectly) certain to be correct. They reject chance, and their thinking is based on an intuitive understanding of the extent to which such an event is improbable. They know they got a bad sample when the cards were dealt—that is, one that was not a purely random expression of a bridge hand.

This same logic underlies the way in which samples are evaluated in modern survey research, psychological studies, and the statistical studies of population experts, economists, and social scientists. If something we might not want to believe is probable, it becomes something we must consider seriously. And if something we might wish to believe has little probability of occurrence we must begin to reconsider its validity.

Although the random sample is usually the best way to proceed in sampling problems, there are situations in which it does not make sense. How do you sample something that is complexly organized—such as a novel? We might take a random sample consisting of every one hundredth word of Mark Twain's *Adventures of Huckleberry Finn.* The sample would be random, and it would have statistical uses. For example, it would allow us to estimate with considerable accuracy what proportion of the entire novel involves words beginning with the letter Z. However, as a sample of the novel itself the idea is preposterous. We could not use a random sample of words to see the essence of the story. In fact, we probably would not be able to see any story at all on the basis of such a sample.

Here is every tenth word from the first page of *Moby Dick,* beginning with the sixth word as a random beginning:[2]

> ago money on and about world the growing drizzly pausing every such strong the I soon and his nothing all very there round surrounds take where by sight circumambulate.

Obviously this is a crazy sample. We cannot see what the first page is about from this ten percent sample of its contents. This illustration suggests a major constraint on sampling procedures as they are used in academic research and in ordinary life: we cannot sample a *structural* system by using elemental random procedures. Even so, to a surprising extent, people persist in making generalizations about structural forms in ordinary life—institutions, societies, the personalities of individuals, books, movies, and so on—based on limited experiences or samples.

A modern form of what we are talking about here appears in the practice of constantly changing television channels with remote control devices—a habit commonly known as channel surfing. In this case, viewers "sample" bits and pieces of a great variety of programs, watching perhaps ten different channels by sampling a few moments of a program at a time. This is an interesting form of sampling because it suggests that sampling is not only common, but that it also has some kind of natural appeal. People would rather sample lots of programs than concentrate on the totality of a single program. However, sampling structured events cannot be done in any reliable manner. When you sample programs by going from one to another with a remote control device, you are likely to miss significant moments; thus, the ending of any single program will be puzzling or enigmatic.

Sampling structured phenomena is a profound matter, and there are further problems that we cannot pursue in detail here. Still, the matter needs to be addressed. Academic sampling statistics, along with hypothesis testing and nearly the entire array of statistical procedures and methods used in social science, are based on the idea that you are dealing with a singular variable or continuum of some kind. Yet the world does not consist of isolated continua. This is particularly true of what might be called the social world.[3]

The real world appears as complex interrelationships between variable conditions. A human being, a novel, a house, or a religion is not a singular continuum. Being intensely devout or not at all devout as a Catholic is not the same as weighing a hundred pounds or six ounces. Being a Catholic is complicated. We get around the problem in statistical studies by lumping Catholics under a single category, assuming that people are Catholic if they say they are, and also assuming that all Catholics are Catholics in the same sense. If they go to church every week they might be considered devout, and if they have not been to church more than twice in the past year, not so devout. These kinds of criteria are called **operational definitions.** Obviously they are highly superficial and arbitrary.

Operational definitions allow counting to take place. Without counting, we cannot use formal statistical techniques. The operational definition transforms social (and physical) realities into quasi-variables at the expense of losing sight of the complex structural forms within which those realities are embedded.[4]

When we begin to see how important sampling is in generating reliable generalizations, we can see why modern social scientists are drawn to operational definitions. Without them, the assessment of sampling procedures and results cannot take place. As you understand the need for such simplifications, you become more sympathetic to the fact that modern social science, when it gave itself over to the idea of operational definitions, had, in fact, made a kind of pact with the devil.[5]

What is discouraging about operational definitions is that they lose information like a sieve. At the same time, however, there is no denying that many fine statistical studies rely on such definitions. Any population study that counts each individual as "one" abandons structural complexities in favor of abstraction. Personalities are bleached out. "One" stands for the president; "one" stands for an infant; "one" stands for a rock and roll star; "one" stands for a homeless person. You lose individual complexity when you ignore structural details. Yet it is important simply to know the numbers of people in various categories. Counting has been a part of political and military concerns since the dawn of civilization. It is not new, and it is not trivial.

The point here is that obtaining information, no matter how you go about it, is an *arbitrary* action. In order to get information, you have to give up a certain amount of other information.[6] Sampling is not just a simple procedure of randomly picking out parts from a whole, as we saw with the sample obtained from the first page of *Moby Dick*. Good statistical awareness is a matter of knowing what you are abandoning as well as what you are getting.

Formal statistical sampling generally ignores the problem of sampling structural systems or integrated units. Yet most of us draw conclusions about novels, cultures, personalities, or other structural forms on the basis of samples. When we consider reading a novel, we scan ten or so pages and either get into the book or put it aside. This form of sampling, though common, falls outside the domain of statistical logic. The adequacy of such samples cannot be established in any precise manner. You might turn down a novel because it begins badly and miss out on one of the greatest conclusions in literature. Sampling a structure is chancy, and we do not yet have a way of accurately assessing confidence in such samples.[7]

THREE FACTORS AFFECTING SAMPLING RELIABILITY

Variability within the Population Being Sampled

Before we get into the technical aspects of sampling, we should develop a general sense of sampling logic. Basically, there are three factors that determine whether a sample is reliable.

The first factor is *the amount of variability in the population being sampled*. If there is a lot of variability, samples come up with different answers each time you draw one. If there is no variability in the population, then samples will always be the same—no matter how big or small they are. If we drew a thousand samples, each involving twenty (or even two or three) cases, they would give the same answer, sample after sample after sample. Actually, if there is no variability in the population, we could rely on a sample of a single case as a basis for generalizing. However, with respect to purely social variables or qualities, there is always considerable variability within the population. As a consequence, samples always vary. We are then less certain with respect to the possibility that any given sample provides us with accurate information concerning the nature of the value for the population.

In the scenarios that follow you are to answer two questions: (1) Would you need a large sample to get reliable results, or could you rely on a small one? and (2) Why? Your answers will depend on your intuitive sense of the extent to which there is variability in the population being sampled.

1. You are visiting a friend who owns a swimming pool. You get up early in the morning and decide to go for a swim. However, you want to find out if the pool is cold before you jump in.
2. You want to determine the sexual make-up of an American prison on the basis of a sample of prisoners. (That is, is it a male or female prison?).
3. You want to determine the average intelligence scores of inner-city kids.

4. As an owner of a publishing firm, you would like to know the average number of books read by the typical American in a year.
5. You are doing a study of the incomes of American university professors. You have to rely on a sample.
6. You want to determine, on the basis of a sample, what language is spoken in a particular country.

These scenarios are meant to be simple. Their intent is to get you to see how variability in the population being sampled affects sample size. Let's go through the six questions quickly.

1. The temperature of the pool is homogeneous. The pool is probably much the same temperature at one place as at another. When this is so, an extremely small sample—such as your toe dipped in the water at the pool's edge—provides all the information you need to decide, with great *confidence,* whether you will go swimming or not that morning.
2. You would only need to determine the sex of one or two prisoners to decide whether it was a male or female prison, as American prisons characteristically have homogeneous populations.
3. Intelligence varies considerably. You could not get by with just one or two people or an extremely small sample here. You would have to get a larger sample to get a sense of what constitutes the typical intelligence of the population.
4. Reading habits vary from those who read no books at all over a year to those who read several hundred. A large sample would be needed to deal with this problem.
5. Professors' salaries vary tremendously. Again, you would need a large sample to get some idea of the typical case.
6. Though there is some variation within nations in terms of languages spoken, the typical case is relative homogeneity. You would only need a few randomly selected individuals to determine what the dominant language is.

At this point we can set forth a general principle:

The more homogeneous a population is, the smaller the sample needed to obtain reliable estimates. The more variable a population is, the larger the sample needed to obtain reliable estimates.

Sample Size

The second factor that influences the reliability of a sample is sample size. Our intuition is generally correct with respect to how the size of a sample relates to sample reliability. The bigger the sample, *assuming it is a random sample,* the better it is in terms of the confidence we can place in its values. The important qualification, of course, is that the sample is randomly drawn. A large biased sample is no better than a small biased sample. It might be worse because we are inclined to think because it is big it is good. A small, carefully randomized sample is superior to a large nonrandom sample. Increasing the size of a biased sample does not do away with the bias or error.

Precision

Finally, the third factor influencing sample reliability is *precision*. Precision was alluded to in the discussion earlier about people feeling more confident when they do not worry about precision. Before going on, what does your intuition tell you to expect with respect to the relationship between sample size and precision? You guessed it—the more precision you want, everything else being equal, the larger your sample is going to have to be. (Don't get too cocky, however. When does making a sample larger not improve on its precision? Hint: When can you rely perfectly on a sample of just one case?)[8]

Samples, in most instances, do not provide precise information. One of the things we want to estimate is the extent to which we can be confident in what our sample tells us. We would like to know the limits of its imprecision. Suppose a random sample taken from a particular political district tells us that 34 percent of the district consists of Democrats and 66 percent Republicans. Obviously, random errors make these values slippery. How slippery? If we wanted to be absolutely certain of our figures, we could say that the Democrats range somewhere between 0 and 100 percent of the district. Such information, though we can place perfect confidence in its truth, is useless. The limits are too broad. While we can be certain the true value is somewhere between 0 and 100, the imprecision of the statement is so total as to make our sense of certainty worthless.

Notice that we are torn between two forces. On the one hand, the probability that we are correct increases as we extend the limits within which the true value, or the **parameter**, might lie. However, as we extend the limits, we lose precision. If we extend them so far that we can be absolutely certain we have contained the parameter, we lose all precision. On the other hand, if we move the limits closer together we get greater precision, but the probability that we have contained the parameter becomes smaller and smaller.

Coming back to the question of the percentage of Democrats in the district discussed above, recall that we were unsatisfied with the conclusion that the true value is between 0 and 100 percent. We might then ask another question: what is the probability that the true value for the district lies somewhere between 30 and 40 percent? If we knew the probability was extremely high, then these limits would define the range of our confidence in our sample—and that is why statisticians call them *confidence limits or intervals*. Sampling procedures are directed toward the goal of establishing the *probability* that given confidence intervals contain the true value of a condition within the population.

These limits (of 30 and 40 percent) are narrower than the range from 0 to 100. If we make our confidence limits even narrower, say from 33 to 35, we would obviously have to have a larger sample to allow us to conclude that we have *probably* limited the elusive population value that we are trying to estimate through our sample. Always keep in mind that we *never* know the precise population or universe value. If we did, we would not need a sample.

We have now covered the three basic factors that determine the amount of variability we can tolerate in the values samples give us as we take sample after sample from a given population. Remember: The more samples are likely to vary from each other, the less trust we can put in what any single sample says. Here, again, are the three factors:

1. Variability or heterogeneity in the population being sampled. If there is considerable variability within the population, then samples will tend to vary a lot. In this case we cannot put as much faith in the values given us by a single sample as we could in a situation in which the population being sampled is homogeneous.
2. Sample size. Generally, if the sample is small, sample values will tend to vary a lot, and we cannot trust what any single sample is telling us.
3. The amount of precision we want. If we do not need much precision, we can tolerate a great deal of variability between samples. However, if we want a high degree of precision, then we want to reduce sampling variability to as low a level as we can manage.

GETTING AT PARAMETERS

Let's consider a specific instance of sampling and see what problems we get caught up in. Say we have taken a sample of the adult population of the United States. It is a random sample.

Before we move on with the example, let's pause a moment to consider how easy it is to say we have a perfectly random sample, but how nearly impossible it is to get one. Think about it. Do all homeless people have a proportionally equal chance of being included in our sample of the adult population of the United States? How about criminals on the run? How about illegal immigrants? How about people living in remote areas? Such small populations are referred to here as *trace elements.* If you say these elements don't count, then you are dealing with a sample that is not a random sample of the total population. It is, therefore, not representative of the total population. Randomness is important only when it promotes representativeness on the part of a sample. It is not, in and of itself, a guarantee of representativeness.[9]

Suppose we get information from the sample concerning the percentage of people who would support a program of total nuclear disarmament by the United States. Let's say that 20 percent of the people in our sample fall into this category. This is a sample value. It is likely to deviate from the real value for the population as a whole. Remember that the real value for the population as a whole is generally referred to as a *parameter.* What we must constantly keep in mind, at this point, is that the parameter is a value that holds for the entire group, population, or universe being sampled. Unless you sample the whole population, you cannot know the exact value of the parameter. You can only estimate it. The question you want to ask of your sample is, how close an estimate do I *probably* have?

We want to know what the parameter is. However, *we can never know the value of the parameter when using sample data.* This is one of the hardest lessons of sampling logic. Nonetheless, we might be able to use our sample to tell us something about what we *believe* the parameter to be. Suppose, for example, we are convinced that 50 percent of the American people favor disarmament. Then consider the following question: if the real value is 50 percent, how likely would we be to get a sample that says it is 20 percent? If we tell you the sample consisted of 10,000 people, what does your intuition tell you? We should have picked up roughly 5,000 people who favor disarmament, and we only got 2,000. It could have happened by chance, but that doesn't seem likely. Therefore, we can reject the hypothesis that the real value is 50 percent.

Suppose we assume that there are no Americans who favor disarmament. The percentage, according to this assumption, is equal to zero. Yet in our sample of 10,000 people we picked up 2,000 people who favor disarmament. A parameter of zero does not seem likely either. In fact, it is impossible. It could be close to zero, but our sample suggests that even this possibility is not likely.

The real value appears to be something lower than 50 percent but higher than zero. So it is, then, that we feel pretty safe in concluding that the real value probably lies somewhere between 0 and 50 percent. Notice that we are thinking in terms of brackets within which the real value is *probably* located. How can we get a good sense of where these brackets should go? How can we establish reasonable limits within which the parameter probably falls?

The notion of limits is the proper way to think about sample data. Indeed, because most of our generalizations are based on samples, it is a good way to think about all of our generalizations. Unfortunately, once a sample value is registered, people tend to lock on it, forgetting that it is an estimate of an unknown parametric value. A sample value should always have upper and lower limits within which it is reasonable to conclude the parameter *probably* lies.[10] Only when the universe (another term for population) is homogeneous or consistent can we conclude that sample values are extremely precise.

USING SAMPLING LOGIC

The logic of sampling is closely associated with the logic of probability, which in turn is closely associated with the idea of percentages.

Let's try our hand at a real sample and see what happens. In *The Statistical Abstract* there is an interesting table on energy use by country.[11] Using data for 1987, we drew a random sample of seven countries. Per capita energy equivalent data are used here. See Table 10.1.

Before we sneak a look at what the mean is for all seventy-three nations from which the sample was taken, let's consider what the sample says it is and what we might anticipate. First, the sample is a small one—only seven cases. We should not expect it to be too reliable. However, it has one very important quality: it is a purely randomly drawn sample. Second, the sample tells us that

Table 10.1. An Illustrative Random Sample, $N = 7$, Obtained from 73 Nations According to Per Capita Annual Energy Use in Kilograms of Coal Equivalent, 1987

Sample Nation	*Per Capita Energy (coal equivalent, kg)*
Bulgaria	5,912
Ecuador	627
West Germany	5,624
Italy	3,570
Pakistan	250
Vietnam	118
Chile	938
Mean = 2,434	
SD = 2,366	

Source: The Statistical Abstract for the United States, *1990, Table 1470.*

we are probably working with a general population that is extremely varied in composition. The range, within the sample, extends from 118 kilograms of coal-equivalent use for Vietnam to 5,912 for Bulgaria. (Bulgaria uses 50 times as much energy, per person, as Pakistan. The United States, incidentally, uses almost twice as much as Bulgaria.) These two considerations should warn us instantly that any generalization from this sample is probably not reliable.

Let's do one more random sample from the same population and see what it tells us. If we suspect our first sample is not reliable, then a second sample is likely to deviate considerably from the first sample. The second sample of seven countries is shown in Table 10.2.

Table 10.2. An Illustrative Random Sample, $N = 7$, Obtained from 73 Nations According to Per Capita Annual Energy Use in Kilograms of Coal Equivalent, 1987

Sample Nation	*Per Capita Energy (coal equivalent, kg)*
Bahrain	14,680
Denmark	5,346
India	275
Kuwait	9,191
Poland	4,810
Sweden	5,004
Chile	938
Mean = 5,749	
SD = 4,572	

Source: The Statistical Abstract for the United States, *1990, Table 1470.*

As we might have suspected, this second sample provides us with a much different estimate—one that is more than twice as large as the value we got with the first sample. If we were to go on drawing sample after sample, we would keep getting different values for each sample. This is an example of **sampling variability.** Sampling variability refers to how much one sample varies from another as we take sample after sample, usually of the same size, from a given population. When sampling variability is great, as it appears to be with our illustration, then we cannot put too much confidence in what any particular sample might be telling us. When sampling variability is low, then our samples are all telling us pretty much the same thing, and we can place considerable confidence in what any particular sample is saying.

Because our population is small, we can find out what the population value is for all the nations involved.[12] When we look at the figures for the whole set of seventy-three countries, we find that the average world figure was 1,921 kilograms of coal-equivalent energy use per person per year. The first sample came fairly close to this parameter, overestimating it somewhat. The second missed it by quite a margin. Now, if we were working with only a single sample and had nothing else to go on, how could we tell whether we were likely to be close to the parameter or not?

Sampling Variability

To answer this question we need to know how much sampling variability we might expect. However, how can we figure out how much sampling variability to expect for a given situation without going to the trouble of taking sample after sample after sample? Obviously, that can become tiresome.

Sampling variability is like any other kind of variability. The key word is *variability.* How do you measure variability? Sampling variability is measured like most other variability; that is, it is expressed in terms of standard deviation units. Remember, sampling variability refers simply to the variability you get in values taken from sample after sample of the same size drawn from the same population. To make the distinction between individual variability and variability between samples, statisticians use the term **standard error.** A standard error is a standard deviation that expresses the variability between measures taken from a relatively large number of samples of the same size.

We see, once again, why scientists are so eager to find constant values. Something that has a constant value can be sampled at any time and in any amount, and the value for the quality being sampled remains the same. Thus, a constant provides a situation in which sampling reliability is perfect. As variability is introduced between samples, however, life becomes more confusing. Things become ambiguous, and it is increasingly difficult to be rigorous in one's thinking and observations. If a population is highly variable in character, we must obtain large samples in the quest for reliability.

Sampling variability goes up as the variability in the population goes up. It goes down as the size of the sample increases. What we need to do is to put these notions together. The following formula does the trick:[13]

$$SE = \frac{\text{Standard Deviation of Population}}{\text{Square Root of Size of Sample}} = \frac{SD}{\sqrt{N}}$$

The formula says simply that the variability of samples (*SE*) is equal to the variability in the population (*SD*) divided by the square root of the size of the sample (*N*). This is an extremely basic formula, and you must understand it in order to gain insight into how samples are evaluated.[14] Suppose we have a population of geniuses, all of whom have an IQ equal to 200. There is no variability in the population. That is to say, $SD = 0$. If we drew a sample of 100 cases from this population, what would we expect to be the amount of variability between samples, measured in terms of standard error? Because there is no variability in the population, we get

$$SE = \frac{0}{\sqrt{N}} = \frac{0}{\sqrt{100}} = \frac{0}{10} = 0$$

There is no sampling variability. Our sample is, therefore, perfectly reliable. If we drew another sample of the same size, it would tell us exactly the same thing.

Let's go back to the energy example above. There we noticed each of our two small samples was warning us about something: there was a lot of variability within each sample, which suggests a lot of variability in the population being sampled. Now, when we are forced to rely on a sample to make a generalization about a population, it means we do not have access to the entire population. We cannot, therefore, determine precisely the variability within the population. How, then, can we find some way to get a sense of the variability of the population? Think about this for a minute before moving along to the next paragraph.

What choice do we have? All we can do is use the sample itself as a basis for estimating variability in the population. As we just saw, both of our samples in the energy example told us that there is great variability among nations with respect to per capita energy consumption. We can, then, estimate the standard deviation for the population by using the standard deviation obtained from the sample. When a sample is small (less than 40 or 50 cases), however, the estimate is likely to underestimate variability in the population.

One of the samples showed a standard deviation of 2,500 kilograms, and the other showed a standard deviation of 4,500 kilograms. Like the mean, the *SD* is subject to sampling variability; that is, one sample is likely to give an estimate that differs from another sample's estimate.

Let's take the higher estimate of variability and assume that our samples come from a highly variable population. Such an assumption means we are not going to be as confident about our sample results. We shall err on the side of caution. Using the formula for sampling variability of the mean, we can at least get a rough idea of how much variability we might expect from sample to sample to sample:

$$SE = \frac{4{,}500}{\sqrt{7}} = \frac{4{,}500}{2.646} = 1{,}701$$

We now have a number—the standard error of a sample mean. But what does it imply? We need to return to our lessons on the standard deviation. The paragraph in the box below is especially important, and you should understand it thoroughly before you move on to the final discussions of sampling in this chapter (it is so important we blocked it off).

If we took an infinite number of samples, where $N = 7$ each time, we could obtain the mean for each of the samples. The standard deviation, *now called the standard error because we are talking about sample values,* would be equal to 1,701. We would then know that approximately two-thirds of all of our samples would lie between 1,701 kilograms above the mean and 1,701 kilograms below the mean (of all the samples). Only a small number of samples would be two standard errors away from the mean. An extremely small number would be three standard errors above or below the mean. *Note:* Although we set *SE* at 1,701, this is a rough estimate. The true value is probably somewhat larger.

The infinite number of samples would be distributed relatively normally, and we would have the situation shown in Figure 10.1.

Figure 10.1. How Sample Values Would be Distributed for a Sampling Distribution with $SE = 1{,}701$

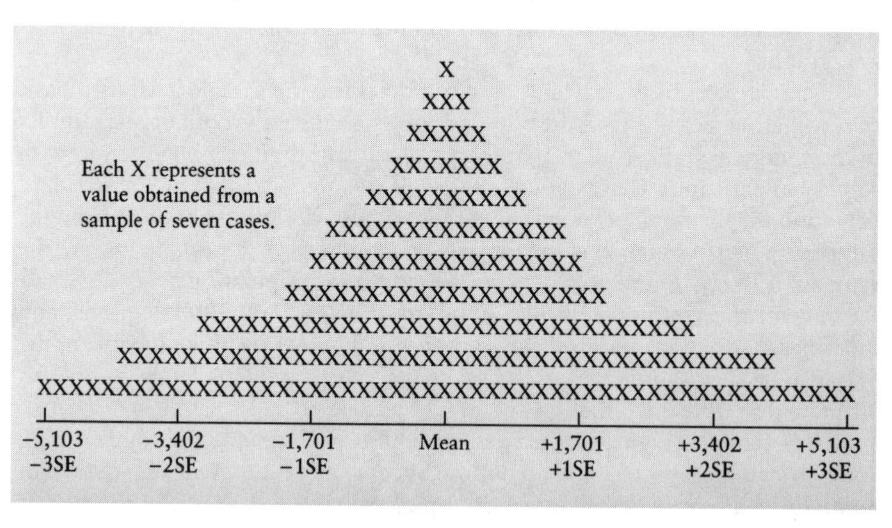

The nature of the standard error should now be relatively clear. If it is not, go back to Chapter 6 and review what was said about measuring variability there (pages 113–120). The major thing to keep in mind is that we are talking about sample values in these distributions. If all the samples gave exactly the same value (something that would happen if the population variability were zero), then the distribution would look like a straight line (see Figure 10.2).

Figure 10.2. How Sample Values for the Mean Would be Distributed for a Sampling Distribution with $SE = 0$

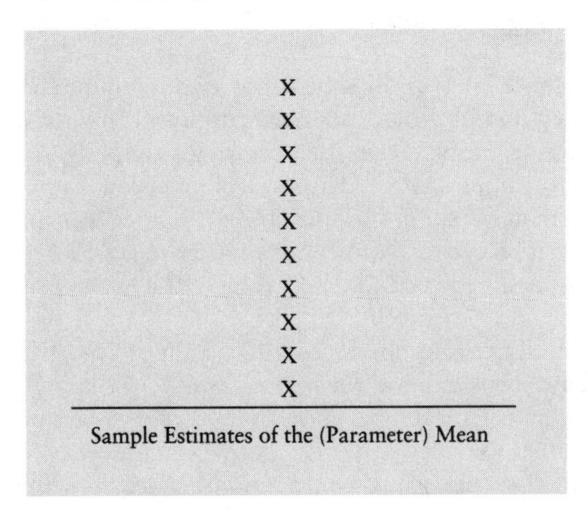

The normal distribution in Figure 10.1 is an example of a logically derived sampling distribution that uses the *SD* of the sample to estimate variability or the *SD* of the population. But the *SD* of any sample of a population usually underestimates the *SD* of the population. For example, the *SD* for the first sample is around 2,400, and for the second it is around 4,600. Note that the sample *SD*'s vary, just as any other sample statistic will vary from other estimates if the samples are drawn from heterogeneous populations. Using a larger estimate of population variability produces larger estimates of sampling variability. If we are erring, it is on the conservative side.

We now have some idea of what sampling variability is. But how can we use this knowledge of sampling variability to estimate what we want to know—the value that exists for the population?

Using the energy example above, let's go two standard errors beyond the mean of our sample (roughly 5,800). The *SE* in this case is much greater; it is roughly 4,600. Two *SE*'s beyond the mean, then, would give us a value around 15,000. If we go two standard errors below the mean we get a value below zero—an impossible value because people cannot use negative energy. The real value probably falls within these brackets. However, the confidence limits are extremely broad, ranging from approximately zero to 15,000. Our samples are essentially useless. We cannot give any kind of precise estimate to world per capita energy use with these samples. We were warned of this possibility, however, by the fact that our samples (a) were very small, and (b) displayed a tremendous range of energy consumption, implying that they were drawn from an extremely heterogeneous population.

The point to be learned here is that *a sample value is not informative unless you have a reasonable estimate of its* SE *to go along with it*—just as a mean is not a very useful statistic unless you have some idea of how much variability is associated with it. In other words, in order to have some idea of the worth of a given statistic, you generally need to know some other values as well as the one you have in hand. Do not get fixated on single statistical values. Always relate them to other values that provide you with a sense of their reliability or unreliability.

Broad confidence intervals like the ones we got with the energy samples are unsatisfying. They do not provide the kinds of precision with which we generally feel comfortable. However, at the same time, sampling logic warns us not to be lulled by *any* samples into a false sense of precision. Sampling logic forces us to recognize that we simply do not know the precise truth, and that our attempts to estimate it can deal only with alternative versions of what the truth *might* be—its probable lower and upper values. The logic of sampling, then, is cautionary philosophy par excellence.

It is worth underscoring this observation. Suppose you had been told that the average energy consumption for nations was 5,749 kilograms. You might have a tendency to lock on to this value, because when we are told something, unless we have other knowledge, we generally begin to believe it.[15] However, a sample value of 5,749 kilograms would seriously overestimate things.

Always be wary of sample values; look at them with a critical eye. While this is easy advice to give, most of us are not generally aware of the extent to which nearly *any* knowledge we have is based on a sample. In other words, we should be wary of nearly every social generalization we know—because it is probably not as accurate as we think.

Why Bother with Sampling Logic?

The logic of sampling is frustrating and difficult for the newcomer to statistical reasoning. Why bother with it? One good reason is that virtually all of our knowledge is based on samples. If we want to understand how we come to think the way we do, we absolutely must gain a sense of what is going on when we sample the world around us.

A second reason is that compared to working with the whole population, sampling is a more efficient, and often cheaper, way to get things done. Modern scientific polling characteristically uses samples of 3,000 or so to get a sense of how the entire population of the United States is responding to a particular political or social issue. It is a lot less costly to use 3,000 people than to poll the entire population.

A third reason for having a better sense of the logic of sampling is that much (in fact, we would be inclined to say nearly all) modern information comes out of samples. If we are to have some comprehension of what our newspapers are saying or what psychologists and other modern experts are telling us, we need to know what the limitations, and the strengths, of sample-based materials are.

For example, when you read in some text that there is a correlation of +0.32 between, let us say, the degree of violence in the home and the consumption of alcoholic beverages, you should ask about the sample upon which the correlation is based. Like any other statistical value, correlations vary from sample to sample. This sample says it is +0.32. Another might say it is +0.23, and yet another might say it is +0.41. What are the appropriate upper and lower limits within which this correlation might be expected to vary for samples of a given size? If the sample were small and nonrandom and the research was badly designed, it could be the case that the correlation of +0.32 is a random aberration. The true correlation might be zero.

What is true of statistics in general is more specifically true of sampling: if it is done right, it gives good and meaningful results. But if it is done poorly, it is like anything else done poorly—it gives bad results.

It cannot be sufficiently emphasized that sample values vary from one sample to another, and that this is what makes samples tricky as a basis for generalization. If we have a sense of how much they might vary, we can begin to see whether we should put a lot or only a little faith in what sample values tell us. In cases where we have reason to believe that a sample is not biased and that other samples of a similar nature would provide much the same results, then, obviously, we can place confidence in the sample. The primary focus of attention in all of this is sampling variability. Keep in mind that if a sample is biased, even if sampling variability is low, the samples obtained in continued sampling would simply go on giving biased information. That is why the assumption of randomness is so important.

THE CENTRAL LIMIT THEOREM

To close off the formal part of this discussion of sampling, it is necessary to mention two interesting features of sampling distributions. First, even when samples are obtained from individual distributions that are highly skewed, the distribution of sample values taken from those skewed distributions tends to become normal as N reaches 20. Even relatively small samples, then, have normally distributed sampling distributions.[16]

A second feature of sampling distributions is that if you draw a large number of samples and determine the mean of the sample values, the mean approaches the value of the parameter as the number of samples is increased. This should be intuitively apparent. If you draw, let us say, five million samples with $N = 50$ in each case, and you obtain the mean for all of the five million sample values, you are, in effect, making an estimate of the parameter based on a sample of 250 million cases. This is an unusually large sample and, if it is properly drawn, its mean should be extremely close to the parameter.

This notion—that sampling distributions approach values equal to the parameter as larger and larger number of samples are drawn—is called the **central limit theorem.** The importance of these observations comes from the fact that sampling distributions tend to be normal, and they are normally distributed around the true mean or parameter. Therefore, once the *SE* is known, it can be used to estimate the probability that a given sample deviates from the true mean.

Suppose you have a sample that gives you a mean of 50. You cannot trust that value because it is taken from a sample. You determine that the *SE* for the mean of 50 is 5. If the parameter were 60, then this sample would be two *SE*'s off. The parameter, then, is probably not as high as 60. If the parameter were 40, the sample would again be two *SE*'s off the mark, again not a very likely occurrence. So, the real parameter probably lies somewhere between 40 and 60—and you can put considerable confidence in these limits.

Intuitive Exercises

Five elementary sampling problems follow. Try to assess them in terms of your intuition and what you now generally know about samples. For each problem, indicate whether you would put a lot of confidence in the result or very little. Then state why. Consider the issues of population variability, randomness, sample size, and implied precision.

1. A social psychologist reports that in a study of 27 undergraduates majoring in psychology at UCLA, she found a correlation of +0.47 between measured intelligence and a concern with environmental issues.
2. A total of 27,346 readers of *Time* magazine returned a questionnaire asking about problems at work. The survey found that over 87 percent were happy with their jobs.
3. A random sample of 2,798 farmers obtained by the U.S. Department of Agriculture revealed that the average farmer owes the banks $18,546.
4. Remarks of an Israeli soldier making a comment before the cameras for a TV documentary on life in modern Palestine: "Your typical Arab is a friendly sort. They can be fiercely loyal. However, they are treacherous. They will thrust a knife into you without warning."
5. "You must try the bean soup at the Teapot Restaurant. I had some there the other day and it was great."

If you have a sense of sampling logic in its general form, then you should begin to see that samples are ruled by just a few basic concerns. Let's go through the five problems above and ask ourselves the following questions:

1. What about the size of the sample?
2. Is the sample possibly biased?
3. Is the population from which the sample was drawn highly heterogeneous?
4. How much precision are we being offered, and is such precision warranted?

Sample 1. Obviously this sample is small. More importantly, however, it is biased. It is based on results obtained from college students—a unique group. We would expect these students to be generally more concerned about environmental issues than less educated people. However, the correlation of +0.47 is set forth as a general statement of relationship. We might wonder what the results would have been with a different group. Certainly we should hesitate before concluding that more intelligent people are, in general, more environmentally conscious.

Sample 2. Here the sample is huge, but it is essentially worthless. It is based on the readership of *Time* magazine, a highly biased group. We would expect this readership to have better jobs in general and, therefore, to be happier with their work.

Sample 3. Department of Agriculture samples are carefully drawn. The sample is large and it is random. We can place some confidence in the results. However, to report the sample value accurate down to the dollar is, of course, to imply greater precision than the sample would permit. While we cannot place much confidence in this highly precise figure, we could place considerable confidence in a figure that indicated the true value as being somewhere between $18,000 and $19,000.

Sample 4. This is a typical stereotypical assertion. It is based on a biased set of observations, and also on a limited set of observations. Although the sample values are reported imprecisely in terms of a wide range of extremes—from loyal friendship to treachery—the impression given is that treachery is a characteristic of all Arabs. The statement gives no hint of the possibility of variability within the Arab population. Each individual Arab is said to vary from being friendly to being treacherous, but this is an easy generalization because it is so imprecise. It probably applies to people almost anywhere.

Sample 5. One of us actually said this to some friends on the basis of a meal he had at a particular restaurant. He went back later and had another dish of their "great" soup, and it was disappointingly below ordinary. In other words, here is another example of a statistician (who should know better) generalizing from a single case and assuming that quality variability was zero. As it turned out, it was not. Even statisticians fall into the temptation to generalize from small samples.

A Final Set of Intuitive Sampling Exercises

Now that you have a sense of how simple, yet devious, sampling logic is, we need to move on to more technical concerns. In the following five problems you can assume that the samples involved are always unbiased or random. What you must do is estimate what you think would be a reasonable set of limits within which the true value, the parameter, lies. Let's define reasonable as a set of limits that are likely to contain the parameter in ninety-nine out of one hundred instances. That is to say, there is only a small probability that the parameter is outside the limits. Make the limits as precise as you can. To get you started, we will go through an example.

Example: You are working for a marketing research firm. You draw a random sample of 360 women who work and find that 67 percent of these women believe (a) that they are badly paid, and (b) that this seriously affects their purchasing power. What might be reasonable limits within which the true value (the percentage of all working women who feel this way) lies?

Analysis: The sample is moderately large. It is random. Obviously not all working women feel that they are underpaid. While we would not expect a lot of variation, we would expect samples of 360 women to vary modestly from each other. Probably the lower limit for the true value is around 64 percent and the higher value might be around 70 percent. (A more precise solution would set the limits just a little broader than this. However, try to first gain confidence in your intuition or estimating ability, and use this ability before running off to the computer for answers.)

1. The standard deviation for a sample of 87 young men on a test of delinquent inclinations is 38. The mean score for the sample is 52. What would you guess the upper and lower limits for the confidence intervals to be?
2. You find that a mistake was made in entering data for the first problem and that the calculated standard deviation for the sample is now 3. What does this do to the confidence intervals?
3. A random sample of 430 counties is taken from the 3,000 or so counties that make up the United States. The percentage of residents who are 65 or over is found for each county. The mean is determined to be 18 percent with a standard deviation of 4 percent. Within what limits would you expect the value to fall for all the counties?
4. You obtain a sample of 2,687 records from the state of New York. The total set of records numbers in the millions. The records deal with, among other things, the length of unemployment of workers after the loss of their first full-time jobs. You find that the range is from one day at the lower end to over twenty years at the upper end. You calculate the *SD* for the sample and find that it is 67 days. You find the mean interval to be 145 days. What would you consider reasonable confidence limits for this finding?
5. Your boss comes to you and says that in order to get a huge contract with Belchfire Industries, your company's research and development department needs some information. They must know, within plus or minus two percent, what percentage of the voting public will vote Republican in your state. How big a sample would you use? (Assume that your boss is the sort of person who, if you waste the company's money, will send you to the branch office in Antarctica.) The thing we are looking for with this problem is how you would go about thinking the problem through. Try to come up with a rough estimate of the sample size needed to achieve the ends you have set for yourself.

Now let's run through a discussion of how the above problems might be addressed before concluding this discussion of sampling.

1. The degree of variability is fairly large. A standard deviation of 38 suggests a lot of variation around the mean. At the same time, the sample is relatively small, consisting of 87 individuals. We shall need to set the confidence intervals rather wide for this one. The true mean might be as low or lower than 40 and as high or higher than 60.
2. This should reduce sampling variability. It would be safe to conclude that the parameter is likely to be found somewhere between 50 and 54.

3. The sample is fairly large. The standard deviation suggests relatively little variation between counties. Therefore, the real value should be close to 18. It would probably not be much below 17 nor much above 19 at the very worst.
4. You have a very large randomly drawn sample. On the other hand, you have to deal with a lot of variability. This one is difficult. The extreme upper limit suggests a highly skewed distribution. It would be wise not to lean solely on the fact that the sample is as big as it is. The true value might be off by as much as two weeks in either direction. The standard deviation of 67 days enables us to make a rough guess and set our limits at around plus or minus three days. Because of the skewed distribution, however, unless there was a need for precision, it might be better to stay with a slightly broader set of limits.
5. It is the demand for precision that makes you want to run out and get the biggest sample you can think of. However, you would probably be safe with a sample of around 700 cases.

More Precise Approaches to Sampling

We shall conclude this consideration of sampling by moving beyond intuition, hunches, and guesses (though these practices should not be ignored or resisted). There are a few fundamental formulas in statistics that are intended to supplement your intuition. Now let's draw on these formulas to deal with some of the problems that we have already struggled with from an intuitive approach.

Let's come back to problem 1, with its sample of 87 men, standard deviation of 38, and mean of 52. We can determine the value of the standard error from this information. We can get a sense of how much we might expect one sample to vary from another. The formula for this, as we saw earlier, is

$$SE = \frac{SD}{\sqrt{N}}$$

Solving for the figures above, we get

$$SE = \frac{38}{\sqrt{87}} = \frac{38}{9.33} = 4.07$$

The standard error for the means taken from samples of 87, in this instance, would be around 4. If the true mean were 44, then our sample would have to be off by two standard errors (two standard deviations), and we already know that things that deviate by two standard deviations are relatively unlikely. If the true mean were 60, we would also know our sample deviated by two standard errors. So a lower limit of 44 and an upper limit of 60 probably contains the true value.

What happens when the *SD* is 3 instead of 38? Then we have

$$SE = \frac{3}{\sqrt{87}} = \frac{3}{9.33} = 0.32$$

If the true mean were 44, then this time the sample mean of 52 would be 25 standard errors away from the real value. It is extremely unlikely that we would get such a peculiar sample strictly by chance. We would, therefore, reject the idea that the real value is 44 and accept a value somewhat closer to 52—a value that is only a few standard errors away. If we decided to let two standard errors be our probable error limits, then we would be willing to settle for the argument that the real average is somewhere between 52.7 and 51.3. The probability of being wrong would be less than one in a hundred.

Problems 3 and 4 are solved in much the same manner. Problem 5 is a little tricky because it does not provide a value or an estimate for the population variability. The best thing to do, in this instance, is to assume the worst case, which would be that the state is fifty percent Republican and fifty percent Democrat. That's the most variability you can have with a dichotomous (two-part) variable. You would then need to go to a statistics text and find the formula that tells you how to find the sampling variability of proportions or percentages:

$$SEp = \sqrt{\frac{p \times (1 - p)}{N}}$$

You want to be extra safe, so you would want a sample large enough to give you a small value for *SEp*. You decided that the value for two *SE*'s should not be greater than 0.02, the limit asked for by the boss (plus or minus two percent). Then *SEp* would be equal to 0.01. To solve you would need to complete the following:

$$0.01 = \sqrt{\frac{(0.50 \times 0.50)}{N}} = \sqrt{\frac{0.25}{N}}$$

If you square both sides of the equation, you get

$$0.01^2 = \frac{0.25}{N}$$

and multiplying by N, you get

$$N \times 0.01^2 = 0.25, \text{ and } 0.0001N = 0.25, \text{ so } N = 2{,}500$$

This more precise solution, though it is still a rough estimate, tells us that a sample in the range of 2,500 cases, randomly drawn, might provide results with a better than a one in a hundred chance of being correct, to within plus or minus two percent of the true value. Our intuitive guess of using a sample of 700 cases was way off.

What would we have come up with if we had used a sample with $N = 700$? Then

$$SEp = \sqrt{\frac{(0.50 \times 0.50)}{700}} = \sqrt{\frac{0.25}{700}} = \sqrt{0.0003571} = 0.019$$

With this value for *SEp*, if we got a sample value for the Republicans of 55 percent, and if we bracketed this value on either side by two *SEp*'s, we would have values of roughly 51 and 59. In this case we would be safe within a fairly broad interval of plus or minus four percent; but the boss wanted the interval to be within half that value.

This example is typical because a sample of 2,500 cases is commonly used in opinion research—a variety of research that generally reports error margins of two percent. Thus, problem 5 illustrates a typical sampling demand. The boss wants two things: she wants precision, and she does not want to spend any more money than is necessary. The whole point of sampling is to save money. If you have infinite resources, time, and energy, then you can go ahead and sample the entire state and not worry about sampling variability. But companies, and individuals, do not have the money or the time for such indulgences. Judgments have to be made economically, and that is the hard bargain that we come up against in sampling logic.

Often, in research for doctoral dissertations or for agencies where budgets are limited and resources are tightly restricted, concessions have to be made. Samples are commonly negotiated (the doctoral student uses a sample that his or her committee is willing to tolerate, for example). Such samples are often samples of convenience, or nonrandom samples, or quite small samples. This is especially true in collecting social data, where simply obtaining a sample, prior to doing any further research, can be costly and demanding.

As a concluding note, it should at least be mentioned that these formulas used for estimating standard errors are crude devices because they estimate population variability in terms of values obtained from the sample itself. As we have seen, samples tend to underestimate variability. The use of erroneous population variability estimations can lead us to attribute greater accuracy to sample values than they warrant. At the very least, however, the *SE* is a warning. It provides some sense of the extent to which any sample value might be subject to variability if similar samples, or samples of a similar size, were drawn over and over and over.

It is now common to see polls presented on television, stating that they have a "margin of error" of two percent, or three percent, or some other figure. Just what is meant by "margin of error" is difficult to determine, because the pollsters never tell us the number of standard errors they are using to bracket their sample values. If they are using 1.96 *SE*'s, then a margin of error of 3 percent would mean that there was one chance in a hundred that the real value exceeded or fell below the reported sample value by more than 3 percent. The margin of error is a probabilistic value, not a mechanical one.

For any statistic that comes out of a sample there is a way of estimating how it varies from one sample to another. These formulas for the *SE* appear in most texts on statistics. It is not necessary to go into all of them here. What we are interested in is the fundamental logic of sampling.

If you have grasped this logic, then you know that if you wanted to find out how much a correlation coefficient (a statistical index that is discussed in detail in Chapter 12) obtained from a sample of 89 randomly selected people varied, you would go to a text to find the proper formula for the *SE* of a correlation coefficient. Suppose the correlation coefficient is $r = -0.43$. The variables being correlated are intelligence, as measured by a test, and number of times the people in the study laughed at a twenty-minute video recording. You would look in the text's index for either standard error or correlation coefficient. Most text indexes will lead you to the formula for the standard error of a correlation coefficient:

$$SEr = \frac{1 - r^2}{\sqrt{N-1}}$$

Solving for the figures you have been given,

$$\begin{aligned} SEr &= \frac{1\,(-\,0.43)^2}{\sqrt{88}} \\ &= \frac{0.81}{9.38} \\ &= 0.09 \end{aligned}$$

Thus there is a probability of less than one in one hundred that the real correlation, assuming the sample is a good one, is less than −0.25 or greater than −0.61. Now we have some idea of how low or how high the real correlation might be. At least we are no longer locked on a value of $r = -0.43$. Instead, we become aware of the fact that the correlation is a sample value and, therefore, something that ranges around the real value. The real value is not likely to be much below −0.25 nor is it likely to be as high as −0.69. If the real value were beyond these limits, we would not have been very likely to get the sample value that we got (−0.43).

The general logic of the *SE* is the important thing. Once you have the logic, standard texts provide the guidance you need with respect to the more specific techniques, as well as their refined use. Once again, the *SE* is a cautionary index. In some aspects of technology and the physical or natural sciences, it can be applied with great precision. In human affairs, however, it keeps us constantly aware of the fact that samples are inconsistent in their findings when they are drawn from variable populations. In human affairs, nearly anything we want to know about is highly variable. Thus, our samples never provide us with absolutely certain information.

SUMMARY

Most, if not all, of our knowledge is based on information obtained from samples. Three factors determine the extent to which we can place confidence in generalizations based on sampling: (1) the degree of variability in the population being sampled, (2) the size of the sample, and (3) the degree of precision we desire. The more variable the population, the larger the sample must be to provide a given degree of precision.

Sample values for any statistical index vary from one sample to the next.

Ten different samples are likely to give us ten different values for the mean of a population when these samples are obtained from a population with a high degree of variability. The true mean value for the population is called a *parameter,* and it is the task of sampling logic to establish how close or how far we might be from the parameter with a given sample value. The best that can be done with sample values is to attempt to bracket the parameter within a set of probable limits called a *confidence interval.*

Sampling variability is estimated by formulas which, in effect, consider two basic influences: (1) the variability within the population, and (2) the size of the sample. As population variability decreases and sample size increases, sampling variability, referred to as the *standard error,* declines. When population variability is zero, as is the case with physical constants in the natural sciences, sampling variability is zero, and a high degree of confidence can be placed in sample findings even when the size of the sample is extremely small. The social sciences are hampered by the fact that social phenomena are highly variable in virtually all instances.

THINKING THINGS THROUGH

1. Suppose some students carry out a research project for a class using all of the names in the first three pages of a city telephone directory as a sample of the people living in their community. Identify as many sources of bias coming out of such a sample as you can think of (there are at least four).
2. You conduct a series of tests on athletes for a campus coach who has hired you to do team research. You record a sample of 27 observations on a particular football player. The data in Table 10.3 are what you put down as your observational data. The data are assumed to have been obtained under relatively constant playing conditions. Is this a reasonable assumption?

   The mean is 5.18 yards per carry and the standard deviation is 16.48. If you use the logic of inductive sampling statistics, what would you estimate as the best confidence limits for the athlete's average carry, assuming

Table 10.3. Fictional Data for the Performance Record of a Football Player

Observation	*Yards Gained*	*Observation*	*Yards Gained*	*Observation*	*Yards Gained*
1	11	10	03	19	−03
2	−02	11	00	20	04
3	03	12	11	21	06
4	02	13	01	22	02
5	−04	14	00	23	−04
6	87	15	10	24	03
7	01	16	01	25	00
8	01	17	01	26	03
9	01	18	02	27	00

confidence intervals that are 99 percent reliable and a sample size of 27? (How else might you discuss the athlete's performance with the coach? For example, what other ways might you use these data to get at what constitutes a "typical" performance?)

In this task the situation has been intentionally set up to provide what generally makes for the best kind of statistical data—purely physical information. This does not mean, however, that solving the problem of the athlete's "real" (parametric) ability is easy. You want to generalize from your sample. The question remains, what is the best way of doing so?

One further hint: we have constantly talked about how peculiar or extreme values, also referred to as outliers, can mess up estimates. What would you do with the 87-yard carry included in the data in Table 10.3? Would you keep it, thus maintaining the integrity of your observations, or dismiss it as a fluke?

3. The standard error for any statistic—such as the mean, the standard deviation, proportions, percentages, and so on—is primarily determined by two factors. What are these factors? What is the basic form they take in equations for the *SE;* that is, what goes in the numerator and what goes in the denominator of these formulas? What determines whether you want a small *SE* or are willing to tolerate a larger *SE?*

4. Using a remote control device, sample all the television channels that come through your system. Put each channel on for just long enough to be able to count the number of men and women on the screen. If no one is on the screen, count it as zero. (You might also be interested in counting the number of people on the screen of different races or apparent age groups.) Obtain one sample during the afternoon and another later in the evening. See if the two values are the same. If they differ, do you think it is because of sampling variability or because television channels offer a different kind of programming in the evening? If you wanted to do a serious study of differences in the categories of people on television using this sampling procedure, how might you go about improving your sampling of television offerings?

5. Although it is mechanical and not directly relevant to social issues, one of the easiest ways to develop a sense of sampling problems is to get a regular deck of cards and play with it for a while. The cards range in value from one (ace) to thirteen (king). Shuffle the cards and draw out five at random. Obtain the mean for the sample. Replace the cards you took out. Continue doing this until you have obtained thirty samples. Calculate the mean for each of the samples. For example, one sample might be made up of an ace, three, seven, jack, and queen; in this case the mean would equal $(1 + 3 + 7 + 11 + 12)/5 = 34/5 = 6.8$. Determine the standard deviation for the thirty values. Then calculate the theoretical *SE* that you should obtain from a deck of cards; see how your observed *SE* compares with the expected value for the mean of the population of fifty-two cards from which the samples were taken.

6. A random sample with $N = 5{,}000$ is obtained from the population of the United States. The sample indicates that the proportion of people who attend poetry readings at least once a month is three percent. Using two

SE's to define your confidence interval, what would be the limits defined for the parameter?

7. Draw a curve showing how the *SE* declines as *N* increases for populations where the $SD = 20$. To draw the curve, plot the values of the *SE* for values of *N* equal to 1, 2, 4, 16, 25, 36, 49, 64, 81, and 100. What happens as *N* becomes larger?
8. You obtain a sample of 200 fraternity members at Western University. You find they have a mean grade point average (GPA) of 2.04 with an *SD* equal to 1. You establish two *SE*'s as your confidence interval. What are the limits within which the parameter possibly lies? In what way must you qualify the assertion that the GPA for fraternity members is 2.04? What can you say about any individual fraternity member's GPA?

Notes

1. In the late 1960s the notion of separate or different realities became popular in social science literature in the United States. We have not seen any theorist tie this in, formally, to a discussion of the logic of sampling. Perhaps the reason for this is that the theorists who promote the idea of separate realities are generally the least statistically oriented social scientists we can think of. Peter Berger, who is trained as a theologian, is associated with this concept. See his *Invitation to Sociology: A Humanistic Perspective* (New York: Pocket Books, 1971) and, with Thomas Luckmann, *The Social Construction of Reality* (Garden City, NY: Anchor Books, 1967). A wildly fictionalized account of what can happen when different samplings of reality clash appears in the works of novelist Carlos Castaneda. In these works Castaneda, an American-trained anthropologist, begins to see the world through the eyes of a Yaqui magician named Don Juan. See, *A Separate Reality: Further Conversations with Don Juan* (New York: Simon and Schuster, 1971).
2. This is technically referred to as a "systematic" random sample (which is a kind of oxymoron). Systematic random samples select one element at random in a series of events and then sample every *n*th event thereafter. While this is not a purely random procedure, it works adequately for most social research sampling problems.
3. One of the original critiques of variable analysis, with its avoidance of structural realities, was made by Herbert Blumer. See Herbert Blumer, *Symbolic Interactionism: Perspective and Method* (Englewood Cliffs, NJ: Prentice Hall, 1969) and Kenneth Baugh, *The Methodology of Herbert Blumer: Critical Interpretation and Repair* (New York: Cambridge University Press, 1990). C. Wright Mills was also critical of surveys that relied on variable analysis. Emphasis on percentages of voters and other measurable quantities tended, Mills argued, to hide the complex political "machines" that brought out the vote. See C. Wright Mills, *The Sociological Imagination* (New York: Oxford University Press, 1959).
4. It is for this reason that purely statistical approaches, in the social sciences, will never prevail completely over nonquantitative approaches. In order to describe structured events, it is necessary to move beyond the more simplified and highly abstracted measures employed by the statistician. Once again we must accept the fact that quantitative and nonquantitative procedures in the study of human affairs are complementary, not antagonistic.

5. Distortion is condemned by modern social science. It is at the very least a sin and, at worst, evil. The artist is aware that distortion is necessary in order to reveal truth—creating a kind of paradox. Social statisticians tend to think that they avoid distortion by concerning themselves strictly with "facts." However, at the very outset of any statistical study, it is necessary to establish definitions. It is obvious that operational definitions are extremely distorting, but still essential. The social statistician, like the artist, is faced with the paradox that only by distorting reality is it possible to get at the so-called facts. It is this paradox that we refer to when we speak of "making a pact with the devil."
6. We can learn something from the ancients here. The God Odin, according to fable, guards the well of wisdom. People asking for a drink from the well are told they may have one at a price—their right eye. The point being made is that the ancients knew that in order to obtain one kind of knowledge, you have to blind yourself to other knowledge. Modern statisticians are not liberated from Odin's bargain.
7. A common way of dealing with the assessment of complex matters such as novels, movies, or personalities is to rely on critics or commentators. The commentator reviews the entire work, assesses it, and then gives us a condensed version, along with a judgment. We can see how capricious this form of sampling is when we notice that six different reviewers usually give us six different images of the event sampled. This problem is illustrated in the ancient story of the blind men who gather their differing conceptions of an elephant by touching the animal's different parts and concluding that the creature is like a tree (from the man touching the leg), a large reptile (from the man touching the trunk), or a huge leaf (from the man who touches the ear).
8. Carl Sagan's *The Dragons of Eden: Speculations on the Evolution of Human Intelligence* (New York: Random House, 1977) contains an essay that praises the consistency of NaCl, or common salt. Every salt crystal is like every other salt crystal. Sagan has reason to praise the consistency of NaCl. If it weren't consistent, trying to deal with it scientifically would be a much tougher problem. In essence, Sagan is praising a typicality that has variance equal to zero—an ideal condition in the creation of solidly factual descriptions of events.
9. Certain sampling procedures attempt to get around this problem by making certain that appropriate proportions of trace elements in a population are, themselves, randomly sampled. This is sometimes called *proportional random sampling*. Suppose you are taking a sample from a district in a large city, and you want to find out characteristics of the Korean population in the district. You know that Koreans comprise only one percent of the total population, and a random sample might not include anyone from this ethnic group. You could compensate for this by randomly sampling the Korean population and then showing how the information you have gathered would affect the total characteristics of the population of the district.
10. Remember, a probability is an average. It is what is expected, on the average, given a large number of replications. So, when we talk about encasing a parameter within given confidence limits with a specified probability, we once again encounter the primary concept of average or typicality. We are saying that, on the average, if we used the same-sized sample, we would contain the parameter ninety-five percent of the time. We mention this only to keep you constantly reminded of the centrality of the idea of average in statistical work. Averages, being the way they are, never allow us, in the individual case, to be certain that we have actually bracketed the parameter.
11. *Statistical Abstract of the United States,* 1990, Table 1470, 852.

12. Although we are dealing with all seventy-three nations listed in Table 1470 of the *Abstracts,* this does not constitute the total population of all nations. In effect, we are taking a small sample from a larger sample which, for purposes of illustration, can be thought of as the total population from which the sample is drawn.
13. Most statistical texts use the formula $SD/\sqrt{N-1}$. This is an attempt to correct for the fact that the smaller the sample, the more likely the sample is to underestimate the variability in the population. This correction, while the logical basis for it is correct, does not work especially well in practical sampling estimation. We leave it to the reader to examine this matter in other texts. Here the primary interest is in the logic underlying the way statisticians go about trying to determine what constitute reasonable limits within which a parameter lies.
14. Virtually all *SE*'s for different statistical indexes (such as means, proportions, percentages, correlations, and so on) are variations on this theme. That is, the *SE* goes up as variation in the population increases, and it goes down as the sample size increases. The *SD* of the population appears in the numerator and the sample size appears in the denominator in calculating *SE*'s.
15. This tendency has been demonstrated over and over by social psychologists and is one of the major empirical findings of modern social science.
16. As sample size increases, sampling distributions quickly converge to normal distributions. Hopkins and Glass show that samples taken from rectangular distributions, with $N = 5$, approximate a normal distribution. See Kenneth D. Hopkins and Gene V. Glass, *Basic Statistics for the Behavioral Sciences* (Englewood Cliffs, NJ: Prentice-Hall, 1978), p. 199.

11

THE STATISTICAL EXPRESSION OF RELATIONS: THE QUEST FOR PATTERNS

". . . the purpose is not to disclose the real essence [of things] but only to track down relations between the manifold aspects of our experience."

—NEILS BOHR

". . . in the investigation of hidden causes, stronger reasons are obtained than from opinions of the common sort."

—WILLIAM GILBERT

". . . nothing ever comes to pass without a cause."

—JONATHAN EDWARDS

The great struggle of human intellectual history lies in the effort to find patterns in the universe, the world, and the domain of human affairs. The general logic of patterns is ancient. At its most fundamental level it is an awareness that the things of this world are related. Prehistoric people knew that the winter solstice is related to the onset of spring and warmer weather. Rain is associated with the growth of crops. Aging is related to physical deterioration and eventual death. Sexual intercourse is associated with having children.[1] Still, stories are told of cultures in which this last association has not, as yet, been established. When you think of it, it takes considerable observation and thought to make the connection. After all, one can have sex and not have children.

Relationships are a matter of knowing that when one thing happens, something else happens along with it. People who insult people have fewer friends. The more you drink, the more intoxicated you become. The more our industrial system grows, the more garbage we have to deal with. The older you are, the less likely you are to have an automobile accident this year—up to a point; after this point you are more likely to have one.

Notice that this last elementary example of a relationship was expressed in terms of probability. Statistical relationships are important because they affect our ability to estimate likely or probable occurrences. We might also add that likely or probable occurrences are closely related to the idea of typicality.

The idea of relationships is easy enough. What is perhaps more interesting is the question of why we bother with relationships. What is so important about knowing that something is related to something else?

WHY RELATIONSHIPS ARE CENTRAL TO HUMAN THINKING

People carry around with them an entire catalog of relationships that they use in their dealings with others. Many of these relationships are actually stereotypes. We know that stupidity is related to playing football. We know that greediness is associated with businessmen. We know that sexiness is related to youth. We know that people who work hard succeed. We know that people who call themselves Democrats are liberal and those who call themselves Republicans are conservative.

Actually, none of the above relationships is supported very strongly by any kind of data we know about. Football players are dumb—and a lot of them are extremely bright.[2] Businessmen are greedy—and many are altruistic. Sexiness is related to youth—yet there are innumerable people who retain an emphasis on their sexuality into advanced ages. People who work hard do succeed—and a discouraging and not generally acknowledged number of them fail (we usually read in the newspapers about hard workers who succeed, but not much is said about hard workers who fail or lazy people who "luck out"). Democrats are liberal—and there are some who make the most conservative Republican look wishy-washy. In other words, where stereotypes depict relationships as hard and fast, the facts generally show that the real relationships are much more complex, if they exist at all.[3]

We need to consider two questions. First, *why* do people go to such lengths to endorse relationships that they will even lean on associations that are, generally speaking, closer to fiction than to truth? Second, *how* do we know whether there is or is not some kind of relationship between things we are interested in?

To get a basic understanding of relationships we need to go back to our fundamental concepts of typicality and variability. Variable things bother us by the fact that they vary. If something varies, it is more difficult to establish what its typical nature is. Now, it should become evident with just a moment's thought that when we have a perfect relationship between two things, when we know the situation for one of these things the situation for the other is invariant. You know, for example, that when you push the accelerator of your car a certain amount, the car will travel at a certain speed—no more, no less. If there were no relationship between the accelerator and the speed of the car, then you would have no idea of how fast your car was going to go when you stepped on the pedal. It might speed up a great deal, or hardly at all.

This introduces another element into relationships. Just as averages are a measure of what is probable, relationships improve one's probability of guessing individual values in a state of ignorance—if there is a relatively high degree of relationship with a known independent variable (see page 201). This is a big "if." It also raises the question of what constitutes a "high degree" of relationship. This question has not been adequately addressed in the social sciences. Ideas about what constitutes a high degree of relationship are relative to particular problems. Sometimes a low degree of relationship is extremely important; at other times a high degree of relationship does not mean much.

We shall put this question off for a while as we consider the basic problems of (a) why we bother with relationships in the first place, and (b) how we go about determining the extent to which they exist. However, we shall eventually pay closer attention to the question of what constitutes a high degree of relationship.

THE FUNDAMENTAL NATURE OF RELATIONSHIPS

To gain a deeper understanding of what is implied in statistical forms of relationship between two or more factors, it is essential to keep an eye on the basic notions of typicality and variability. In our discussion of averages and variation we noted that when there is no variability around a mean value, we can generalize to all cases. The importance of this simple observation cannot be overemphasized. We also noted that when this is the case, we reduce the likelihood of making errors in estimating or guessing individual values to zero. For example, if we know that the mean level of trustworthiness is high for all members of a corporation, *and that there is no variation around this value,* then we can assume that any individual is trustworthy and we will not be in error. As variability is reduced, there is less likelihood of error.

The catch in all of this, of course, is that things *do* vary. Constants are the exception rather than the rule on this planet—especially in the realm of human social affairs. How can we cope with this variability?

At the outset, let's get firmly in mind the fact that variability, though it might be the spice of life, is vexing to anyone who is trying to exert control over some situation. It suggests disorder. It suggests that the situation might not be predictable. Whenever you need an image of variability, think of a toy balloon that has been filled with air and then released to fly about, this way and that, in an erratic and unpredictable manner. You just don't know what it is going to do. We do not want people to act like toy balloons, just as we do not want our machines or anything else in our lives to act too capriciously, or to get "out of control."[4]

The quest for relationships, statistical or otherwise, is basically a quest to curb the problems raised by variability. When we find relationships, we begin to see the controlling factors behind things that had previously seemed capricious or unpredictable. That is all there is to relationships, but it is enough. It forms the motivation behind not only our science but also our art and our literature. Astrology, a form of magical practice, finds patterns between the scattering of the stars and the personal affairs of individual human beings walking around on a small and remote planet in a minor galaxy millions of light years away. The stars, in their paths, are knowable. That fixed knowledge can be used to deal with the variable nature of human fate—or so the astrologer would like us to think.

Suppose astrology were true in its claims. We know that people vary considerably in terms of their personalities and experiences. Some are unorthodox in nature; others are extremely conservative. Without any other knowledge, we have no way of knowing whether strangers are conservative or unorthodox when we first encounter them. If, however, we knew that all Sagittariuses are unorthodox and all Leos are conservative, we have a situation in which, for either of these groups, variability is zero. If we meet a stranger and find out, early in our conversation, that she is a Sagittarius, we know, without going further, that she is unorthodox. The problem of variability is solved. It is solved by the fixed relationship between one's astrological sign and one's character. What was once a matter of guessing is now known.

Astrology makes up the patterns it uses for predicting individual characteristics. If nothing else, this suggests how powerful the desire for control and reduction of variability is. We are prone to turn to almost any kind of theory, speculation, or quackery to help us find some kind of pattern behind the ebb and flow of fortune and misfortune that make up human experience.

The basic point here is that as relationships become stronger or more marked, the problem of variability declines. Our primary concern in establishing relationships is to overcome variability. In the social realm this is not as easy to do in any precise manner as we might hope. Before we get further into the mechanics of relationships, one important thing must be kept constantly in mind in assessing whether or not we are really accounting for variability. We must worry about whether any obtained relationship has been influenced by the way we defined our variables; once again, then, the problem of categorization rears its head. If, for example, you correlate the percentage of the population of each county of the United States that is white with the percentage that is nonwhite, you will get a perfect negative correlation. The reason for this is that the category "percentage white" also defines the category "percentage nonwhite"; once you know one value, you know the other. In such instances you are correlating something with a slightly different version of itself. Such correlations are known as **spurious correlations.** Spurious correlations are largely a problem of categorization of variables. If you get a perfect correlation between two social variables, it is practically certain to be the result of correlating a quality with a differently stated version of itself.

DEGREE OF ASSOCIATION

Formal approaches to statistical relationships are characterized by two distinct qualities. First, statisticians know that some relations are strong and others are weak—and there are ways of establishing how strong or weak a relationship is. Second, relationships must be defined in terms of variables that can be identified. We can call this the "no ghosts" rule. If you think there is a factor associated with something you are interested in, that factor must be identifiable. It cannot remain mysterious.

There really is little more to good reasoning, scientific or otherwise, than this. Here, then, is the process: first, we find something puzzling in its variable nature; second, we attempt to understand that something by finding what it might be related to.[5] Some students, for example, succeed in school. Others fail. In other words, there is variability in levels of academic success. We might, then, conduct a statistical study and find that success is associated with parental guidance. Parental guidance then becomes a factor that helps us understand variation in student achievement.[6]

Something in which we are interested as a problem for research is commonly referred to in statistical work as a **dependent variable.** The term itself helps order our thinking. A *dependent* variable is something that we suspect is dependent on something else.[7] Something we think might *explain* a dependent variable is referred to in statistical work as an **independent variable.** We want to find out just how much our dependent variable might depend on some independent variable.

In ordinary, day-to-day living we commonly order our thoughts in this fashion. For example, suppose we are interested in the fact that some people are good and some are bad. This is a variable quality and one that generates confusion. If we are in the company of bad people and are not aware of it, we can suffer as a consequence. It would be nice to know who is bad and who is not. What independent qualities might tip us off to the character of strangers or others with whom we relate?[8]

Statistical relationships are concerned with reduction in variability. Perfect relationships reduce variability to zero. (Remember our previous example: if all Sagittariuses are unorthodox, then variability within the class "Sagittarius" is zero.) Moderate relationships reduce variability moderately. Nonexistent relationships leave us with as much variability as we had before the relationship was tested. (In reality, Sagittariuses are as mixed in their character as the general population; the independent sun sign does no better job of predicting character than guessing would.)

We have spent considerable time now building to this major point: when we have a perfect relationship, knowledge of the independent variable provides us with precise knowledge of the dependent variable. There is no more variability to cope with. The tricky aspect of variability has, so to speak, been tamed.

It is useful to determine how much we have tamed variability in a dependent variable by finding out precisely how it is altered under different conditions. The degree of association we find between variables—high, medium, or low—suggests how much faith we can put in a particular pattern. For example, we might believe goodness is associated with church membership. However, it could be the case that there are good people who do not belong to churches and bad people who do. In scientific or statistical studies (or anywhere else), exceptions weaken the rule. Exceptions, then, are simply another way of saying that variability is still in effect; no exceptions implies no variability.

BASIC FEATURES OF RELATIONSHIPS

When we seek to uncover the extent to which aspects of the world are related, we need to consider four fundamental concerns.

1. *Degree of relationship*. We want to know how closely this and that—*X* and *Y*, or the dependent and independent variables—are related. If they are closely related, we can rely on *X* to tell us something about *Y*. If not, then an awareness of *X* leaves us no better off than we were before when it comes to an awareness of *Y*. When these variables are closely related, the variability in *Y* is reduced for any value of *X* we might consider. Statistical analysis generally establishes the degree of relationship by letting 1.00 represent a perfect correlation (or relationship) and letting 0.00 represent no correlation. Values in between represent imperfect degrees of association. There are, however, many indexes of relationship used in statistics that range between perfect unity and zero relationship. Interpreting these values calls for a careful understanding of the specific index being used. Many of these indexes can be misleading or difficult to interpret.[9]

   In the next section of this chapter (beginning on page 204), one of these indexes, the Pearsonian correlation coefficient, will be discussed. This coefficient is the clearest measure of relationship used in statistical analysis. Even so, its meaning is easily misinterpreted.[10]

2. *Direction of relationship*. It is important to know whether *Y* increases or decreases as *X* increases in value. For example, the longer you drive your car, the less gas there is in the tank. The correlation between time on the road and gas in your tank, then, is a negative one. The more pressure you put on the gas pedal, the faster the engine goes. The correlation between gas pedal pressure and speed, then, is a positive one. There are thousands of mundane examples of positive and negative relationships—most of them quite important in dealing with the ordinary affairs of daily life.

3. *Form of relationship*. This problem of relationships is not as extensively dealt with in the social sciences as it is in the physical sciences. In the social sciences, most relationships are dealt with as simple linear relations. As *X* increases, *Y* increases in a constant increment. Figure 11.1 illustrates such a simple linear relationship. Figure 11.2, on the other hand, depicts a **curvilinear relationship**.

Figure 11.1.

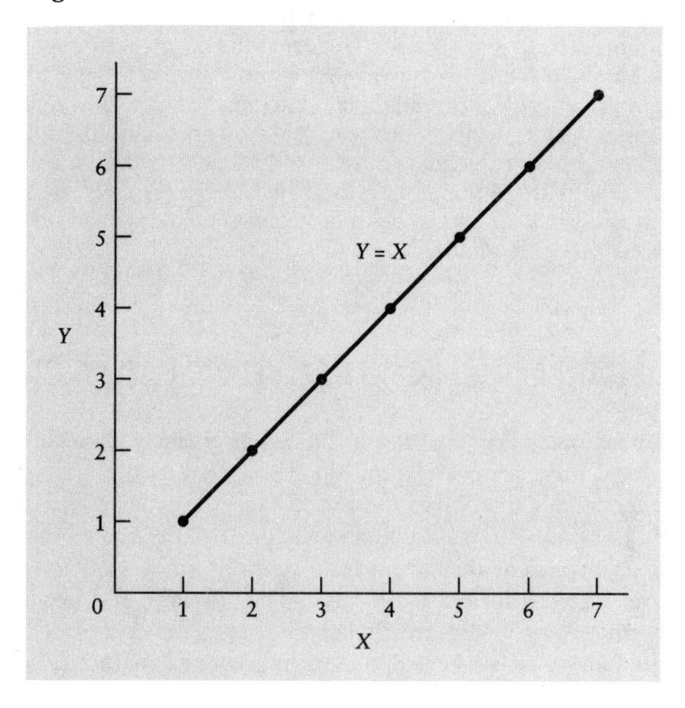

Figure 11.2.

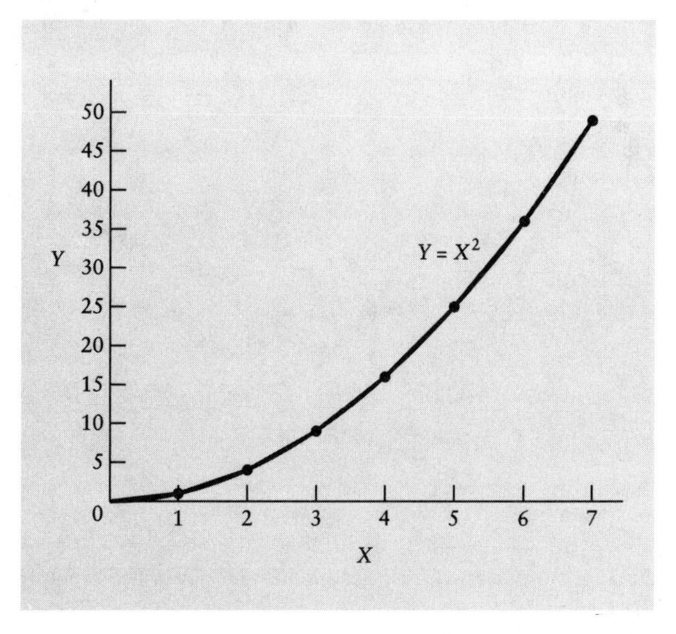

4. *Significance of relationship*. A relationship can happen just by chance. Suppose you have a test of ability that you administer to one hundred people, correlating it with their performance on a specific task later on. You find a slight tendency for people who did well on the test to perform well at the task. Is the test really related to performance, or did the relationship happen just by chance? A great deal of work in biological, sociological, and psychological statistical research is concerned with this problem. This is perhaps the dominant research **paradigm** (or framework of questions) in statistical studies in psychology and sociology: can it be said that a relationship exists at all?

THE PEARSONIAN CORRELATION COEFFICIENT

The **Pearsonian correlation coefficient** (named after the statistician Karl Pearson) is an ideal measure of relationship. It requires that whatever is being related be expressed in interval or ratio scales. It is not only an ideal measure of relationship, but it is also, we think, an ideal place to begin a discussion of the nature of relationships as they are seen in a more formal statistical sense. The logic of **correlation coefficients** is the basic logic of relations. Once the correlation coefficient and its logic are understood, it is easier to understand the strengths and limits of other ways of trying to deal with the problem of whether one variable is related to some other variable.

Let's begin with a highly idealized and artificial set of values. The following set of data, along with a **scattergram** that gives a picture of the data arranged along a vertical and horizontal ordinate (shown in Figure 11.3), are instructive.

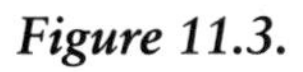

Figure 11.3.

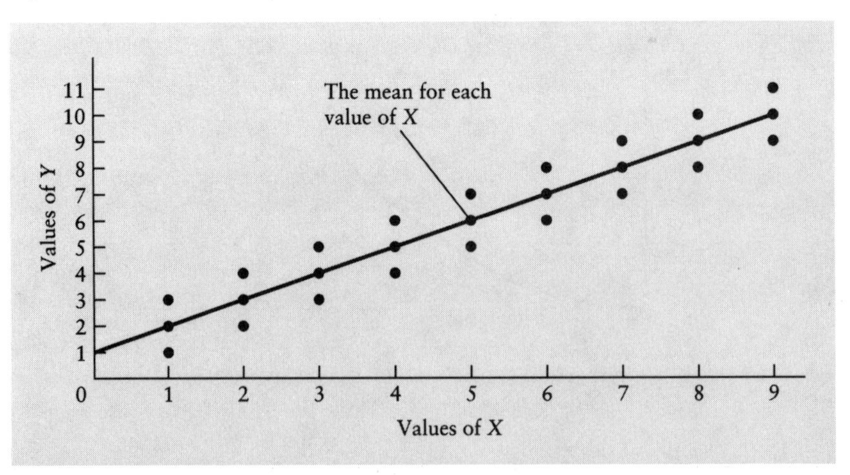

X	Y	X	Y	X	Y
1	1	4	4	7	7
1	2	4	5	7	8
1	3	4	6	7	9
2	2	5	5	8	8
2	3	5	6	8	9
2	4	5	7	8	10
3	3	6	6	9	9
3	4	6	7	9	10
3	5	6	8	9	11

Notice that for each specific value of *X* there are three different values of *Y*. Also, as *X* increases, so does the *average* value for *Y*. However, *there is variability around that average.* So, even though *X* provides us with some good information about *Y*, it does not provide perfect information.

What is important, in diagrams such as the one in Figure 11.3, is the arrangement of variances. We are concerned with what is happening to *Y* on the basis of what we know about *X*. What we discover is that for any value of *X*, the variability in *Y* is a specific amount. Can you determine the standard deviation of *Y* for the case when $X = 4$? There are three cases of *Y*. What is the mean for these three cases? What is the standard deviation? It is equal to the square root of $\frac{2}{3}$. Do not move on until you see clearly how this was obtained. Do another standard deviation for another value of *X*, say the case where $X = 8$. What is it? Why are all the standard deviations the same? What is the *average* standard deviation of the values of *Y* for any value of *X*? What is the variability around this value for the average standard deviation? It is, of course, zero.[11]

So, in this instance we have a distribution such that the *average* standard deviation for *Y* for any value of *X* is 0.8165. We still have some variability in *Y*, even after we have knowledge of *X*, but it looks relatively small compared to the amount of variability we had before we introduced *X* into the picture. How much overall variability is there in *Y*? To get the standard deviation for *Y*, we find the root of the mean squared deviations for all the cases of *Y*. It turns out that it is equal to 2.7081.

In working with correlations in statistics, it is easier to work with the square of the standard deviation, which is technically referred to as **variance.** So, let's square the two standard deviations. The square of 0.8165 is 0.6667. The square of 2.7081 is 7.3338.

The larger value is the variance in *Y* before we know anything about *Y*'s relationship with *X*. The smaller value is the variance in *Y* after we know something about *Y*'s relationship with *X*. The larger value is the **total variance.** The smaller value is the **residual** or **unexplained variance.** It is the variance that *X* was not capable of wiping out. Remember, we would like to get rid of all variance, if we are going to know *Y* in terms of *X*. But *X* only allows us to get a pretty good guess at *Y*, because it has lowered the variance in *Y* from 7.3338 to 0.6667. That is a substantial reduction in variance.

If the value 0.6667 represents unexplained variance and 7.3338 represents the total variance in *Y*, then 7.3338 minus 0.6667 would represent the amount of variance we got rid of—the variance we have explained. Accordingly, it is called **explained variance.** It would be nice to know what percentage of the total variance we explained; the following computation gives us that value:

$$\frac{(7.3338 - 0.6667)}{7.3338} = 0.9091$$

X explains about 90 percent of Y, and about 10 percent is still unaccounted for (notice that we are working here with the simple and ordinary idea of percentages). This value, 0.9091, is the value of the correlation coefficient (r) squared. The squared correlation coefficient tells you how much of the total variance is accounted for by the dependent variable.[12] Ideally, any research using correlation coefficients would include a statement of the value of r^2. To find r, the correlation coefficient, all that is required, then, is to obtain the root of 0.9091: $r = 0.95$.[13]

This approach to the correlation coefficient reveals the close relationship between three forms of variance: (a) the *total variance* in the dependent variable, (b) the *explained variance* after the variable is related to an independent variable, and (c) the *residual* or *unexplained variance*. Such an approach provides a good look into the logic of relationships.

FORM OF RELATIONSHIPS

So far we have been concerned with the degree of relationship. That means we have had to concern ourselves with the extent to which our explanatory factor gets rid of variability in the variable we are trying to understand.

We also want to know something about the form of the relationship. In the natural sciences, where relationships commonly have a mathematical precision, there is little interest in the degree of relationship, which is usually extremely high. The relationship between pressure and temperature in contained gases is a nearly perfect relationship and can be represented by a well-defined curve. The nature of the curve becomes important in such studies. In the social sciences, however, relationships are difficult to establish, and the primary concern becomes that of determining the extent to which a relationship exists at all. (Which of the two ways of dealing with relationships—the scientific or the social scientific—generates greater descriptive utility?)

If two variables are related, one of the most fundamental questions we can ask about the form of the relationship is whether Y increases or decreases in value, on the average, as the value of X changes. The correlation coefficient provides an answer to this question by displaying a negative value for negative relationships and a positive value for positive relations.

The Pearsonian correlation assumes linearity in relationships. If two variables have a strong curvilinear relationship, the Pearsonian coefficient will tend to underestimate the degree of relationship.

A REAL CORRELATION

The illustration that began this discussion of correlation coefficients is extremely artificial because it provides a simple way to get an average value for the variance in *Y* across all values of *X*. What do you do about this when you have a more complicated pattern? Table 11.1 presents real data arranged in two rows, *X* and *Y*. The *X* row represents the density of lawyers in the various American states—that is, the number of people living in a state for each lawyer residing there. Washington, D.C. has 22 people for every lawyer. Arizona has 628. The *Y* column is the number of people in the state under custody for a criminal activity for every 10,000 people. It is immediately obvious from the data that ascertaining a pattern between *X* and *Y* is not a simple task.

Table 11.1. Population per Lawyer (Variable *X*) and Number of People in Correctional Systems per 10,000 Population (Variable *Y*) by States of the Union and the District of Columbia, 1988 (Source: *Statistical Abstract of the United States,* 1990.)

State	*X*	*Y*	*State*	*X*	*Y*	*State*	*X*	*Y*	*State*	*X*	*Y*
AL	597	123	IL	304	138	MT	411	87	RI	381	135
AK	286	152	IN	579	165	NE	374	123	SC	657	145
AZ	405	145	IA	471	86	NV	400	170	SD	574	80
AR	628	116	KS	439	138	NH	457	70	TN	537	137
CA	312	179	KY	531	71	NJ	323	147	TX	410	341
CO	284	107	LA	422	173	NM	450	87	UT	479	98
CT	277	203	ME	457	74	NY	244	143	VT	386	160
DE	384	249	MD	338	282	NC	665	177	VA	421	96
DC	22	510	MA	262	232	ND	532	47	WA	378	209
FL	361	210	MI	444	209	OH	395	136	WV	689	51
GA	328	316	MN	367	132	OK	397	152	WI	462	105
HI	373	148	MS	608	105	OR	362	159	WY	458	93
ID	493	91	MO	143	134	PA	428	159			

When dealing with tables of this complexity, look for the extreme cases. For example, Washington, D.C. has a very low value for *X* and a very high value for *Y*. North Carolina (NC) has a very high value for *X* and a relatively low value for *Y*. Arkansas (AR) has a high value for *X* and a low value for *Y*. Missouri (MO), however, has a low value for *X* and a relatively low value for *Y*. Obviously, X and *Y* are not perfectly related.

This complicated tangle of information can be made simpler by drawing a graph. The graph produced (in Figure 11.4) is another example of a scattergram.

Scattergrams are commonly ignored in social and psychological research when people hurry to determine the degree of relationship.[14] This can be an error, as our example shows. The scattergram suggests that the relationship is one where as the value of *X* goes up, the value of *Y*, *on the average,* goes down.

Figure 11.4. Scattergram for Values in Table 11.1

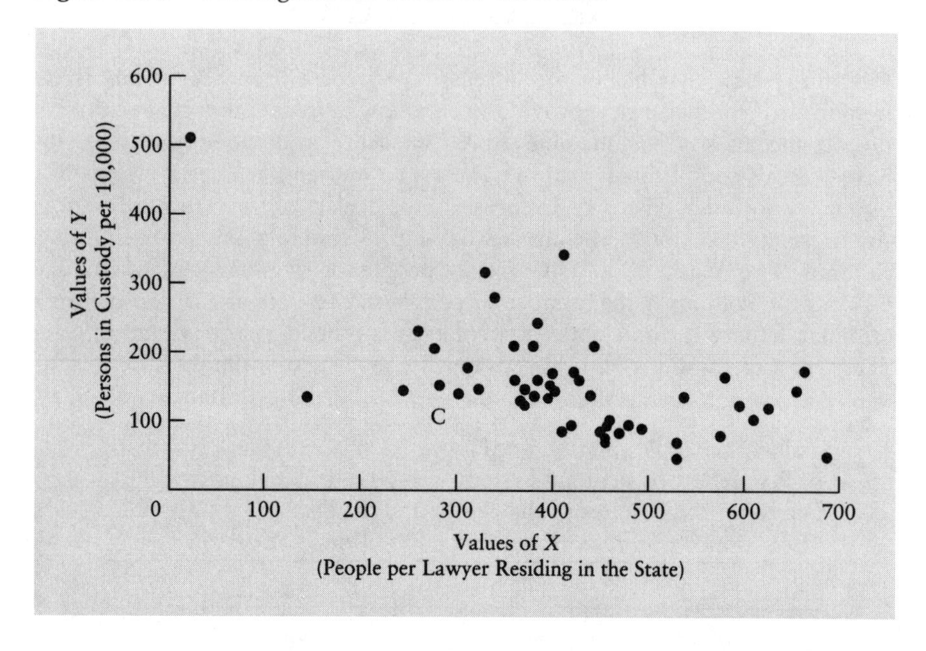

In Figure 11.4 we plot the value of *X* along the horizontal axis and the value of *Y* on the vertical axis. The state of Colorado (CO), for example, has 284 people for every lawyer residing in the state (the *X* value). It has 107 people per 10,000 population under correctional supervision (the *Y* value). The proper point to place Colorado is the one where these *X* and *Y* values intersect. In Figure 11.4 this is represented by the C. The other symbols represent values for other states.

The scattergram provides us with a picture of a relationship. What we see in the diagram is a tendency for the *average* value of *Y* to go down as the value of *X* increases. States that have a lot of people for every lawyer (that is, relatively few lawyers) also have relatively few people under correctional supervision. Those that have a high incidence of lawyers have a large number of individuals under supervision. Notice that all the states with more than 550 people for every lawyer have less than 200 people per 10,000 under correctional supervision. All the states with fewer than 300 people per lawyer have relatively high proportions of people under supervision.

A lot of statistical work ignores the scattergram and relies directly on indexes of relationship, particularly the correlation coefficient. This is unwise because the correlation coefficient can provide overestimates or underestimates of relationship, depending on how the *X* and *Y* variables are distributed. In Figure 11.4, for example, the District of Columbia is quite unusual in that it has an extremely large proportion of lawyers (one for every 22 people in the district) and an extremely large proportion of people under supervision. This one observation, alone, carries a lot of weight in creating a relationship. To find out how much, let's calculate the correlation coefficient, first with the District of Columbia datum included and then with this datum excluded from our figures.

The correlation coefficient, with the District of Columbia included, turns out to be -0.55—a fairly high relationship for social data. Notice, here, that the negative sign tells us that the relationship is a negative one. Again we must be careful. It is negative because states with higher densities of lawyers have lower index values. In preparing this finding for publication it would be proper to warn the reader of this.

If we remove the District of Columbia from the picture, we get a much lower degree of relationship. Now the correlation coefficient is equal to -0.37. This is most important to notice. *The inclusion or exclusion of a single extreme case alters the correlation coefficient radically.*

We know that the correlation coefficient should always be squared before we interpret it. The high relationship, $r = -0.55$, tells us that about thirty percent of the variance in the proportion of people in correctional institutions is taken into account by using X to make predictions. The low correlation, $r = -0.37$, tells us that about fourteen percent (less than half of what we got before) of the variability is removed by using X. So a single case can radically alter the extent to which we think two factors are related. If nothing else, this warns us that unless we have some idea of what a relationship looks like, through use of a scattergram or some other device, we cannot be certain what a correlation coefficient is telling us.

Notice, as we go along, that extreme cases that distort distributions in the study of relationships operate in much the same way as extreme cases do in the study of averages. One extreme case can lower or raise the estimate of an average all by itself. We are left with the problem of determining whether the average is better with or without the extreme case. Do we have a right to throw away information because it makes the data "look better"? On the other hand, should we keep it, even though it is obvious that the extreme case is biasing our results? In the above example, what should be done with the data for the District of Columbia: should it be left in or tossed out?[15]

The primary information provided by the correlation coefficient has to do with whether there has been any reduction in the amount of variation in the dependent variable. The dependent variable, in the example above, is the variability in the number of individuals under correctional supervision per 10,000 members of the state population. This number ranges from a low of 47 for North Dakota to a high of 510 for the District of Columbia. That is a considerable amount of variability. The District of Columbia has more than ten times the incidence of people under correctional supervision as North Dakota. The unique character of the District of Columbia as the center of federal governmental operations doubtless has a lot to do with this extreme value. At the same time, the District of Columbia has a tremendously large number of lawyers per capita. There are only 22 people for every lawyer in the District. West Virginia, on the other hand, has 689 people for every lawyer—roughly thirty times more than for the District of Columbia. Again, there is tremendous variability in these figures.

Because the basic computational formula for a correlation coefficient involves large numbers and is cumbersome, it is illustrated in an appendix to this discussion (on pages 211–213) using the data in the table on page 205. Anyone interested in a more detailed understanding of correlations should undertake the calculations of at least a few examples in this appendix before using computers to do the job.

SUMMARY

Why are we interested in relationships? We can gain a sense of "explanation" when we discover Y is related to X. If we know that cool people dress cool and dorks dress like dorks, then if we see someone dressed like a dork, we know he's a dork. We don't have to guess what he's like. Now we *know*. Clothes, in this instance, offer a partial "explanation" of the person's character. One statistical index of relationships is the Pearsonian correlation coefficient (r). When the assumptions underlying its use are met, the Pearsonian correlation coefficient can tell us the degree, direction, form (linear), and (in conjunction with other indexes) significance of the relationship (is it, for example, just a matter of chance?).

The correlation coefficient, when squared, tells you how much you have reduced variance in the dependent variable by co-varying it with some other variable. A correlation coefficient of $r = +0.60$ means you have reduced the variance in Y, on the average, by 36 percent, or about a third. Two-thirds of the variance in Y remains as a random element. This might, in nontechnical and slightly distorted terms, be viewed as meaning that in such an instance, on the average, when you use X to predict Y, you will be close to the mark one-third of the time. You will be off the mark roughly two-thirds of the time. As is the case with any average, however, you cannot tell for the individual case how close or far from the mark you are. But your guessing ability with respect to individual cases has been improved.

THINKING THINGS THROUGH

1. The following table presents suicide rates (suicides per 100,000 population) for thirteen countries for men and women. Is there a relationship between the rates for men and women? How would you interpret your findings?

Country	*Men*	*Women*
United States	20.6	5.4
Australia	19.1	5.6
Canada	22.8	6.4
Denmark	35.6	19.9
France	32.9	12.9
Italy	12.2	4.7

Japan	25.6	13.8
Netherlands	13.9	8.2
Poland	22.3	4.7
Sweden	27.1	10.1
United Kingdom	11.6	4.5
West Germany	26.7	11.8

Draw a scattergram from the above data. Estimate the degree of correlation before you calculate it precisely.

Is the correlation a spurious one? Is it a causal one—that is, do you think female suicide rates "cause" male suicide rates or vice versa? If you were trying to get at the causes of suicide, what are some other variables you might want to relate to the above data? List at least ten.

2. Movies can be seen as patterned events in which a variety of things are related. For example, in the film *Dick Tracy,* all the bad people are ugly and all the good people are attractive. In political campaigns serious efforts are made to establish relationships in such a way that the opposition appears totally negative while "our side" is perfectly positive. How are such relations asserted? What degree of relationship is usually presumed?
3. Below are the data for murder and rape rates (per 100,000 population) for geographical regions of the United States. Do the data suggest a relationship? Interpret your findings. A scattergram will help a great deal.

Division	*Murder Rate*	*Forcible Rape Rate*
Northeast	3.8	28
Middle Atlantic	8.8	29
East North Central	7.3	43
West North Central	4.3	27
South Atlantic	10.4	39
East South Central	8.6	34
West South Central	11.1	44
Mountain	6.6	36
Pacific	9.2	43

4. Using either *The Statistical Abstracts* or some other official source of data, put together a short essay in which you discuss the relationship between capital punishment, as a policy in the United States, and issues of race. Use statistical information to make your point. You do not need to lean mechanically on the correlation coefficient, but you should constantly keep in mind how you are dealing with such matters as degree, direction, form, and chance with respect to the relationships you talk about in your essay.

APPENDIX: HOW TO CALCULATE THE CORRELATION COEFFICIENT

Some fictional numbers were used earlier in this chapter to show that the squared correlation coefficient is an estimate of the average amount of variance in the dependent variable that is "absorbed" by Y's relationship with X. These numbers are repeated below for convenience, along with other relevant figures. Note that these numbers are "unreal"; one is usually not able to deal directly with residual variances. We need another way of computing the correlation coefficient (r) when we have nothing but raw data pertaining to values of X and Y. Note that the additional necessary values for solving the correlation for these hypothetical, unreal numbers (X^2, Y^2, and XY) have also been provided here.

	X	Y	X^2	Y^2	XY
	1	1	1	1	1
	1	2	1	4	2
	1	3	1	9	3
	2	2	4	4	4
	2	3	4	9	6
	2	4	4	16	8
	3	3	9	9	9
	3	4	9	16	12
	3	5	9	25	15
	4	4	16	16	16
	4	5	16	25	20
	4	6	16	36	24
	5	5	25	25	25
	5	6	25	36	30
	5	7	25	49	35
	6	6	36	36	36
	6	7	36	49	42
	6	8	36	64	48
	7	7	49	49	49
	7	8	49	64	56
	7	9	49	81	63
	8	8	64	64	64
	8	9	64	81	72
	8	10	64	100	80
	9	9	81	81	81
	9	10	81	100	90
	9	11	81	121	99
Sums	135	162	855	1170	990*

*The sum of XY is called the sum of cross-products and is obtained by multiplying each value of Y by X and then summing across all products.

Here is the computational formula for r, or the Pearsonian correlation coefficient:[16]

$$r = \frac{[N\Sigma XY] - (\Sigma X)(\Sigma Y)]}{\sqrt{[N\Sigma X^2 - (\Sigma X)^2] \times [N\Sigma Y^2 - (\Sigma Y)^2]}}$$

Note that ΣX = the sum of X. The Greek letter sigma (Σ) usually stands for summation in statistical texts.

Work through this example. It calls for summing the values of X and Y. The sum of X is 135, and the sum of Y is 162. The next step is to square each X and Y and get the sum of X^2 and Y^2. The sum of X^2 is 855, and the sum of Y^2 is 1170.

The next step is to get the value for X times Y and then sum the products of XY. The sum of XY is 990.

The final step is to substitute these values in the formula for r. When we do that we get:

$$
\begin{aligned}
r &= \frac{[27(990) - (135)(162)]}{\sqrt{[27(855) - 135^2)] \times [27(1170) - (162^2)]}} \\
&= \frac{26730 - 21870}{\sqrt{(23085 - 18225) \times (31590 - 26244)}} \\
&= \frac{4860}{\sqrt{4860 \times 5346}} \\
&= \frac{4860}{5097} \\
&= 0.95 \\
r^2 &= 0.91
\end{aligned}
$$

Our calculated value for this relationship provides a correlation coefficient of +0.95 (the same value we found in the introductory discussion of correlation), indicating an extremely high and positive association. The squared correlation indicates that slightly over ninety percent of the variance in Y is accounted for by X. Keep in mind that although this is an extremely high degree of correlation, ten percent of the dependent variable is still varying randomly or in ways that are not accounted for. In sum, ten percent of Y is still operating within the domain of chance.

Notes

1. Harold Garfinkel demonstrated that people are quite capable of finding "sensible" patterns in nonsense communications. See his *Studies in Ethnomethodology* (Englewood Cliffs, NJ: Prentice-Hall, 1967). Projective psychological tests, such as the Rorschach, are further evidence of the extent to which people seem naturally inclined to find patterns in things. The examination of relations is not something specific to statistical analysis. Statistical analysis is a refined form of what comes close to being an instinctive, and certainly a universal, response on the part of people to their environment.
2. James S. Coleman's research on the American high school found that football players, at least at the high school level, were generally superior in both social and academic skills. See James S. Coleman, Thomas Hoffer, and Sally Kilgore, *High School Achievement: Public, Catholic, and Private Schools Compared* (New York: Basic Books, 1982).
3. Throughout this book, it will be argued that the primary contribution of modern social science has not been the creation of a mechanical science but, instead, the undermining of the "mechanical" mentality that is characteristic of ordinary social discourse. Where extremely high relationships are often thought to exist between

intelligence and ethnicity, for example, modern psychology and the social sciences find virtually no relationship. The social sciences, instead of generating a "mechanistic" science, replaced the mechanistic thinking of folklore with the probabilistic thinking of the empiricist. If the social sciences have had a profoundly revolutionary influence on modern thought, this is it.

4. This discussion of relationships is relevant to one of the most profoundly cherished moral concepts in the United States: the idea of freedom. We mention this because people sometimes think that statistical and ethical discussions are completely separate things. However, if we can establish perfect relationships, then we have shown that freedom does not exist. For example, if there were a perfect relationship between getting good grades in school and getting a good job, then a person who got poor grades would not be free to get a good job (nor would a person who got good grades be free to get a poor one). Any person's life would be determined by a relational pattern. Pure freedom, like the balloon going this way and that, would mean that human patterns were perfectly unpredictable. The fact that they are statistically predictable—to some degree—undermines the argument that they are perfectly free. Obviously freedom is something you want to get rid of when you are trying to understand what is happening. A purely free event, by definition, exhibits a variance that prevents any kind of association. In any event, the question of whether human actions are associated with causal forces, and therefore determined, or are free and therefore removed from causal forces, remains a powerful philosophical issue in our time. Statistical analysis, in the social sciences, is never far removed from moral, ethical, and ancient humanistic issues. Because it is based on the assumption that causes can be found for human actions, it tends to run counter to prevailing ideology. On the other hand, the fact that all of its findings reveal an inherent element of chance tends to leave the issue open.
5. Two powerful concepts in human discourse are expressed in the terms *understanding* and *explanation.* This discussion of relationships argues that understanding or explaining things is only a matter of relating this to that. We understand disease as we discover that it is associated with germs. "Real" understanding—trying to find out what is behind reality, or what the meaning of life might be—is a religious problem, not a scientific one. Science only describes how reality appears; it cannot go beyond that. The problem of what we mean by terms such as *understanding* and *explanation* is a serious one, and it is generally deceptive. In a disappointing sense, we can never understand the world; all we can do is describe it.
6. We can, like a child, go on and on. We can ask what is related to parental guidance. Suppose it is ethnic background. We can then ask why certain ethnic groups emphasize education more than others. There is really no end to it and, ultimately, no final explanation or understanding. This, combined with the fact that all social relationships are imperfect and therefore probabilistic rather than mechanical in nature, implies that social issues are always open systems. Social facts cannot be conclusive because they do not permit closure in the same sense that some physical systems permit closure. For example, in designing an airplane you are dealing with a closed system—the airplane and nothing but the airplane. If you are dealing with, let us say, the American banking system, however, you are dealing with something that is associated with not only the social and economic system of the United States, but the economic systems of the entire world. In that sense, it is not a closed system but an open one.
7. It is worth noting that the simple logic of dependent and independent variables promotes a sense of "cause" and "effect." The social sciences are "scientific" to the

extent they seek to find natural relationships. A dependent variable must be accounted for by some other observable condition.

8. This problem has been examined by social psychologists in a large number of articles since World War II. Hitchhikers who do not "look right" are less likely to be given a ride than someone wearing a business suit and carrying a briefcase. Teenagers are constantly looking for and assessing signs that indicate "coolness" or "being in." When they do this, they are doing essentially the same thing the social statistician does in assessing relationships. If you are *dressed* (independent variable) "cool," you should *be* (dependent variable) "cool." We mention this only to point out that the logic of relationships is profound and penetrates nearly every aspect of a person's social life.
9. The value r^2 tells us the percentage of variance in the dependent variable that is, on the average, accounted for by the independent variable. Many indexes of association do not do this and should, therefore, be interpreted with great care. See Chapter 13 for a further discussion of this issue.
10. The most common misinterpretation is to think that the Pearsonian correlation coefficient (r) gives you a proportional explanation (to think, for example, that an r of -0.70 offers you an accounting for seventy percent of the dependent variable). This error is made even by experts who should know better. Generally you must square r before talking about proportional explanation.
11. It is easy to see in this case that all of the variances or standard deviations are the same across all values of X. This is called *homogeneity of variance* and is an important assumption in both correlation analysis and analysis of variance. The term *homoscedasticity* is sometimes used to refer to homogeneous variance. It is an extremely severe restriction on the proper use of correlation and the analysis of variance.
12. There are some quite complicated exceptions to this. When the measures of the dependent and independent variables are not connected by some hidden or third condition, then r^2 is reliable. However, where such hidden connections exist, r^2 is distorted. For example, identical twins, reared apart, are still connected by a perfectly associated common biology. Therefore, IQ tests performed on identical but separated twins have a correlation of roughly $+0.70$. The square of this figure is not an appropriate measure of variance. Can you see why? What assumption must be made about environmental influences in this case? Resolving such difficulties is extremely important in a society like our own, where much is made of the question of whether or not certain ethnic or racial groups have superior or inferior "intelligences."
13. While this is relatively clear, it is not an appropriate way to determine the value of r because it will always give positive values. Other formulas for computing r enable the relationship to be evaluated as negative or positive.
14. Richard J. Herrnstein and Charles Murray's *The Bell Curve* (New York: Free Press, 1994), a book that has created a storm of debate, presents IQ data, correlated with racial and other so-called independent social factors, without using scattergrams. A false sense of precision is thereby created. This is an astonishingly bad practice, but an all too common one in social statistics.
15. Statisticians are aware of the problems that idiosyncratic values pose in a study. Should you leave them in or take them out? A recently coined term in statistical analysis is *robustness* of estimate. Robustness has to do with making estimates that in various ways take into account "peculiar" or "out-of-the-way" data that creep into a survey or set of observations. We do not want to enter into the problems of

curving and smoothing observations in order to attain such robustness here—one cannot cover everything in a brief book. This is too bad because here, particularly, is an opportunity to discuss another place in statistical analysis where arbitrary treatments of information lead to quite different consequences. Any advanced studies of statistics should give some consideration to the problem of attaining robustness. One of the major problems with dismissing outliers is that one can begin to find excuses for getting rid of other items in a data set because, for some reason or other, they might "contaminate" an aggregated index. The literature on robustness is extensive but technical. Interested readers might want to look at Peter J. Huber, *Robust Statistics* (New York: Wiley, 1981), or Moti Tiku, *Robust Inference* (New York: M. Dekker, 1986). The latter work contains a bibliography on the subject.

16. This formula can be found in just about every statistical text ever written. You should browse about in other books to find out more about various statistical techniques and the approaches endorsed by different writers. In this instance, the formula for r was obtained from Kenneth D. Hopkins and Gene V. Glass, *Basic Statistics for the Behavioral Sciences* (Englewood Cliffs, NJ: Prentice-Hall, 1978). The discussion of the derivation of r by Hopkins and Glass is more thorough than that found in most social statistics books.

12

THE STATISTICAL EXPRESSION OF RELATIONS: THE BASIC LOGIC OF ANOVA

"Really, universally, relations stop nowhere, and the exquisite problem is . . . to draw . . . the circle within which they . . . appear to do so."

—WILLIAM JAMES

". . . order in variety we see."

—ALEXANDER POPE

Throughout this book we have emphasized the importance of reliable measures of central tendency. Variability reduces the utility of measures of central tendency. Therefore, a central concern of most statistical work has to do with eliminating the effects of variability. In Chapter 11 we considered an ideal approach to controlling variability: the use of an extrinsic variable that has a high degree of relationship with a variable condition in which we are interested. If X is strongly related to Y, then we can use X to predict values of Y. When X and Y are ideally categorized in terms of ratio or at least interval scales, the correlation coefficient provides a relatively good index of the degree and the direction of association—though it should be used with caution. In the case of the data for attorneys and number of people in custodial institutions, for example, we saw that a single case can seriously distort the estimate of the degree of association—especially when the correlation is drawn from small samples.

It is also worth remembering that a correlation coefficient is just another *average*. It is the average amount of variance that is accounted for by some explanatory factor. The same things that we learned about the concept of typicality in general apply to the statistics of correlation. For example, distortions caused by such factors as skewedness or bimodality make the assessment of average degree of relationship more difficult.

THE ANALYSIS OF VARIANCE

The discussions in this book are concerned with the elemental forms of statistical reason. Less emphasis has been placed on techniques and specific mathematical procedures in order to make clear the logical forms underlying statistical analysis. The following consideration of a technique known as *the analysis of variance,* or **ANOVA,** concentrates on the basic logic underlying this technique.

If you have reviewed the earlier discussions of typicality, variability, and the logic of relationships, then the analysis of variance should not be especially difficult for you. In a sense, the entire statistical quest, whether it takes the form of a formal study or simply our day-to-day coping with ordinary problems, is a matter of dealing with variability, variance, or deviations—things that refuse to settle down into consistent typical values. In a way, it would not be a bad idea to use the term *analysis of variance* to refer to statistical methods in general, because that is largely what they are. However, this term is reserved for a particular statistical technique. The general logical concern of this technique is the same as that of the concept of relationships in general.

With correlation, you have two or more measures involving ratio or interval scales. However, how do you measure relationships when one of the measures you are interested in is expressed as a ratio or interval scale, while the other measures you are considering are ordinal or, worse yet, nominal? It is necessary to keep in mind at this point that nominal measures are merely nouns such as doctor, tennis player, student, and so forth. Ordinal terms involve some kind of ranking or ordering. Interval scales not only order, but provide equivalent scales that tell us how much more or less we have of some quality in any given case. Finally, ratio scales provide a zero point along with equal scales. Correlation provides a way of measuring relationships when we have two or more variables measured in terms of equal intervals or ratio scales. What do we do, however, if one of our measures is more crude—an ordinal or nominal scale? ANOVA deals with this problem. ANOVA asks: Can variance in a given variable or set of variables be broken into components (such as nominal categories, for example) such that, when the components are taken into account, variance in the dependent variable is significantly reduced?

Let's consider the analysis of variance by using a simple illustration. Table 12.1 displays thirty values. Let's say these values represent the scores of thirty people on a test of knowledge of ways to preserve the environment. A score of 10 indicates a lot of knowledge, a score of 0 indicates no knowledge. The test is an interval scale (an important presumption).

The thirty people taking the test were asked to indicate the extent to which

Table 12.1. Fictional Data Illustrating a Problem in the Analysis of Variance

	Frequency of Newspaper Reading		
	Every Day	*Once a Week*	*Hardly Ever*
	10	7	7
	10	7	6
	9	6	5
Score	9	6	5
on	8	5	5
Sensitivity	8	5	4
to the Environment	7	5	3
	7	4	2
	6	4	1
	6	4	0

they read newspapers. Their answers were registered as (a) every day, (b) once a week, or (c) hardly ever. This is an ordinal scale but a purely nominal scale could be used as well. The test scores are distributed by readership.

First of all, the *total* variability, across all groups, ranges from 0 to a high of 10, and the *SD* is 2.41. The mean is 5.7.

Within each group, we get the following mean values and *SD*'s:

	Every Day	*Once a Week*	*Hardly Ever*
Mean	8.00	5.30	3.80
SD	1.49	1.16	2.25

The variability in test scores (when we are ignorant of how newspaper reading might affect things), as measured by the standard deviation, is 2.41. The first thing we notice is that if we know a person's readership level, we have reduced variability. For the every day group, the standard deviation of 1.49 is smaller than the standard deviation for the entire population of thirty people. The once a week readers have a standard deviation value of only 1.16. Notice that we are now talking about variability within each group or the **within-group variability.** The average variability, within the groups, unweighted, is 1.63. This variability is about 67 percent of the total variability. We reduce variability by about 33 percent when we consider the effect of newspaper reading, if we use *SD*'s as our measure of variability.

There is also variability *between* the groups, or **between-group variability.** We can see this when we compare the averages of the groups. The *total variability* of the test scores can only come from two sources: within and between group variability. *The greater the variability between the groups, the less variability there will be within groups.*

The analysis of variance begins to look like the basic logic of relationships. If there is little variance within the groups, then the relationship is high. If there is no variance within the groups, then it follows that variance between the groups will have to be high. Variability in the dependent variable is the sum of the within- and between-group variabilities.

To compute variability as the sum of squared deviations, or variance, we make the following calculations:

1. The total sum of squares = sum of (Y − mean of Y)^2—where Y is any dependent variable.
2. The within-category (or group) sum of squares = sum of (Yc − mean of Yc)^2—where Yc is the values of Y within each category.
3. The between-category (or group) sum of squares = sum of Nc(mean of Yc − mean of Y)^2—where Nc is the number of cases in the category.

The total sum of squares for our hypothetical values is 168.3. The within-category sum of squares is 77.7 (check this yourself). The between-category sum of squares is equal to 90.6 (again, check this yourself).

Notice in this calculation that the total sum of squares, a measure of variability, is equal to the sum of the within-group variability and the between-group variability. It is here that we should focus our attention. The analysis of variance illustrates the logic of relationships once again. That logic stipulates that as a relationship approaches unity, the degree of variability on the part of a dependent variable approaches zero for any given value of an independent variable. This would be the case, for ANOVA, if all the values *within* each category were the same; in this case we would have the ideal of categorical constants.

Notice that the between-category sum of squares is 54 percent of the total variance. When we use the more technically correct variance values, we find that newspaper reading accounts for over half the total variance in the scores in the test population.

Let's take another look at ANOVA, this time with a tiny set of hypothetical values that illustrate the fact that as within-group variability decreases, the variability between groups increases. Here are nine values:

10 20 30 10 20 30 10 20 30

The mean for this group of values is 20, and the sum of squared deviations is equal to 600. (Why should one be careful about interpreting or using the standard deviation in this situation? Hint: the distribution on which it is based is not normal.)

Suppose we arrange our values by some independent factor and find the following distribution of observations:

A	*B*	*C*
10	20	30
10	20	30
10	20	30

What is the within-group variability? If we use the sum of squared deviations, it has to be zero because there are no deviations from the *within-group* means. The categories A, B, and C give us constant values. If you know that an individual falls within category A, for example, then you will know, without exception, that his or her value will be 10. Within-group variability has been totally eliminated, and all individuals in each category are the same.

The *between-group* means of 10, 20, and 30 show variability. For group A the group mean is 10 units below the average for the group as a whole. For group C it is 10 units above the average. Group B does not deviate. The contribution of the means to the sum of squares is equal to the deviation of the mean of the group from the total mean, weighted by the number of cases in each group.

Group A, then, gets −10 deviation points for each observation, and there are three observations. The square of 10 is 100. We weight this by the number in the group for a squared deviation value of 300. The same thing happens in group C. The total squared deviations, based on the group means, is equal to 600, which is the same as the sum of squares for the total distribution.

The total sum of squares is 600. The within-group sum of squares is 0, and the between-group sum of squares is 600. In this instance, between-group variability accounts for all the variability in the dependent variable.

WITHIN GROUPS AND BETWEEN GROUPS

In the social sciences the analysis of variance is primarily used to establish the extent to which there is a significant degree of association between events. In other words, is the relationship under consideration one that could have happened accidentally or is it "real"? This is known as *hypothesis testing,* and it is an approach to studying social issues that relies heavily on the concept of probability. A separate chapter (Chapter 14) is devoted to a consideration of hypothesis testing and the problem of significance. We will continue the discussion of the analysis of variance in that chapter, because most uses of ANOVA in psychology and sociology are concerned with estimating the probable significance of a relationship.[1]

However, before considering the problems of significance, it is a good idea to examine the general character of the analysis of variance. What is it doing? How does it deal with the problem of relations? Again, the answer is that it deals with the problem of relations by examining the extent to which original variability in some condition is altered when you apply an explanatory set of categories to it. For the moment, what is important is to see that total, within-group, and between-group variabilities are all there is to the analysis of variance.

Before rushing to a computer to have it crank out significance values using analysis of variance programs, pause and take a look at how your data are distributed within the categories you have assigned them. In the first hypothetical case presented in Table 12.1 above, we can see that the group that did not read newspapers had a low mean score on the test, and this group also displays a tremendous degree of variability—almost as much as in the population as a whole. The basic difference, in the mean values, lies between two groups, the newspaper readers and the nonreaders.

In our discussion of the correlation coefficient, we noted that it is a good idea to examine scattergrams to get a more solid sense of how your variables are related. The same is true with the analysis of variance. It is a good idea to see what is going on within and between groups. To bring this discussion to a close, then, we will present a real example of what an examination of within- and between-group variances offers. The data in Table 12.2 were obtained from *The Statistical Abstract of the United States* for 1990. The data are divorce rates for selected states, grouped into four divisions: the New England, Middle Atlantic, Mountain, and Pacific states.

Table 12.2. Divorce Rates (per 1,000 population) for Selected Divisions of the United States, 1985

	New England	*Middle Atlantic*	*Mountain*	*Pacific*
	5.20	3.80	5.20	5.70
	4.90	3.90	6.20	5.40
	4.40	3.40	7.50	4.80
	3.70		5.90	6.90
	3.80		9.10	4.60
	3.50		6.60	
			5.30	
			14.10	
Unweighted Mean	4.25	3.70	7.49	5.48
SD	0.69	0.26	2.96	0.91

Mean for all twenty-two states = 5.63, and *SD* = 2.37.

What do we see? First of all, there is not much variability in the Middle Atlantic states; there are only three of them, and their rates are low. There is tremendous variability in the Mountain states, and they seem to have relatively high rates overall. There is a great deal of variability among the means for the divisions, suggesting that divorce rates are associated with divisional categories of states.[2]

Before rushing to any conclusions, however, look at the figures. Don't just rely on an index of significance, a correlation value, or a hasty rejection or acceptance of *the null hypothesis*. (The null hypothesis, which is discussed further in Chapter 14, is the argument that any obtained relationship is attributable to chance. In social research the acceptance or rejection of this idea is fundamental to the scientific validity of a researcher's findings.)

We see quickly that one state has an extremely high rate for divorces. It is in the Mountain division. Can you guess which state this is? How is this one case going to affect our analysis of variance?

Actually, this figure affects ANOVA considerably. It jacks up any value we get for variability for the total. We have three categories in which the within-category variability is relatively low. For the Middle Atlantic states it is almost one-tenth the value for the overall variability. We begin to think that the divisional classification of states is associated with divorce rates—and it probably is, but this one case is doing some particularly funny things with variability. It is making the assessment of average variability difficult to judge. The "peculiar" state is Nevada. Should we keep it in or toss it out? This is the same problem we encountered in the discussion of correlations.

Watch what happens to total variability, as measured by the standard deviation, when you remove Nevada from the calculations. The standard deviation drops from 2.37 for all twenty-two states to a value of 1.46. The within-category standard deviation drops from 2.96 to 1.37.

This single case, then, has a lot of influence on how we are going to interpret our data. For this reason techniques such as correlation and ANOVA require relatively large samples for reliable results. These techniques also require a very limiting assumption: that variance across the range of the independent variable remains relatively constant. This is called **homoscedasticity,** or the assumption of a common variance for all cases within the different categories of the dependent variable. The illustrative case used here has too few cases to be reliable, but it does reveal, extremely effectively, what happens if you have one category in which variance is distorted by an extreme case.

It is also important to note that even with large samples involving several hundred cases within each group being examined, a few extreme cases can distort results. This is what is meant by the assumption of homogeneity of variance or variability within groups in ANOVA. As is the case with correlation, unless you take a close look at your data, you cannot be certain that your values are being distorted by skewed distributions. Again, this is because these techniques are dedicated to finding *average* changes in variability, after we take an explanatory variable into account; and we have seen that averages are affected by skewedness in a distribution.

SUMMARY

This brief discussion of the analysis of variance has been concerned with two elemental points. The first is the importance of assessing the extent to which variability is reduced when we are studying relationships. In the analysis of variance this assessment is done by looking at what happens to variance within categories. The variable under consideration within the categories must be measured on an interval scale, i.e., it must be a measure that provides equal intervals (such as a yardstick, where any inch is equal to any other inch).

The second point addresses the problem of skewed or distorted cases within categories. The analysis of variance presumes that the variances within categories are not affected by skewed distributions—that the variability from one group to the next is relatively homogeneous. Considerable variability between groups suggests that the samples were drawn from populations with different variances. This makes comparisons problematic. In the example of the divorce rates above, we saw that variability, measured by the standard deviation, varies considerably from one category to the next. The division with the greatest variability had a standard deviation about ten times larger than the division with the smallest variability.

Using the analysis of variance to obtain a test of significance, in this instance, would be inappropriate. However, comparing the variances within and between groups can be informative—and this is essentially what the analysis of variance seeks to do.

The discussion of relationships in general and ANOVA in particular is not complete until the issue of *hypothesis* or *significance testing* has been considered. The more elaborate applications and developments in ANOVA are dedicated to the primary objective of testing significance. This topic is so central to the social sciences that it merits a separate discussion (in Chapter 14). Usually ANOVA is discussed primarily in terms of significance testing. In this chapter, however, we have suggested that before you get caught up in the problems of establishing the significance of relationships, it is important to have a sense of how the degree of relationship is established.

THINKING THINGS THROUGH

1. You are studying alcoholism and you have a ratio scale that measures alcohol consumption in terms of liters of alcohol consumed per month. You want to know if people who live in houses as opposed to apartments have a different average level of alcohol use. You find the following:

	Type of Residents	
	Home Dwellers	*Apartment Dwellers*
	1.0	1.5
Alcohol	2.3	6.2
Consumed	1.8	5.4
per	5.7	4.2
Month	2.1	1.9
(in liters)	3.1	3.6
	0.0	5.0
	5.3	7.5

   How much does the between-group variance account for the total variance in consumption among all sixteen consumers? Show that the between-group variance and the within-group variance account for the total variance.

2. This problem is much the same as the one above, except it uses real data and asks you to do more than simply account for how divisions affect variability in child abuse. For example, is there some kind of pattern in the data beyond that indicated simply by dealing with variability between divisions?

Table 12.3. Reported Child Neglect and Abuse Cases, by Division, United States, 1980 to 1987 (Rates reported per 1,000 population.)

Year	*NE*	*MA*	*ENC*	*WNC*	*SA*	*ESC*	*WSC*	*MT*	*PA*
1980	3.9	2.5	2.5	4.5	4.2	4.0	2.9	3.3	4.4
1981	3.1	2.8	2.6	5.0	4.6	4.5	3.6	3.2	4.4
1982	4.0	2.9	3.0	5.6	4.8	5.1	3.8	3.4	4.6
1983	4.7	2.7	3.4	5.4	5.9	4.2	4.0	4.0	4.8
1984	4.4	3.9	4.4	6.2	5.2	6.2	4.4	4.8	5.5
1985	4.7	4.1	5.0	6.1	5.6	8.0	4.8	6.1	6.2
1986	7.5	6.1	7.5	11.8	7.3	8.1	7.0	7.6	11.7
1987	7.8	6.5	8.0	10.2	7.0	8.3	6.5	9.7	12.1

NE = Northeast; MA = Middle Atlantic; ENC = East North Central; WNC = West North Central; SA = South Atlantic; ESC = East South Central; WSC = West South Central; MT = Mountain; PA = Pacific.

This is an intrinsically interesting array of data. It is offered as a problem in the analysis of variance because it suggests that one should not leap immediately into measures of within-group and between-group variance to see how much of the total variance is accounted for by the independent categories (divisions). There is also obviously a relationship with the year in which the data are reported. How would you go about describing that relationship? What do you think accounts for it? How might you change the table so that the rates are reported *as if* they were all obtained in a standard year, say 1987?[3]

3. Relationships in the social sciences never perfectly eliminate variance. It is virtually always the case that huge amounts of variance remain. Provide an example that illustrates this argument and write a five-page essay in which you argue for or against the following statement: Eventually social scientists will achieve statistical precision, in analyzing purely social variables, that will eliminate all but five percent of the original variance in the dependent variable.
4. Ten teachers receive teaching evaluations before and after their supervisor implements a teaching merit policy that awards good teachers a special bonus. The teaching evaluation scores, for both before and after the merit policy went into effect, are listed below. How much influence did the merit policy have?

	Evaluation	
Teacher	*Before Merit Policy*	*After Merit Policy*
A	3.28	2.96
B	1.97	2.40
C	2.57	2.40
D	1.33	3.92
E	3.97	3.93
F	2.61	3.40
G	1.67	0.98
H	2.85	2.80
I	1.68	2.75
J	2.50	3.01
Mean	2.44	2.86

5. Given the above list of evaluations, assume that a teacher is selected at random and you have to guess whether his or her evaluation simply improved or declined. What would be your ratio of errors? The difference between means in these data suggests that the policy was a success. But the ratio of errors suggests it did not accomplish much. Which set of data is correct? How do you make your decision with respect to the success of the merit program on the basis of the information you are given here?

Notes

1. The dominance of hypothesis testing in the social and behavioral sciences must be carefully considered. It has taken on a nearly ritualistic quality. For a critique of this approach see Stanley Lieberson, *Making It Count* (Berkeley: University of California Press, 1989).
2. People determine causes of social actions casually and almost invariably incorrectly. Elaborate techniques have been devised to tease out causal connections in complex social systems. The value of these techniques is controversial. For a discussion of the quest for causation in human affairs see Hubert M. Blalock, ed., *Causal Models in the Social Sciences* (Chicago: Aldine Atherton, 1971) and David R. Heise, *Causal Analysis* (New York: Wiley, 1971).
3. We are beginning to touch on more complicated forms of ANOVA with this example. For technical discussions of these more elaborate forms see works such as Richard S. Bogartz, *An Introduction to the Analysis of Variance* (Wesport, CN: Praeger, 1994) or Lynne K. Edwards, ed., *Applied Analysis of Variance in Behavioral Science* (New York: M. Dekker, 1993.) The literature on analysis of variance is voluminous and highly technical. However, the basic logic remains the same: as variance is diminished, the ability of central tendency values to predict individual values is heightened.

13

THE STATISTICAL EXPRESSION OF RELATIONS: ASSOCIATIONS

"The shrewd guess, the fertile hypothesis, the courageous leap to a tentative conclusion—these are the most valuable coin of the thinker at work. But in most schools guessing is heavily penalized and is associated somehow with laziness."

—J. S. Bruner

When we can calculate standard deviations or the sum of squared deviations, we have a base for determining the degree of a relationship. The variability that exists prior to establishing a relationship can be compared with the variability that exists after the relationship is determined. If the relationship is strong, the original variability is greatly reduced. Determining standard deviations and other relatively precise measures of variability requires at least an equal interval scale.

Our concern with the statistical forms of reason now continues with a consideration of the concept of associations—a form of relationship that can prove to be logically imprecise or mathematically elusive. The question we shall consider at this point is, what problems do we encounter when we try to determine relationships between purely *nominal* conditions? (Remember, anything that is measured in terms of, let us say, a ratio scale—the ideal measure—can always be nominally categorized. For example we could measure height in terms of millimeters; we could also refer to individuals as falling into one of two height groups, A or B, without identifying which letter designated the taller group, thereby avoiding bias that might come from ordering the two categories.) The classes A and B would be nominal.

In Chapter 11 we saw that we can use the ideal measure of degree of relationship, the Pearsonian correlation coefficient, when we have variables measured in equal intervals or in terms of ratio scales. As we learned in Chapter 12, we can use ANOVA when we have one set of variables that is interval data or ratio-scaled and another set that consists of nominal or ordinal categories. Since most social categories are nominal categories, we are commonly faced with the problem of trying to relate two or more sets of data that are nominal in form. For example, we might want to know the relationship between being Protestant and being ambitious.[1] These are both nominal categories.

Nominal measures are the simplest and crudest measures available to a researcher. They consist merely of the classification of names of things: Protestant, drug user, banker, housewife, clerk, and so on. The very crudity of such measures makes the problem of assessing relationships precisely more difficult than is the case with equal interval or ratio scales. The heart of the problem lies in the fact that nominal scales force us to deal differently with variability. You begin to see, as you work with nominal data, that trying to get a precisely determined measure of the degree of relationship between two or more nominal sets of data is surprisingly difficult.[2] More important is the discovery that when working with nominal measures, we are led into a different logic of relationships—one with many implications for social research.

Although nominal data do not readily provide us with simple indexes of degree of relationship, we have to consider the fact that nominal data appear surprisingly simple with regard to prediction—and prediction is one of the main reasons we bother with relationships. Moreover, there are many instances in which we can predict nominal qualities with nearly perfect reliability. For example, a teacher can reasonably anticipate—that is, predict—that any student about to enter his or her classroom will be wearing clothes. While this is not an intellectually interesting prediction, it is worth noting because it suggests that there are aspects of the social order that are highly predictable. Correlation coefficients never come close to this ideal in behavioral research.

To illustrate various aspects of relationships that use nominal information, let's imagine a situation in which some researchers are interested in the spending habits of college students in American universities. A journalist finds out about the research, and an article appears in the papers under the heading, "Engineering Students Big-Time Spenders." This headline, in itself, is interesting because it suggests that all engineering students are big spenders—whatever that nominal category might mean. If this were the case, we would have perfect predictability. That is to say, we would know an individual was a "big-time spender" if we also knew the individual was an engineer. The journalist, however, is distorting what was, in fact, observed. (We like to refer to this as an example of the "rhetorization of fact." Facts are sometimes dramatically enhanced by journalists, and often by researchers themselves. This might be an unconscious response to the common desire to exaggerate the details of a story, or it might result from an effort to promote interest in some finding. Beware of this tendency when reading the newspapers.)

Suppose the article goes on to tell us that 55 percent of engineering students spent fifty dollars or more on a date, while only 45 percent of other students spent as much. This looks like a fairly significant relationship—a matter of a 10 percent difference between the two groups. Let's assume a very large sample was surveyed. We want to know the degree of relationship between being a big spender and being an engineer. Just what does the 10 percent difference between engineers and nonengineers mean? One way of thinking about this is to ask how much better off (in terms of knowledge of students' spending habits) are we in knowing about the engineers than we would have been if we didn't know about the engineers? If you said 10 percent, you would be wrong. The reason you would be wrong has to do with the fact that, in the case of our fictional study, engineers make up a minority of the student population.

This illustration relies on a statistic referred to as **lambda.** In itself, lambda is not especially imposing. It is an ordinary sort of statistical index. However, in a broader sense it is unusually evocative. It is a statistic that forces us to reexamine the logic of correlations. If nothing else, that makes it worth our time. Lambda relies on an interesting logic that adds new perspectives to the question of what we mean by things being related to each other. This logic is commonly referred to as **proportional reduction in error,** or **PRE.**

PROPORTIONAL REDUCTION IN ERROR

PRE calls for an answer to the following question: in percentage terms, how much better off are we after we know about a relationship than we were before we knew about the relationship? Think about that for a moment. How well off are you with regard to predicting events before you know something is related to something else? Then, after you know the relationship, how much better off are you? (We are reminded of Ronald Reagan's famous query in the debates preceding the 1980 presidential election, when he ran against President Jimmy Carter: "How much better off are you today than you were four years ago?" The logic in this case goes, if you aren't better off, then perhaps Carter did not have an effect.) The logic of PRE is extremely basic. We now need to see how it is put into effect.

Let's say, for purposes of illustration, that the engineers made up 20 percent of the student population of the university where the observations were made. If we took 20 engineers, we would find that 55 percent of them, or 11, fell into the big spender category, while 9 would fall outside that category. If we took 80 nonengineering students (giving us a total sample of 100 students), 45 percent of them, or 36 individuals, would be big spenders and the remaining 44 would fall outside that category. Make certain you see how this is established before going on. We can illustrate these numbers in the following simple table:

	Engineers	*Others*	*Total*
Big Spenders	11	36	47
Other Spenders	9	44	53
Total	20	80	100

Now we see that the big spenders constitute 47 percent of the total and other spenders constitute 53 percent of the total. If we had to guess the most predictable event, using just the marginal total, we would guess "other" spender and only be right 53 percent of the time. We would make mistakes or errors 47 percent of the time if we always guessed "other."

If we used our information about engineers, however, we would make 9 errors in the engineer column, and 36 errors in the "others" column by guessing the mode or most likely event. We would make a total of 45 errors. Our first prediction produced 47 errors, the second 45. The difference is 47 − 45 or 2. Proportionally, we reduced the error level 2/47ths or about 4 percent. This is the proportional reduction in error or lambda.

Data presented in this fashion need to be looked at carefully. As we shall see in a moment, it is possible to have a relationship and lambda will tell you no relationship exists. Here we have moved from an apparent considerable difference of 10 percent to a point where the relationship is extremely modest. Is a 4 percent reduction in error all that great? It is definitely not as dramatic as the 10 percent difference that led the reporter to declare that engineers are big spenders. Suppose one student decided to go out on one hundred random dates, while another student dated one hundred engineers. Using the knowledge of this research to enhance the probability of going out with a big spender, the second student would have, on the average, only two more dates who were big spenders than the first student.

This illustration tells us several things. First, it suggests that when we are told that several groups differ in some respect, we have to take into account the relative sizes or weights of the groups in assessing the meaningfulness of that difference. Second, it forces us to think further about the logic of relationships. Although it seems at first that we have a fair association between engineers and other students when it comes to spending on dates, a brief consideration shows that the degree of relationship might not be all that great. Third, it forces us to consider the possibility that we might have had a relatively high degree of predictability all along. Finally, we have to pay attention to what we want from a statement of relationship. Do we simply want to know that there is a relationship? Do we want to make a prediction? Do we want to improve our knowledge? What specific use do we intend to make of our findings?

Suppose the relationship were such that all engineers were big spenders and all the other students were not. Then, in a group of 20 engineers and 80 nonengineering students, we would get the results in the following table:

	Engineers	*Others*	*Total*
Big Spenders	20	0	20
Other Spenders	0	80	80
Total	20	80	100

If we were going on a blind date, without any knowledge of the spending habits of engineers but with the knowledge that big spenders were only 20 percent of the students, what would be our best guess with regard to whether we were going out with a big spender? We would probably anticipate not being with a big spender, and we'd be right 80 percent of the time. Notice that the simple percentage provides us with much more than a fifty-fifty break. If we knew we were were going out with an engineer or someone who was not an engineer, we would know what to expect, and we would be right 100 percent of the time. We have improved things only 20 absolute percentage points by having this knowledge. The reason is that even in a state of ignorance, we had a relatively high predictive efficiency: 80 percent. The perfect association gave us a 100 percent predictive efficiency. Thus we increased our predictive efficiency by 20 absolute percentage points. We reduced our error by 25 percent, proportionally. This figure,

$$100 - 80 \text{ divided by } 80 = \frac{20}{80} = 0.25$$

is the proportional reduction in error, or PRE, or lambda.

We have gone through this exercise to show that a perfect association does not necessarily mean a 100 percent improvement in our ability to deal with the world. In fact, we might have been doing fairly well to begin with. Remember, if we have dichotomous, nominal categories as our concern, the worst-case situation is one in which things are 50–50. That means we have at least a 50 percent chance of being right when we guess. In correlation analysis, the situation is more complicated. We are likely to be wrong nearly every time at guessing a dependent value on the basis of a correlated independent value, but not as wrong as we would be without the correlation. Take time to compare the logic of PRE with that of correlation.

Now let's turn to real, rather than fictional, data to conclude this discussion of lambda. Table 13.1 contains data pertaining to the distribution of religions throughout the world. The data are, in themselves, of some interest. Here we will concentrate on the problem of determining the extent to which regional locales throughout the world are related to various religious groupings. We want to know the *degree* of relationship. We will use lambda as a means to an answer.

Table 13.1. Religious Populations of the World (in millions), 1993

Religion	*Africa*	*Asia*	*Latin America*	*Northern America*	*Europe*	*Eurasia*	*Oceania*	*Total*
Christians	341	300	443	241	410	112	23	1,870
Muslims	284	668	1	3	14	43	0+	1,013+
Nonreligious/ Atheists	3	888	21	26	74	137	4	1,153
Other	75	1,435	10	12	4	3	1	1,540
Total	703	3,291	475	282	502	295	28+	5,576+

Source: Statistical Abstract of the United States, *1994, modified version of Table 1354.*

Pause before you begin calculating a measure of degree of association. Eyeball the table. Does there seem to be a relationship? Is there some kind of pattern? Even though the data are moderately complicated, we can see that Latin America, Northern America, Europe, and Oceania have relatively large Christian populations. Asia has a large population of "other" religions. Africa is relatively balanced between Christians and Muslims. Eurasia has a fairly even division between Christians and those categorized as nonreligious or atheists.

To the extent a particular religious category dominates a region, variability is reduced in the region. When variability is reduced, we have a higher degree of relationship. This is the basic logic of relations. We anticipate, then, a value for lambda indicating that reliance on a knowledge of the world's regions enhances the predictability of religious populations.

If we knew nothing about a particular region, we would predict Christian and we would be right 1,870 million times out of 5,576 million times, or 1,870/5,576 × 100, or about 34 percent of the time. Our total number of errors is 5,576m − 1,870m, or 3,706m. (Don't let the large numbers intimidate you.) Notice we are relying on a simple percentage to tell us our error ratio. Notice, also, that we had about 33.5 percent predictability without knowledge of locale. That is not bad, considering correlations rarely bring us to a level of 34 percent predictability. If we use our knowledge of how religious populations are distributed through the world, we get the following set of guesses, errors, and total errors (all values are in millions):

World Region	*Guess*	*Correct Guesses*	*Wrong Guesses*	*Total Population*
Africa	Christian	341	362	703
Asia	Other	1,435	1,856	3,291
Latin America	Christian	443	32	475
Northern America	Christian	241	41	282
Europe	Christian	410	92	502
Eurasia	Nonreligious/Atheist	137	158	295
Oceania	Christian	23	5+	28+
TOTAL		3,030	2,546+	5,576+

The total number of errors is obviously reduced when we rely on the world regions. Our correct guesses amount to 54.3 percent of the total. Before knowledge of the world's regions and their association with religious populations, we were 33.5 percent effective when we simply guessed Christian. Thus we have improved our predictive ability by about 31 percent.

When we compare this with the results we obtained in our fictional example about the engineers and spending, we must be careful to keep in mind how likely we would have been to guess correctly *before* we examined the relationship. With the engineers, we had 80 percent initial predictability; with the world religions, we had just 33.5 percent. Our improved predictability is always relative to the initial amount of predictability.

We can carry out the calculation of lambda more simply by taking the number of errors we have before we have knowledge of the relationship, subtracting the number of errors we have after gaining knowledge of the relationship, and then dividing by the total errors before we knew the relationship:

$$\text{Lambda} = \frac{\text{Errors before} - \text{errors after}}{\text{Total errors before}} = \frac{3{,}706 - 2{,}546}{3{,}706} = \frac{1{,}160}{3{,}706} = 0.313$$

Lambda, then, equals 0.313, or 31 percent.

ASYMMETRY

As a rule, **asymmetry** does not occur with correlation, but it is a disturbing feature of lambda. Although lambda's logic appears elementary and reasonable, we get different degrees of relationship between two qualities when we reverse the relationship. For example, let's return to the data in Table 13.1 and ask the question, to what degree does the religious composition of a region enable us to identify the region? In this case we are asking the reverse of our earlier question. To solve this problem we can refer to Table 13.2. If we were predicting a random draw, knowing only the regional totals, we would predict Asia and we would be right 59.0 percent of the time. That's impressive, considering that we are drawing from seven different regions. If we rely on religion to help us predict a region's identity, we get the following errors (again, all values are in millions):

Religion	*Guess*	*Correct Guesses*	*Wrong Guesses*	*Total Population*
Christian	Latin Am.	443	1,427	1,870
Muslims	Asia	668	345	1,013
Nonreligious/Atheist	Asia	888	265	1,153
Other	Asia	1,435	105	1,540
TOTAL		3,434	2,142	5,576

Table 13.2. Regions of the World by Religious Populations (in millions), 1993

Region	*Christians*	*Muslims*	*Nonreligious/ Atheists*	*Other*	*Total*
Africa	341	284	3	75	703
Asia	300	668	888	1,435	3,291
Latin America	443	1	21	10	475
Northern America	241	3	26	12	282
Europe	410	14	74	4	502
Eurasia	112	43	137	3	295
Oceania	23	0+	4	1	28+
TOTAL	1,870	1,013+	1,153	1,540	5,576+

Source: Statistical Abstract of the United States, *1994, modified version of Table 1354.*

We had 5,576m − 3,291m, or 2,285m errors without knowledge of religion. With knowledge of religion we have 2,142m errors—not much of a reduction. Solving for lambda, we get

$$\frac{2{,}285 - 2{,}142}{2{,}285} = 0.06$$

or about a 6 percent reduction in error. The value for PRE changes as we reverse the data.

Most unfortunately, lambda has the bad habit of telling us no relationship exists between two qualities when, in fact, a relationship does exist. Consider the following case:

	English Majors	*Social Science Majors*	*Total*
Big Spenders	512	422	934
Others	16	306	322
Total	528	728	1,256

Continuing with our fictional "big spenders" example, we would, in this case, guess, in a state of ignorance, that we would be more likely than not to get a big spender on a random draw. We would be right 934 times out of 1,256 total cases or 74 percent of the time. Our total errors would equal 322 out of 1,256, or about 26 percent of the total. If we use our knowledge of the majors, we would predict "big spender" in both cases, and we would get 16 errors for the English majors and 306 errors for the social science majors—a total of 322 errors. This is the same number of errors we had before we relied on our knowledge of the majors. If we solve for lambda, we will get zero.

Notice however, that there is a relationship. Social science majors have a much heftier proportion of "others" than the English majors. So, lambda is telling us that no relationship exists when, in fact, there is one. We are led, once more, to an awareness that attempting to establish the degree of relationship for the case in which we are relying on nominal data is not as simple as it might, at first, appear.

Before leaving this elementary discussion of PRE we want to state once again that what is important is the underlying logic of any relationship of concern to us. It is also extremely important to first eyeball basic data before relying on an index of relationship. Indexes should never take the place of serious thought about a set of data.

If nothing else, this review of PRE demonstrates that establishing an easily interpreted, precise, and consistent measure of the degree of association is difficult. So far no index has been developed that is capable of providing a meaningful, symmetrical index of association that is not affected by extraneous considerations, such as how many columns and rows the data are divided into or unusual distributions in the marginal totals.

CHI-SQUARE

We cannot conclude this discussion of association without at least a quick look at yet another response to the question of what constitutes a proper logic of relationships. In this instance the problem is solved by (a) setting up data showing how things would look if there were absolutely no relationship, and (b) seeing what the actual observed values look like in contrast. If the data in both cases are similar, then there is no relationship. If they differ, then there is more likelihood of an association. What we are interested in, then, is the difference between an expected, purely zero association and some actual obtained values.

Suppose we found the following data for essays judged "brilliant" and "ordinary" for a sample of 1,000 students in a literature class:

	Essay Evaluation		
	Ordinary	*Brilliant*	*Total*
Used Computer	400	100	500
Noncomputer	100	400	500
Total	500	500	1,000

It is obvious that there is an association here. What is less obvious is how the data would appear if no association existed, retaining the same total values. You should be able to figure this out intuitively. If there were no association between these categories, we would have the following distribution:

	Essay Evaluation		
	Ordinary	*Brilliant*	*Total*
Used Computer	250	250	500
Noncomputer	250	250	500
Total	500	500	1,000

This is what we would expect *if* computer usage and essay quality were not associated. Notice that the row and column totals remain the same. We expected only 250 ordinary essays from the computer users and we observed a larger number. Something is obviously going on. The tough question is, how much is going on?

The logic we have been sketching here is expressed in a workhorse statistic called **chi-square.** Chi-square is used so commonly in dealing with nominally ordered information that it is necessary to outline a few of its essential features before leaving the topic of measures of association. Chi-square is especially pertinent because it relies on a different logic of associations. It approaches the problem by asking how different what we observe is from what would be the case if there were no association between our variables. Thus, chi-square is a general test designed to evaluate whether the difference between what we observe and what we might expect if there were no relationship at all is a significant difference—that is, one not likely to occur by chance.

In Table 13.3 we might be interested in whether there is a relationship between time and the extent to which marital status relates to legal abortions. It is evident from the table that in the twelve-year interval from 1973 to 1985, the number of abortions for the unmarried category more than doubled while those for the married category only slightly increased.

Table 13.3. Legal Abortions (100,000s) by Marital Status of Women for 1973 and 1985, United States

	1973	*1985*	*Total*
Married	216	266	482
Unmarried	528	1,323	1,851
Total	744	1,589	2,333

Source: Statistical Abstract of the United States, *1990, Table 101.*

If we calculated lambda to get a sense of the degree of relationship, we would predict unmarried for each of the two years, and we would be wrong 482 times out of 2,333 guesses. If we know the year, we would guess unmarried in 1973 and be wrong 216 times, and we would guess unmarried in 1985 and be wrong 266 times. The total number of errors, taking time into account, is the same as it is when we use the marginal total. Lambda tells us, in this instance, that there is no relationship. Obviously, lambda is leading us astray.

Chi-square approaches the problem of measuring the degree of association with a different, but persuasive, logic. It asks the question, how much does what we observe differ from the case of no relationship? Determining what the case of no relationship should look like is a matter of prorating the values within the cells of the table so that they are distributed according to the marginal totals.

The total percentage of abortions is 20.7 percent for married women and 79.3 percent for unmarried women. If this proportion held in 1973, then 21 percent of the 744 abortions in 1973, or 154 abortions, would be among married women. The number among married women in 1985 would be 328. Thus, the data would look like the following figures if there had been no association whatsoever between date and marital status for number of abortions:

	1973	*1985*
Married	154	328
Unmarried	590	1,261

In sum, the variation between cells is maximized, assuming the marginal totals remain constant. If there is a relationship, this variation will be reduced.

To obtain chi-square:

1. Establish the null expected frequency for each cell by multiplying the row and column totals for that cell and then dividing the product by the total number of cases (N).
2. Find the difference between the expected and the observed value in each cell.
3. Square the difference.
4. Divide by the expected frequency.
5. Sum the dividends across the cells.

For our abortion data, then, chi-square equals:

	1973	*1985*
Married		
Observed	216	266
Expected	154	328
Unmarried		
Observed	528	1,323
Expected	590	1,261

In the first cell, the married women for 1973, the difference is 216 − 154 = 62. This difference is squared, equaling 3,844. When we divide 3,844 by 154, we get 25. We continue through the other three cells doing the same thing:

$$[(216 - 154)^2/154] = (62^2/154) = 3{,}844/154 = 25$$

$$[(266 - 328)^2/328] = (-62^2/328) = 3{,}844/328 = 12$$

$$[(528 - 590)^2/590] = (-62^2/590) = 3{,}844/590 = 7$$

$$[(1{,}323 - 1{,}261)^2/1{,}261] = (62^2/1{,}261) = 3{,}844/1{,}261 = 3$$

When we have the values of chi-square for the four cells, we sum these values to obtain the final value for chi-square for the entire contingency table:

$$25 + 12 + 7 + 3 = 47$$

An increase in the value of chi-square means an increase in the difference between what we observed and what we would have expected if there were no relationship. The larger chi-square is, the larger the degree of relationship is between the qualities we are studying. If chi-square is zero, it means that our observations conform to what we would expect in the case of no relationship whatsoever.

One problem with using chi-square to estimate the degree of relationship is that the value of chi-square can be greater than one. It can be three, five, sixteen—even thirty, forty, or fifty. If there are a lot of cells in a table, the value for chi-square can be quite large.

There is, however, a class of measures of association that use chi-square to get an index of relationship that has a lower limit of zero and an upper limit of one. One such index is *phi.* Phi is appropriate, however, only for tables with two rows or two columns. Its calculation is simple:

$$\text{Phi} = \sqrt{\frac{\text{Chi-square}}{N}}$$

For the abortion data, then, phi equals

$$\sqrt{\frac{47}{2{,}333}} = \sqrt{0.02} = 0.14$$

The degree of association indicated by phi is low here, suggesting a limited relationship. At the same time, an examination of the table suggests that something peculiar was going on between 1973 and 1985. The unmarried category more than doubled its number of abortions during this interval, while the married category increased only slightly. This example suggests that the use of various indexes of association should be carried out with great caution. Do not rely on the indexes exclusively. Instead, look carefully at what the data are saying as they appear in the basic tables from which the indexes of association are derived.

SIGNIFICANCE

In our extended discussion of relationships, ranging from a consideration of correlation coefficients and analysis of variance to indexes of association, we have not talked about the problem of whether a relationship is significant.

The analysis of variance and chi-square are used primarily to determine whether an observed relationship can be attributed to purely accidental or chance occurrences. So strong is the emphasis on tests of significance in the social sciences and psychology that discussions of statistical methods tend to bypass the much more meaningful problems of (a) the *degree* to which a correlation exists, and (b) the *form* of the relationship. The natural sciences pay attention to form; the social sciences are concerned less with form and more with the question of whether a relationship exists in the first place.

The question of significance of relationships will be considered in the next chapter. There we will once again come back to the analysis of variance and chi-square. In this chapter we were concerned with the logic of relationships and the question of how we can determine the extent to which several variables are related. If we can measure variability or variance in some precise way, ascertaining relationships is a matter of seeing how that variance is affected by a particular relationship. If we cannot measure variance precisely, then it is more difficult to develop a sense of the degree of relationship between variables. Because nominal data do not permit precise measures of variability, finding an index that measures the degree of relationship between nominally presented data is a difficult task.

A NOTE ON THE LOGIC OF RELATIONSHIPS

Organisms existing within a given environment must be related to that environment in some systematic way. This means that humans, if they are to create, sustain, and interact with their social environments, must have some way of relating to that environment. This means that people must be able to establish various relationships. If they cannot establish relationships, then social order cannot be sustained.

Within the research of social scientists, correlation analysis appears to reveal low-level predictability. Correlations are often on the order of less than ± 0.40. This suggests that 84 percent of the variance in whatever it is we are concerned with (our dependent variable) is residual variance, or variance that is not accounted for. How can people relate to their social milieu when it is so "chancy," so loaded with heavy elements of probability?

The answer might lie in a trade-off, as seen in the nature of associations. Nominal events, though much more crudely defined and vaguer in character, are more highly predictable. This is readily seen, for example, in the fact that sports analysts are not at all good at predicting the score of, let us say, a single football game. However, given certain information, they can do a fair job of predicting who will win or lose.

What do you have when you determine the degree of a relationship? We must raise this question because so many people believe that any degree of relationship suggests a *total* relationship. This is as misguided as believing that if a baseball player has a batting average of .235, he or she will get a hit every time at bat. Correlations, percentage reductions in error, and phi coefficients offer hints when there is nothing better to go on. In the social sciences these measures of relationship always caution us by reminding us that our relationships, in one way or another, are burdened by imprecision.

SUMMARY

The primary concern of this chapter has been to draw attention to the fact that attempting to establish the degree of relationship between variables that are nominal in character is quite difficult. The problem centers, at least in part, on the fact that nominal measures do not allow precise measures of variability. One measure of degree of association for nominal measures, lambda, provides an example of this difficulty. When row and column headings are reversed, lambda can give different values for the degree of relationship—something that, ideally, should not happen. It can also indicate no relation in situations in which a definite relationship exists.

The statistical index known as chi-square relies on evaluating the extent to which observed values deviate from what we would expect if there were no relationship between our variables whatsoever. A measure of association known as phi, derived from chi-square, is sometimes used to measure the degree of association between two related variables. But its obtained degree of association can be misleading.

The general conclusion of our discussion, then, is that measures of association must be dealt with most carefully. The solution to measuring association when variables are nominal in character remains a difficult and unresolved problem.

Thinking Things Through

1. A researcher notices that sports announcers generally attribute momentum to teams as the score changes in the team's favor. Assume that an announcer ascribes momentum 100 times during a season, and every time this announcer does so, it happens as the score changes in favor of the team to whom momentum is being attributed. A researcher quickly jots down a two-by-two contingency table showing the association, and it looks like the following:

	No Mention of Momentum	*Momentum Is Mentioned*
No change in score	100	0
Change in score	0	100

What is the degree of association in this case? What kind of relationship is this? What's going on? Is momentum real? (This problem, incidentally, is based on a story told by a social psychologist who reported that a student of his made a similar real finding.) Calculate the values of phi and lambda.

2. Observe an old television program or film in comparison to one that was made recently. (Define "old" as a film made before 1945 and "recent" as one made since 1985, for example.) Over a fifteen- or twenty-minute interval, determine the length of shots that make up the program. A shot is any clearly defined sequence of action that is completed with the entrance of the next shot showing a different sequence of action. Define any shot that is less than three seconds long as short, a shot that extends between three and ten seconds as medium, and any shot that is longer than ten seconds as long.

   Draw up a contingency table showing the results. Is there an apparent relationship between length of shot and whether the program or film is an old or new one? Would the relationship be altered by the manner in which you defined the length of the shot? How would you go about showing the existence or lack of relationship between your variables? If there is a relationship, how would you interpret it?

3. What degree of relationship is generally given to social relations by folk knowledge? For example, how does folk knowledge see the relationship between, let us say, being a liberal or a conservative and accepting or

rejecting capital punishment? What do you think the real degree of relationship is? How might you go about finding out? Suppose you obtained the following findings by interviewing people more or less at random:

Capital Punishment	*Liberal*	*Conservative*
Percent in Favor	25	52
Percent Opposed	37	43

Do these data support or undermine folk wisdom? What does this suggest with respect to the impact that the statistical findings of social scientists has had on folk knowledge? Which do you think is superior, social science knowledge or folk knowledge? (This is, by no means, a simple question. All we can offer is the suggestion that it depends on the situation. But in what ways does it depend on the situation?)

Suppose the data above are obtained from two areas. One area is 90 percent conservative. The other area is 90 percent liberal. How does this affect your interpretation of the percentages in the above table?

What is the percentage of exceptions to the folk idea that all conservatives will be in favor of, and all liberals will be opposed to, capital punishment? What statistical concept is the calculation of exceptions similar to?

4. Why do people go to great trouble to create games in which the outcome is uncertain and then pay sports analysts large sums to try to predict the outcomes of the games? How would you go about trying to set up a statistical model for predicting the outcome of a single National Football League team's season? How would you first go about predicting wins and losses? Then, how would you go about predicting the score of each game? At which effort do you think you would be more successful? Why?

   What lessons lie in this exercise for the social scientist? Has any statistician proved better at predicting the outcomes of sporting events than well-informed sports analysts?

Notes

1. One of the major theorists in social science, Max Weber, suggested that religious affiliation has a great deal to do with ambition. See, for example, *The Protestant Ethic and the Spirit of Capitalism.* But there are tremendous problems when it comes to establishing what is meant by these categories. Furthermore, even if these problems can be resolved, there are still extremely difficult technical problems that arise in trying to find a simple index that provides a meaningful measure of the extent to which the categories are associated with each other. As a result, scholars have argued over Weber's contention for decades.
2. Despite the problems that are inherent in establishing indexes of relationship between nominal variables, textbooks in statistics commonly discuss as many as a dozen or so different ways to measure such relations. Most of these measures are nearly impossible to evaluate in any direct manner. For a discussion of the analysis of nominal data see H. T. Reynolds, *Analysis of Nominal Data* (Beverly Hills, CA: Sage Publications, 1977).

14

SIGNIFICANCE: THE LOGIC OF HYPOTHESIS TESTING

"What we have to do is to be forever curiously testing new opinions and courting new impressions."

—WALTER PATER

"A wise man will not leave the right to the mercy of chance. . . ."

—HENRY DAVID THOREAU

When asked why he had not mentioned God in his work Mecanique Celestes: *"Sire, I have no need of that hypothesis."*

—PIERRE SIMON DE LAPLACE

By far the greater part of statistical research in psychology and social science involves **hypothesis testing.** Hypothesis testing is largely a matter of using the logic of probability to deal with the problem of relationships. However, there are cases in which a hypothesis consists not of a statement of relationship but a statement of a presumed value. In such instances probability is used to determine the likelihood that the presumed value is acceptable or unacceptable. The logical form, in any case, essentially goes back to the logic of sampling probabilities.

A common example of testing a presumed value occurs when a political survey is undertaken to find out if a particular candidate has a majority of the vote. In such an instance the hypothesized value might be stated as $H_1 > 50\%$. The researcher would then be called on to survey the population involved and obtain a value for the proportion voting for the candidate. Suppose the researcher found that sixty percent of the voters polled said they were going to vote for the candidate. Is this necessarily good news? It depends. On what does it depend? Has the hypothesis $H_1 > 50\%$ been tested?

Suppose further that the pollster had only polled ten people, all of whom were friends. In this case the sixty percent figure does not appear to be much of a test. The sample is small and biased. In such an instance you would not be able to reject the notion that the sample was a fluke of some kind (notice how this is worded). The candidate probably does not have a majority.

Suppose we told you now that the pollster interviewed ten people but carefully selected them at random. The sixty percent figure still is not especially impressive because it is evident that the pollster could quite easily have selected a sample that just happened to be in favor of the candidate by a ratio of six to four. We cannot reject the idea that the good news coming from the sample is not a matter of chance or random fluctuations.

Notice the way this is put: *we cannot reject the idea that chance is not at work*. This means we must consider the possibility that chance is at work. In statistics the acceptance of chance is acceptance of what is called **the null hypothesis,** or H_0.

Suppose we told you that the pollster interviewed ten people carefully selected at random, and all ten expressed a preference for the candidate. Do you now think that there is a probability that something other than chance is at work? Would you reject the null hypothesis, H_0? The null hypothesis, in this case, says that what you observed was a consequence of chance rather than some alternative possibility. However, the possibility of this happening, just by chance, is about one in a thousand—*if the public is actually evenly divided.* Is that a low enough probability for you to begin thinking that perhaps something other than chance is at work—that perhaps the candidate has a real majority?

Notice that a very small sample, $N = 10$, *if it is randomly selected,* can enable us to make decisions even where variability is assumed to be high. In this case we are assuming an equal division (the most variable the situation can be between the candidates), and our small sample says this assumption is *probably* wrong.

TYPE 1 AND TYPE 2 ERRORS

Hypothesis testing sounds modern and scientific—and it is. However, people have been creating hypotheses of various sorts and testing them for thousands of years. In what we might call the "folk form" of hypothesis testing, it is common for people to create the evidence that is used to sustain the hypothesis.

This happens a lot in films, literature, and theater, in gossip, and in ordinary stories of the day. For example, one hypothesis developed in the movie *Batman* is that evil people are ugly, and that good, law-abiding citizens are attractive. The story sustains its hypothesis by creating bad people who look odd and good people who look nice. This is an ancient literary device—one that is still popular. Hypotheses are always being tested in one fashion or another. However, once you see the logic of hypothesis testing as it is used in psychological and social research, you begin to appreciate the extent to which folk forms play fast and loose with the problem of providing good evidence to back up their hypotheses.

Hypothesis testing in psychological and social research rests on probabilistic thinking. It requires a good sense of the meaning of probabilities. Here is a world where everything becomes a matter of probability values. Nothing is absolutely certain. There is always the possibility you might be wrong.

When you play with probabilities, two basic things can happen: you might be right or you might be wrong. Unfortunately, when you are expressing yourself in probabilistic terms, you cannot say for sure which side of the truth you happen to be on. You can only say that you are probably right. But being probably right implies that you might, to some degree, probably be wrong. The more probably right you are, the less probably wrong you are. However, unless you have reduced the probability of being in error to zero, there is always a possibility that you have reached a conclusion that is not, in fact, valid. This is true in social research, and it is also true in day-to-day living.

Consider the problem faced by a teacher who has a student who misses an exam. The student tells the teacher that she missed the exam because her grandfather died. Then she misses a second exam and says it was because her grandmother died. Then she misses a third exam and excuses herself by saying that it was because her car broke down on the way to school. The teacher is faced with a problem. One folk hypothesis that might come to mind is that the student is lying. The null or chance hypothesis would be that she got caught in a series of improbable events. Her misfortunes were merely chance or random happenings. Notice that either of these possibilities might be true.

Suppose the teacher assesses the probability of a grandfather dying during the semester at 1 in a 100, or $P = 0.01$. He does the same for the grandmother dying. The probability of the car breaking down on the way to school is given a probability of 1 in 50, or 0.02. The teacher feels he is being extremely generous. He knows that the probability of these individual events happening to the same person in a single semester is the product of the individual probabilities: $0.01 \times 0.01 \times 0.02$, or one in five hundred thousand. That each would occur just at the time of an examination makes things still more improbable. The probability that these events happened by chance is extremely low.

In this example, the teacher does what people do all the time—he makes a judgment on the basis of casually estimated probabilities. It just is not "reasonable," we might say, for such a series of disasters to happen all at once and, moreover, always at a time when an examination is being given. Suppose, however, that it turns out that the student was telling the truth. The teacher made a wrong judgment, even though it was a highly reasonable one. He made what statisticians call a Type 1 error. He rejected the idea that chance was at work and concluded, instead, that what was really going on was not a matter of chance but a matter of the student engaging in a deception. But actually, chance *was* at work.

Here is another ordinary example. Chuck makes a date with Sally, and she stands him up. Chuck concludes something went wrong and that an accident or chance happening prevented Sally from showing up for the date. But actually, Sally snubbed him intentionally. In this incident Chuck accepts the null hypothesis, or the idea that fate or chance or happenstance was behind what he experienced, when, in fact, he should have rejected it. Chuck just committed a Type 2 error.

If nothing else, a knowledge of Type 1 and Type 2 errors makes us more sensitive to the extent to which we are vulnerable to error, even when we are very careful with evidence.

SETTING ALPHA

Sometimes, in dealing with the events of the world, we do not need to think much about the probability, or "reasonability," of an event. The fact, for example, that in the United States, there are twenty-two men in prison for every woman could perhaps be a chance event, but the probability is so low as to be absurd. There is obviously something going on here. Now, what is important is to notice that while we can reject chance, we cannot be certain what constitutes the alternative explanation. It might be that men are naturally more inclined to commit crimes. It might be that the police and the judicial system are more lenient with respect to women. Or it might be that men are encouraged to be aggressive, and aggression is associated with criminality. While we cannot be certain of the alternative hypothesis, we can be relatively certain that the random hypothesis is unacceptable. We do not have to engage in fancy probabilistic arguments to see this.[1]

There are other times when circumstances are not at all clear, and we have to decide when we are going to give up on the idea of chance and accept, instead, some alternative idea. Suppose you are the director of a nuclear power station. A local group protests that the plant is causing illness among elderly people. It is generally known that the number of people suffering from illness within a given population varies randomly from year to year. Sometimes the number is high and at other times it is relatively low. At the moment, let's say, the number of sick elderly people in the area of the power station is high. Is this a result of chance or is it due to something other than chance? Should you accept H_0 or H_1, the alternative hypothesis? (Once again, note that if you reject H_0, it does not necessarily mean that the power plant is the cause of the illness; but the case against the power plant is certainly a little stronger.)

Now, the first question we have to deal with is how high or how low we are going to set the value for determining if chance is at work. Note that if we set the value extremely low, then we lower the chance of rejecting H_0. If we set it extremely high, then we increase the likelihood of accepting the null hypothesis.

The local protest group puts together some information on the number of sick elderly people before the plant was built. They find that there were 27 cases of illness per 1,000 people 65 years of age or over. After the plant has been operating for fifteen years, the same statistic is 31 cases per 1,000. There is an increase, but is it a significant increase? Is it one that probably did not happen by chance?

What do you need to know to answer the question? First, and most simply, *you need to know what you are going to call a chance happening.* We need a cutoff point. We might say that if the difference is one that could not have happened more than once in a hundred times on a pure chance basis, then we will reject the idea of chance. In this case we run the risk of being wrong one in a hundred times. We might set the limits more loosely and say that if the difference could not happen by chance more than once in twenty times, then it was a nonchance event. Now we run the risk of being wrong one out of twenty times.

Setting the level at which we are going to decide whether something is happening by chance or not happening by chance is called setting the value of **alpha.** Ideally, alpha should be set before you begin testing whatever ideas you are out to test. Notice that alpha is purely arbitrary; it can be set low or high. You can set alpha equal to or less than 0.40, 0.30, 0.20, 0.10, 0.05, 0.01, 0.001, 0.0001, or 0.0000000000001. You can set it wherever you like.

Some care should be taken with alpha. In psychological and social research, the tradition has developed of setting alpha at either 0.05 or 0.01. That is, H_0 is rejected if the observed value could not have occurred by chance more than one out of twenty or one out of one hundred times. If alpha is set at the 0.05 level, there is the possibility that out of every one hundred rejections of the null hypothesis reported in the literature, five are wrong (unfortunately, you can never know which five are in error). Notice that alpha, in this instance, is established by a purely arbitrary convention.

Sometimes, however, alpha should be thought about more deeply. If, for example, you have a close friend with whom you play backgammon on a regular basis, you want to set alpha levels very low before you conclude that this friend is cheating at the game. Here you might want an alpha level of as low or lower than 0.001, or even 0.00001, before you are forced to conclude your bad luck with your friend is not really a matter of luck. (It might also be true that your friend is simply a better player.)

On the other hand, suppose you are in a bad neighborhood where you have reason to think local merchants are not always completely honest. In this case the least indication of cheating will probably set off alarms. Here you might want to set alpha levels much higher, perhaps at the 0.10 or 0.20 level.

If you were the director of the nuclear plant, how would you want alpha set, assuming that you wanted the plant to continue in operation? Obviously you would want it to be set very low, because then it would take a lot of evidence to dispute the notion that what was really going on was the result of chance. If you were a local protester, what alpha level would be convincing for you? In this case you would want alpha to be set relatively high, because then a modest amount of evidence might lead to rejection of H_0, or the idea that the higher illness rate for the elderly was only a random aberration.

Suppose you are a meteorologist, and you are responsible for warning people of an impending hurricane. There is a probability it might strike a large region. Then again, it might not. If you predict a hurricane and none takes place, you caused needless preparations, expense, and worry. If you do not predict one and one does happen, then you are responsible for leaving people vulnerable to the storm. Where would you set alpha in such a case? Would you set it relatively high or relatively low? (If you thought statistical decision-making was purely logical and mechanical, thinking about this example should change your mind in a hurry.)

Notice that in ordinary life there is an analogue with the logic of alpha levels. People who see cheating in every setting and are suspicious of nearly anything, have, in effect, set alpha high. They accept nearly any evidence as incriminating. On the other hand, some people are extremely trusting. It takes an overwhelming amount of evidence to convince these people that something might be behind a run of bad luck. Such people set alpha very low. What the statistician does formally with alpha levels, then, is done informally all the time as people go through their daily affairs.

SAMPLING ERRORS

Most hypothesis testing in social research deals with problems arising out of sampling errors. In discussing sampling in Chapter 10, we saw that if samples vary a great deal from each other, we cannot put much faith in what a single sample tells us. There is room for chance to influence things.

Let's get back to the nuclear power station. Cases of illness went from 27 to 31 per 1,000 elderly people. Could this be due to random fluctuations in the samples being used? Say the protest group sets the value for alpha at 0.01, and they feel they are being relatively lenient (some of the protesters want alpha set at 0.05).

Let's suppose that the samples involved are fairly large. The protest group bases its case on a telephone survey of elderly people completed just before the nuclear station was built. A similar survey is done fifteen years later. Both samples involve 1,000 respondents aged sixty-five and over. The sample is meticulously random, with the only bias being that it involves people listed in telephone directories. In the first sample, 27 people mention an illness. In the second, 31 people mention an illness.

We would like to know to what extent we would get no difference, modest differences, or big differences when we compare two samples. Pause for a moment and think about this problem. Suppose there is an infinite population of elderly people in which the annual number of illnesses is 30 out of a thousand. Most of the time, assuming large samples, our findings will be fairly close to thirty, and the difference between any two samples will be small (on the *average* the difference will be zero). Once in a while we will get one sample at the low end of the population and another sample at the high end. However, on the average this will happen relatively rarely—but it still can happen just by chance.

There is a formula in statistical analysis that allows us to estimate how much the differences will vary from each other. This is the formula for the **standard error** of the difference in proportions. If we can expect a big standard error for differences between samples, then a large difference between two samples could easily happen by chance. If the standard error for differences is very low, then a modest difference between our sample findings could be significant (that is, not accidental or due to luck or chance). Here is the formula for the standard error of the difference between sample proportions. (There are also ways of estimating the *SE* of differences between means and any other commonly used statistical index. Because we are more concerned with the logic of this procedure, these other indexes have not been included. Elaborate discussions of these indexes can be found in most standard texts or statistical references.)[2]

The following is the formula for computing the standard error of the difference between sample proportions:

$$SE_{p1-p2} = \sqrt{\frac{(p_1q_1 + p_2q_2)}{n_1 + n_2}}$$

- SE_{p1-p2} is the standard error for a *difference* between proportions, assuming that the mean difference for all pairs of samples is equal to zero;
- p_1 is the proportion in the first sample, and $q_1 = 1 - p_1$;
- p_2 is the proportion in the second sample, and $q_2 = 1 - p_2$; and
- n_1 and n_2 are the sample sizes for the respective samples.

In our problem it is obvious that the proportion of illnesses for each sample is relatively low—0.027 for the one sample and 0.031 for the other. Although the difference is low, is it one that did not happen by chance? The logic is what is important; the moderately complicated computations can be mastered with a little patience. The logic says that if we had every reason in the world to expect no difference, then any difference would be significant. What if we could presume that we might expect a difference of 0.002 to be very unlikely, and we found a difference of 0.004. In this instance, we would suspect something.

The trick, then, is to find the number that leads us to suspect the difference is larger than a random fluctuation. That is what the *SE* of a difference between proportions provides us. It is a standard deviation, and we know that two standard deviations is getting into the realm of the peculiar, odd, or unusually deviant. So, if we set two standard errors as our "suspicious number," then anything bigger than that might make us begin to think we are dealing with more than mere random sampling errors. Our standard error for the difference, after we solve for the equation above, is 0.016. We would need a difference about twice as large as this *SE* value for it to be unusual or improbable. Our observed difference was less than half of a percent. We needed a difference of about three percent to match what we would have expected if the samples were unusually different from each other. The difference we found was not especially significant, statistically. It could easily have happened by chance.

Such a procedure allows us to use a more careful logic to examine issues that are highly emotional in character. In this instance, there is no realistic basis for closing down the plant. The differences in the incidence of illness among the elderly could have happened accidently.

There is one more thing that needs to be pointed out in this example. The protesters are arguing that the nuclear plant is dangerous and, in doing so, have suggested that illness should increase. If illnesses had remained the same or declined, their hypothesis would immediately be negated. However, we know that although the change was small, it was in the predicted **direction of relationship.**

We need to remember the character of the normal curve in considering this situation. We know that if we are 1.96 *SD*'s above the mean *and* below it, then we have taken up a total of five percent of the curve in the two tails. Each side involves 2.5 percent of the curve. However—and this is a big however—if we are concerned with only one side, either above or below the mean, then we do not have to go as far to take up five percent of the curve. We only need to go 1.64 *SD*'s to get into the five percent region. If we are going to define anything that is so unique as to be a one-out-of-twenty-event as improbable, then we only have to go 1.64 *SD*'s to get into that domain. Where before we had to go nearly two *SD*'s to get something that unusual, we now only have to go a bit over one and a half *SD*'s to get something that unusual—*if we are able to anticipate the direction that events are expected to take.*

We can say that something is going to be different from something else. Or, we can be more specific and say we expect something to be bigger or smaller than something else. If we can do the latter—that is, if we can give direction to our expectations—we have more of a case if, in fact, the evidence goes in the direction we anticipated.

To return one last time to the power plant example, we expected illnesses to increase and they did. If they had decreased, then there would be absolutely no case against the power plant, no matter how small the decrease might have been. The fact that illnesses did increase is suspicious. To be more careful with our logic, we can now multiply our obtained *SE* by 1.64. The *SE* of a difference in proportions was 0.016. If we multiply that by 1.64 we get 0.03. However, the difference we got, though in the right direction, was only about one half of one percent. The critical value for rejecting H_0 at the 0.05 level called for a much higher difference between the two samples of elderly people. We still cannot conclude that the difference is anything other than an accidental one. We have to accept the null hypothesis.

Being able to state an argument more precisely by indicating direction makes it easier to reject chance as the reason for some observed value or difference. It is another reason we like to have at least an elementary idea of the form of a relationship because this gives direction to our hunches or guesses.

Suppose we have two groups taking a test in mathematical ability. One group is from the United States, the other from Japan. We know, from our general knowledge, that there is a relationship between nationality and mathematical skills and that Japanese students tend to do better. Without this knowledge we might predict a difference, but we would not be able to predict which group would get the higher score. With this knowledge, we predict the Japanese will get the higher score. If they do, the fact that the data are in the direction we anticipated strengthens our argument.

If the data go even slightly in the other direction, then it is obvious that the directional hypothesis is not sustained by the facts. However, testing H_0 for the opposite case is still necessary if you now believe that the relationship exists, but is reversed. In sum, then, directionality enables you to reach relatively solid conclusions with more tenuous formal standards.

SIGNIFICANCE AND LARGE SAMPLES

One of your authors participated in a court trial in which the defendant, in effect, was a correlation coefficient. The correlation coefficient was $r = +0.14$, a correlation for the relationship between a personnel test for candidates for the job of firefighter and how well the candidates later performed when rated by their supervisors in the field. The correlation, of course, accounted for about two percent of the variance in evaluated performance.

Why was such a poorly correlated test being used? (As it turned out, the paper and pencil test correlated rather nicely with ethnic and racial background—allowing the test to function relatively well as a disguised discriminatory device.) The answer is that the test had been evaluated in terms of whether it was able to do better than pure chance in predicting performance. The hypothesis that the test would do no better than chance was tested and ultimately rejected. The test did better than chance at the 0.05 level.[3]

The reasoning here was fundamental hypothesis-testing reasoning. It was assumed that if you checked sample after sample after sample, you would get sampling variations in obtained correlations. What would the standard deviation—that is, the standard error—of the obtained sample correlations be? As it turns out, the standard error of a correlation coefficient is determined largely by the size of the sample used to obtain the correlation. Here is the formula for the standard error of *r:*

$$SEr = \frac{1 - r^2}{\sqrt{N-1}}$$

In the case of the fireman's test, the sample size used was 525 cases:

$$SEr = \frac{1 - 0.14^2}{\sqrt{524}} = \frac{0.98}{23} = 0.043$$

How many *SE*'s from a correlation of zero would a correlation of +0.14 be? It would be 0.14, 0.042, or 3.33 *SE*'s from a correlation of zero. How many *SE*'s do you have to go out to in order to be in the domain of five percent probabilities? You only have to go out 1.96 *SE*'s. Obviously, this is a significant correlation. We can reject the idea that it happened by chance. At the same time, it is a correlation so low in value as to be worthless for prediction.

So dominant is the logic of hypothesis testing that tests that correlate at low levels with **criterion variables** (such as workers' performance in the field, grades in college, or the effectiveness of a given medication) are used if they pass the significance test. However, keep in mind that something that is significant at a probabilistic level might not be significant at a substantive level. This is not an easy issue to deal with. After all, science is a matter of uncovering subtle relationships, and a test of significance can often determine that a relationship exists between variables when that relationship is extremely subtle. The relationship between the personnel test for firefighters and later competence in the field was amazingly subtle.

Whether subtle relationships are worth further thought is a theoretical issue and not a statistical one. A statistical test might reveal that something could not have happened by chance. Just what this might signify depends on the intellectual content of the relationship. *What is of value with respect to intellectual content is determined by theory or by the relationship's humanistic significance, not by its probabilistically established significance.*[4]

Suppose we found that the number of deaths attributable to a nuclear plant was a significant figure. Let's suppose that the number of deaths attributable to the plant is 2 per 100,000 population per decade. This might be a statistically significant finding, but is it substantively significant? Is this a large enough value to warrant closing down the plant? Some people might think it is. Others might argue that it is a reasonable cost. After all, we pay with our lives all the time for the advantages offered by modern technology. Automobiles are an obvious example. They kill and maim people in large numbers and threaten global destruction through their polluting effects. Nonetheless, we love them. The point here is that a significant difference might or might not be—in a theoretical, intellectual, philosophical, or practical sense—a significant difference. Statistical tests do not mean that you should give up thinking about other theoretical or practical issues—although statistical methods are sometimes used for that purpose.

Be wary of tests of significance based on extremely large samples. Remember that if samples are infinitely large, then even infinitely small differences are going to take on significance. Infinitely small differences are extremely subtle. They are also, by definition, substantively meaningless. This, perhaps more than anything else, constitutes the Achilles' heel of hypothesis testing. Extremely large samples are more likely to provide an opportunity to reject the null hypothesis. This implies that significance is a simple function of the size of the sample used; but of course we would like to think that whatever it is we are interested in is not relative to sample size.[5]

CHI-SQUARE AND SIGNIFICANCE

In our discussion of contingency and measures of association in the last chapter we introduced a commonly used statistic known as *chi-square*. In that discussion, we used chi-square to deal with the question of how much association exists between several variables, or the *degree* of association. Chi-square is more often employed, however, to measure the significance of an association.

Chi-square asks, does an observed distribution deviate significantly from the distribution we would expect if there were no relationship whatsoever between the variables being observed?

Consider the set of observed frequencies for joggers by age and swiftness in Table 14.1.

Table 14.1. Frequency of Fast, Medium, and Slow Jogging Speeds (Data are fictional)

	Young	*Middle-Aged*	*Old*	*TOTAL*
Fast	25	20	15	60
Medium	20	20	20	60
Slow	15	20	25	60
TOTAL	60	60	60	180

What would a totally even or perfectly "smeared"—or chance—distribution look like for the figures above? We can see that a preponderance of the old category also fall in the slow category. A preponderance of young fall in the fast category. If we evened things out, each cell would have twenty cases. One way of calculating this is to multiply the row total by the column total and divide by the total *N*. In this instance it would be a matter of multiplying 60 by 60 and dividing by 180 = 3600/180 = 360/18 = 20. This is the value you would expect in each cell if there were absolutely no relationship between age and speed. To what extent do the observed values deviate from this expectation? Chi-square is essentially the summation of these differences. As with the standard deviation, chi-square squares the differences and then extracts the root:

$$\text{Chi-square} = \sum\left[\frac{(fo - fe)^2}{fe}\right]$$

where *fo* is the value of the observed frequency in a cell and *fe* is the expected value for a cell assuming no relationship. These figures are calculated for each cell in Table 14.2.

Table 14.2. Figures for calculating chi-square (based on data in Table 14.1)

	Observed (fo)	*Expected (fe)*	$fo - fe$	$(fo - fe)^2$	$\frac{(fo - fe)^2}{fe}$
Young–Fast	25	20	5	25	1.25
Young–Medium	20	20	0	0	0
Young–Slow	15	20	−5	25	1.25
Middle–Fast	20	20	0	0	0
Middle–Medium	20	20	0	0	0
Middle–Slow	20	20	0	0	0
Old–Fast	15	20	−5	25	1.25
Old–Medium	20	20	0	0	0
Old–Slow	25	20	5	25	1.25
					Sum = 5.00

Chi-square, then, is the sum in the last column, or chi-square equals 5.

Determining what this means requires the introduction of a statistical concept known as **degrees of freedom** or *df*. We do not want to enter into an elaborate discussion of degrees of freedom other than to point out that it should be obvious, after a while, that the more opportunity (or freedom) there is for cases to fall in different boxes or cells, the more opportunity there is for chi-square to have larger values. The number of cells in a contingency table determines the degrees of freedom. We determine df for interpreting chi-square by the formula

$$(r - 1) \times (c - 1)$$

where r represents the number of rows in a table and c represents the number of columns. In the above table we have three rows and three columns. The value for *df* is $(3 - 1) \times (3 - 1) = 2 \times 2 = 4$.

We now have all the information we need to answer the following question: Could this large a value of chi-square have happened by chance? Suppose we set alpha at 0.05. (Is this a good alpha level? What are the implications of not being able to answer this question in any noncontroversial manner?) We then go to a table of chi-square values.

Table 14.3 is an abbreviated chi-square table. The column headings provide different alpha values. The rows give different values for *df*. For example, to obtain a significant chi-square in a table with four degrees of freedom and alpha (P) set at equal to or less than 0.10, chi-square would have to be 7.78 or higher.

We set alpha at 0.05. In doing so we determined that if what we have observed might have happened by chance only once out of twenty times, or less, then we would reject H_0. We have four degrees of freedom. Therefore, we

Table 14.3. Values of Chi-Square Arrayed by Degrees of Freedom and Probability of Occurrence Ranging from $P = 0.90$ to $P = 0.01$

df	*P = 0.90*	*P = 0.70*	*P = 0.50*	*P = 0.30*
1	0.16	0.15	0.46	1.07
2	0.21	0.71	1.39	2.41
3	0.58	1.42	2.37	3.67
4	1.06	2.20	3.36	4.88
5	1.62	3.00	4.35	6.06
df	*P = 0.10*	*P = 0.05*	*P = 0.02*	*P = 0.01*
1	2.71	3.84	5.41	6.64
2	4.60	5.99	7.82	9.21
3	6.25	7.82	9.84	11.34
4	7.78	9.49	11.67	13.28
5	9.24	11.07	13.39	15.09

are interested in the value of chi-square at the point where the alpha (P) = 0.05 column and df = 4 row meet. The table tells us that only when chi-square is equal to or larger than 9.49 can we conclude that our observed distribution was a one in twenty probability. We conclude that we cannot reject H_0 because our chi-square value was equal to 5.

We said earlier that larger samples can generate larger values for chi-square even when the degree of association remains constant. We leave the calculations for you to check in Table 14.4. Do notice, however, that the table is identical to Table 14.1 with the only difference being that the sample has been increased by a factor of ten. For this table, chi-square turns out to be equal to 50. The value of chi-square is obviously related to sample size. This is awkward.

Table 14.4. Frequency of Fast, Medium, and Slow Jogging Speeds (Data are fictional)

	Young	*Middle-Aged*	*Old*	*Total*
Fast	250	200	150	600
Medium	200	200	200	600
Slow	150	200	250	600
TOTAL	600	600	600	1800

We can see that tests of significance are strongly influenced by sample size. The same degree of association can be attributed to chance according to one sample while in another H_0 is rejected. This is troublesome. It raises some serious questions about what we are getting at when we seek to establish significance through purely statistical operations. If the task is difficult when we use extremely sophisticated devices such as statistical tests of significance, think of the problems created when we try to do the same thing on a purely intuitive basis.

ANOVA AND STATISTICAL SIGNIFICANCE

In our discussion of relationships in Chapter 12, we looked at a procedure called the analysis of variance. We saw that variance can be divided into within-group and between-group variances. When between-group variance is high, we have evidence of a relationship. When between-group variance is low, we do not have a relationship. The question once again becomes, if we have some evidence of between-group variance, is it enough to let us reject the idea that the variance between groups is not occurring simply as a result of random fluctuations?

Consider the data in Table 14.5. The hypothesis suggested by the table is

Table 14.5. Females as Percentage of Total Civilian Employment for Selected Countries for 1980 and 1987

	Females as Percentage of Civilian Employment	
Country	*1980*	*1987*
United States	42.4	44.8
Australia	36.4	39.8
Canada	39.7	43.4
France	39.5	42.0
Italy	31.7	33.8
Japan	38.4	39.6
Netherlands	30.3	36.0
Sweden	45.2	47.8
United Kingdom	40.4	42.5
West Germany	38.0	39.6

Source: Statistical Abstract of the United States, *1990, Table 1662.*

that the female percentage of the civilian labor force was significantly affected by the seven-year passage of time from 1980.

We really do not need to use analysis of variance to see that 1987 differs significantly from 1980. There are ten nations involved, and each one increased its percentage of employed women between 1980 and 1987. If we used the analogy of a coin toss, arguing that a nation was as likely to increase as decrease on a random basis, then this distribution obviously differs from what we would expect on a chance basis. The likelihood, under this assumption, that all ten nations would increase would be 0.50 raised to the tenth power. H_0 can be rejected at the 0.001 alpha level.

Because huge populations underlie percentages or proportions such as those listed in Table 14.5, probability statistics are not generally used with such information. Let's suppose, however, that our test is a test of two independent samples, one drawn in 1980 and the other in 1987, with each sample consisting of ten randomly drawn cases. We could then consider whether the between-group variation is great enough to suggest that the two samples are significantly different.

If there is no difference between 1980 and 1987, then we would expect the means for the two periods to be similar. Any differences could be attributed to chance. The greater the difference between group means, the less variability we should find within groups.

As was the case with chi-square, ANOVA requires taking into account the extent to which our values are free to vary. The more cases there are, the more freedom there is. The more categories we have, the more freedom there is. The degrees of freedom, in ANOVA, are calculated by the following basic formula:

$$DFw = N - k$$

where DFw is the degrees of freedom for the within-group variance,

N is the total number of cases, and

k is the number of categories.

For our example, we have twenty cases and two categories, providing a value for *DFw* of 18.

The degrees of freedom for the between-group variance is determined by the following formula:

$$DFb = k - 1$$

We have seen that k equals 2; therefore *DFb* is equal to 1.

The logic of ANOVA was outlined in our discussion of relationships. We want to know how much variance within categories has been reduced, relative to overall variance. Basically, overall variation is the sum of variation within groups and variation between groups. The more variation there is between groups, the less there is within groups.

We begin by determining the variation between groups using the sum of squares (*SSb*):

$$SSb = \text{Sum of number of cases in category} \times (\text{the mean of the category minus the overall mean})$$

The mean for 1980 is 38.20, and the mean for 1987 is 40.93. The mean for the group as a whole is 39.56. So the *SSb* for 1980 is 39.56 − 38.20 times 10, or 13.6. For 1987 it is 39.56 − 40.93 times 10, or 13.7. The total *SSb* equals 13.6 + 13.7, or 27.3.

The total sum of squares is equal to 372.51. Notice that we already have an indication that there is not much of a relationship. The influence of the change in time is modest. The reduction in the total sum of squares by the variation between means is only about seven percent.

The *SSw*, or within-group sum of squares, is obtained by subtracting the between-group sum of squares from the total sum of squares: 372.51 − 27.3 = 345.21.

We now estimate the within-group variance by dividing the within-group sum of squares by the appropriate degrees of freedom:

$$\frac{345.21}{18} = 19.18$$

The between-group variance is obtained by dividing the between-group sum of squares by the appropriate degrees of freedom:

$$\frac{27.3}{1} = 27.3$$

All of this has been done to attain a ratio, the ratio of between-group variance to within-group variance. In this case the ratio is equal to

$$\frac{27.3}{19.18} = 1.42$$

This is called, in ANOVA, an *F ratio.* As with chi-square, to determine whether this ratio indicates statistical significance it is necessary to find out what constitutes a significant value of *F* relative to the degrees of freedom granted our data. Table 14.6 provides an abbreviated table of *F* ratios.

Table 14.6. *F* Values for Alpha Levels Equal To or Less Than 0.05

Between-Group df	*Within-Group df*				
	1	*2*	*3*	*4*	*5*
1	161.00	200.00	216.00	225.00	230.00
4	7.71	6.94	6.59	6.39	6.26
8	5.32	4.46	4.07	3.84	3.69
12	4.75	3.89	3.49	3.26	3.11
16	4.49	3.63	3.24	3.01	2.85
18	4.41	3.55	3.16	2.93	2.77
30	4.17	3.32	2.92	2.69	2.53
40	4.08	3.23	2.84	2.61	2.45
120	3.92	3.07	2.68	2.45	2.29
infinity	3.84	3.00	2.60	2.37	2.21

Between-Group df	*Within-Group df*				
	6	*8*	*12*	*24*	*infinity*
1	234.00	239.00	244.00	249.00	254.00
4	6.16	6.04	5.91	5.77	5.63
8	3.58	3.44	3.28	3.12	3.93
12	3.00	2.85	2.69	2.51	2.30
16	2.74	2.59	2.42	2.24	2.01
18	2.66	2.51	2.34	2.15	1.92
30	2.42	2.27	2.09	1.89	1.62
40	2.34	2.18	2.00	1.79	1.51
20	2.17	2.02	1.83	1.61	1.25
infinity	2.10	1.94	1.75	1.52	1.00

The following discussion deals with values of *F* that must be reached for given values of the degrees of freedom to attain significance at the 0.05 level. (If significance at the 0.01 level is being sought, then *F* will have to be higher. Other tables provide values of *F* for alpha levels of 0.01 and 0.001.)

With 18 degrees of freedom for the within-group variance and 1 degree of freedom for the between-group, we see that we would need an *F* ratio of 4.41 to attain significance at the 0.05 level. Our obtained ratio was 1.42, so we accept the null hypothesis.

However, this example is worth thinking about because it is obvious that the shift from 1980 to 1987 is a significant one. Every country shifted upward. The huge numbers from which the percentages were determined would give us quite different results if, for example, a test of the significance of difference in proportions were used.

The example is also interesting insofar as an analysis of variance points out that the relationship really is a modest one. The shift is consistent and it comes out of huge populations. However, the shift is only a few percentage points upward.

In sum, do not apply analysis of variance to a problem simply because you have found that you have a situation in which total variances, between-group variances, and within-group variances can be determined. Look carefully at your data and let them tell their own story. Do not push things unless you are relatively certain of your techniques. A simple, clear presentation of information can be more useful and valid than an improper application of a complex technique. Go for content, not form. (Try not to be swayed by the unfortunate fact that journal editors, for example, are often more impressed by complex, scientific-looking mathematical indexes than by simpler, and usually more informative, frequency distributions.)

SUMMARY

Determining statistical significance for social data is not a purely mathematical exercise. Nonetheless, this approach has come to dominate Western quantitative social research. The good thing about tests of significance is that they provide a kind of alarm. They tell you there is a possibility that a relationship might, in fact, exist. The bad thing about tests of significance is that if your data are so confusing that they do not readily reveal the existence of a relationship, then the degree of relationship is probably so low as to be relatively useless for practical work—or even intellectual work, for that matter.

Keep in mind that if you relied on significant relationships to get your car from here to there, you would be in deep trouble. Suppose the steering wheel were significantly correlated with the front wheels, with the degree of relationship indicated by a correlation coefficient of $r = +0.20$. You would be able to steer the car better than you could under pure chance conditions over the long run, but you still would not be able to steer the car effectively in any individual instance. Ninety-six percent of the time the steering wheel and the front wheels would be at odds with each other.

Tests of significance in modern sociology and in psychology have become somewhat like Egyptian temple art. The form of the presentation might be more important than the content. We quickly hasten to add that this discussion does not deny the utility of tests of significance. At the same time, however, this is an area where those who intend to do professional work should think long and seriously about the meaning of their findings.[6] Certainly one should not simply calculate chi-squares or standard errors by the dozens and feel elated simply because a few turn out to be significant at the 0.01 level.[7]

If, for example, you find that there is a significant relationship between watching television and violence, it is important to look at a variety of things. Are there a few extreme cases that might be producing the effect? Were extremely large *N*'s involved? How strong is the relationship? Is the degree of violence that is a product of watching violent programs sufficient to justify a response?

This last question is not an easy one to deal with. For example, numerous violent incidents have occurred during and after football games in the United States and soccer games in Europe. These sports generate a small and possibly significant amount of violence—where significance is defined in statistical and probabilistic terms. However, few people would see this as a justification for doing anything other than admonishing fans to be more considerate. There is possibly a significant relationship, but not one of sufficient degree to lead to a serious effort at control. In other words, statistical tests are merely the beginning, not the end, when it comes to thinking about what we should do.

One last word: a number of researchers in the social sciences have developed a kind of statistical bigotry with respect to statistical significance. If a relationship is significant, it is considered an important finding. Researchers are commonly disappointed by not finding a significant relationship and do not publish their results when nothing seems to be happening. The consequence has been a bias resulting from researchers publishing their findings when they are able to reject H_0, but not publishing their findings when they are unable to reject it.

Good research should find that support of H_0 is as intellectually interesting as its rejection. Much of the significance of psychological and social research in the twentieth century has come not so much from what it has found to be significantly related as what it has found *not* to be significantly related. For example, although a lot of controversy still rages around the subject, it is certainly important to establish, finally, that there is no significant correlation between race and intelligence. If there is a relationship, then that is interesting. If we accept the null hypothesis, that is also interesting. Given the injury that the alternative hypothesis, in this instance, has managed to do in its long history, accepting the null hypothesis is, in our opinion at least, much more interesting in its implications than rejecting it.[8]

THINKING THINGS THROUGH

1. To keep you aware that statistical logic is unified and holistic rather than a variety of separate and distinct techniques, here is a problem you encountered earlier. Solve it as a problem in hypothesis testing. That is, define H_1 and H_0, set the value of alpha, and make a decision.

   You are playing cards with a man named Doc. You play two games, and twice in a row he comes up with four aces and a wild deuce. What do you conclude? Give precise values to your very strong intuitive sense that H_0 should be rejected. What does H_0 mean in this instance?

2. You notice in a book that reviews films that are currently available for video rental that old films seem to get high ratings more often than new films. You do some research and get the following array of films by rating:

	Rating (One star is low; four stars is high.)			
	One	***Two***	***Three***	***Four***
Old Movies	15	22	18	9
New Movies	17	33	16	5

From this information can you conclude that H_0 should be rejected?

3. Find something in the real world where you can get a count.
For example, you might notice how men and women arrange themselves in a classroom by rows from the front of the class, compiling data like the following:

		Men	*Women*
	1	2	6
	2	3	5
	3	3	5
Row	4	4	4
	5	6	2
	6	5	3
	7	6	2
	8	7	1

Once you have made your observations, test H_0. "Eyeball" your data before you get into the math of the problem. Get a sense of what your technical result will be. Use your imagination with this problem. Don't just mechanically set up some obvious relationship. Provide an example that requires X^2 and one that requires ANOVA for testing H_0.

4. A very large sample of students is obtained. It is found that students who entered college with well-defined professional aspirations did better than those who entered with no defined professional goals. The data look like this:

	Well-Defined Goals	*Undefined Goals*
Above Average	5,327	4,722
Below Average	4,673	5,278

Is this a significant finding? (Consider how, in one sense, it is significant, while in another sense it is not.) What is the proportional reduction in error (PRE) for these data?

5. Evaluate each of the following in terms of substantive or intellectual significance (Int. Sig.) and purely probabilistic, or statistical, significance (Stat. Sig.).

   a. The difference between mean IQ of African Americans, Asian Americans, and Caucasians is significant with $P > 0.672$.
   Stat. Sig.: High_____Low_____ Int. Sig.: High_____Low_____

b. The correlation, r, between eating raw onions and having bad breath is $+0.87$, $P < 0.0001$.
Stat. Sig.: High_____Low_____ Int. Sig.: High_____Low_____
c. The relationship, r, between recycling of waste done by a city and mean education of the city's population is $r = +0.16$, $P < 0.0001$.
Stat. Sig.: High_____Low_____ Int. Sig.: High_____Low_____
d. The mean difference between percentage of cures of AIDS victims using medicine A and medicine B is seven percent, $P > 0.05$.
Stat. Sig.: High_____Low_____ Int. Sig.: High_____Low_____
e. The difference between military distinction and sexual orientation of the enlisted person or officer is found to have a P value equal to 0.13. (How does direction of the difference affect how you judge this finding?)
Stat. Sig.: High_____Low_____ Int. Sig.: High_____Low_____
f. The grade point average for students who study compared with that of students who goof off could only be different by chance with $P < 0.00001$. (What if we tell you the mean GPA is higher for the goof-off gang? How does this affect things?)
Stat. Sig.: High_____Low_____ Int. Sig.: High_____Low_____

APPENDIX: A NOTE ON USING PRE TO GAIN PERSPECTIVE ON HOW REJECTIONS OF H_0 CAN BE MISLEADING

It is possible to have situations in which even though it is known that one group tests higher than another at a significant level, guessing individual values is not especially enhanced by this knowledge. The following small example is included here to underscore, once again, that tests should not be applied mechanically.

Consider a comparison of men and women on a test of some kind. Let's say it is a leadership test. The numbers in each column show the scores.

	Men	*Women*
	3	3
	4	4
	5	5
	6	8
	7	7
	8	8
	9	9
Sum =	42	44
N =	7	7
Mean =	6	6.29

We see immediately that the women's mean value is 6.29 and the men's is 6.0. This is reported to the papers in a small article with the headline, "Women Get Higher Scores in Leadership Test." However, the difference could be accidental. With the above data, the relationship would not be significant. (We can, of course, keep the same proportional differences and achieve significance just by boosting *N*, so the illustration is reasonable.)

Suppose the relationship were significant at the 0.05 level. We are then inclined to conclude that women are generally higher in leadership. However, let's ask some questions. Suppose you are in the company of a man and a woman, randomly selected from the test groups. You have no knowledge of their scores. You are asked to bet which one has the higher score. Your best bet, of course, is that the woman will have the higher score. Here is the way such bets go over a repetitive course:

Man has higher score	*Woman has higher score*	*Tied*
19	21	6

There are forty-six things that can happen. Out of the forty-six, the favored event—that the woman will have the higher score more times than the man—will occur on twenty-one occasions, or forty-six percent of the time. In other words, although the statistical significance test implies that your best bet is to wager that women are superior (and they are on the average), in the individual case you are definitely better off betting that when you encounter any pairing of a man and woman, the woman will *not* be superior to the man. The reason, of course, is that there are a fair number of tied cases.

This little example is included to make the point, once more, that the application of aggregated values can lead to some wrong-headed understandings when applied to individual cases. The leap from what is true for large numbers to what is true at the individual level should be made with great care—it can be a dangerous leap. This example also suggests that proportional reduction in error (PRE) is used to determine just how much relationship has, in fact, been established. Quite often, rejecting H_0 at the 0.05 level might involve a negligible difference between groups, despite the general sentiment among social researchers that it is cause for rejoicing.

Notes

1. One of the reasons the *Statistical Abstract of the United States* is such a powerful source of statistical information is that it is based on *N*'s so large that *any* difference or relationship is significant.
2. The indexes of most statistics texts contain a section under the heading "Standard Error"; beneath this heading are references to the formulas for the standard errors of various indexes such as the mean, percentages, proportions, standard deviations (which vary from sample to sample), correlation coefficients, and so forth.
3. The U.S. government has published a small monograph with the title *Federal Guidelines for Personnel Testing*. According to this monograph, the primary criterion for the adequacy of a personnel test is its ability to distinguish between two groups—let's say high and low performers on some task—at the $P < 0.05$ level.

This criterion permits nearly any test one can imagine to be acceptable for use in personnel matters if it is tested on large enough populations.

4. Anything and everything can be described statistically. Statistics is just another way of talking about things, and we are free to talk statistically about anything under the sun. We are reminded of the movie quiz that asked how many black spots were used in Disney's *101 Dalmatians* (the answer: nearly 1,600,000). The fact that this is a statistic however, does not make it intellectually worthwhile. It is trivial information because it is not relevant to any significant practical or theoretical concern. We mention this because there is a growing sentiment that one is *either* a statistician *or* a theorist. The consequence has been a spate of irrelevant and trivial statistical studies on the one hand, and theoretical literature that has little sense of the empirical problems involved in testing its assertions on the other.
5. This is an extremely disturbing problem. If you want statistically significant findings, all you have to do is get big enough samples. We would like to think that significance is not going to be affected by a mechanical factor such as sample size. Worse yet, there is the implication that the null hypothesis exists only under the most limited conditions; in the general case, it is always, in principle, rejected. This suggests that the entire logic of hypothesis testing is profoundly flawed. For a devastating critique of this dominant research paradigm in the social sciences, see Jacob Cohen, "The Earth is Round ($p < 0.05$)," *American Psychologist* (December 1994): 49, 997–1003.
6. There is a tendency to transform probability findings into mechanical findings. We are reminded of a case in which a very slight difference between two groups who were given a test was found to be statistically significant. This meant that if you drew an individual at random from one group and then the other and guessed which individual had the higher test score, you would make a lot of mistakes. However, the people responsible for the findings began speaking as if *all* the members of one group scored higher than all the members of the other group. We might refer to this as the "fallacy of mechanistic loading." That is to say, in a case like this one, probability findings are inappropriately loaded with mechanical implications. This fallacy should be avoided like the plague. Probabilistic findings should always be seen probabilistically.
7. It should be obvious that if you test one hundred relationships you will, on the average, find five that can be rejected at the 0.05 level—simply by chance. In other words, you will make 5 Type 1 errors out of 100 investigations with alpha set at 0.05. If these get published and the others are neglected, then the journals are burdened with a larger proportion of findings that are Type 1 errors than would be the case if findings were reported without a bias in favor of rejecting H_0.
8. All we are saying here is that good research has serious implications regardless of whether H_0 is accepted or rejected. Most of the findings of modern social science have been significant for what they have not found rather than for what they have found. For example, most of the claims of folk knowledge—such as claims about the weakness or mental inferiority of women, for example—have been shot down by modern statistics. What was found was that in many important ways, women do not differ significantly from men.

15

A BRIEF NOTE ON MODELS AND STATISTICAL CONTROL

". . . nature has all kinds of models. What we experience of nature is in models, and all of nature's models are so beautiful."

—R. BUCKMINSTER FULLER

"Models are intellectual toys and should never be confused with the reality they are supposed to represent."

—KENNETH BOULDING

"The controlling intelligence understands . . . what it does, and whereon it works."

—MARCUS AURELIUS

The concept of models is now a part of everyday speech. We hear of computer models that will, we hope, inform us of the extent of global warming. Models are used to test aircraft designs, construct buildings, predict death rates, and perform a wide variety of other essential tasks. We also hear of models who pose for magazines. A *model* is a representation. Just as the women and men who pose for fashion magazines are an idealization of style, intellectual models are idealizations of reality. Models are extremely useful and we cannot, in this day and age, get along without them. At the same time, we should develop a sense of the limitations, as well as the strengths, of models.

To think that models are the same thing as the reality they attempt to describe is to make the mistake of thinking that a map of the Grand Canyon is the same thing as the Grand Canyon. *Models are necessarily distorted.*[1] The only perfect map or model of some aspect of reality would be a perfect duplicate of the reality it was describing. To have a perfect map of the Grand Canyon, you would have to have a duplicate of the Grand Canyon itself. Obviously, such a map would not be especially helpful.

Models reduce the complexities of the reality they describe to a set of manageable dimensions. In physics, there are only several dozen basic variables or factors that provide the complex descriptions of reality that come out of the natural sciences—mass, charge, velocity, temperature, pressure, length, gravitational attraction, and magnetic attraction, among others.

In the social sciences, several variables have acquired special prominence in nearly any kind of research undertaking; these include status, sex, education, income, occupation, race, age, residence, and a few others. The variables used in social research do not, however, have the same reliability and referential precision that primary physical variables have. This is important to remember, because models are precise only to the extent they are capable of providing precise analogues for the reality they attempt to represent.

THE PURPOSE OF MODELS

The principal purpose behind models is to generate accurate expectations when a given set of conditions is defined. Meteorologists working with global climate models try to determine what we can expect if a given amount of industrial and automotive pollutants is released into the biosphere. An aeronautical engineer might want to know what to expect if the center of gravity of an aircraft design is altered. A criminologist might like to know what to expect if a policy of capital punishment is endorsed.

As we have suggested, a model is much like a map. A map is designed to fulfill your expectations when you go on a trip. A map might show you that if you head south for fifty miles and then turn east for another forty, you will be in Topeka, Kansas. You *expect* to arrive at a certain place because the map is a model of the terrain you are moving across.

The analogy of the map offers insight into the problems encountered when we attempt to model social reality. Physical reality seems to lend itself nicely to a variety of maps. Maps never perfectly reproduce the reality they represent; it is not the task of maps to do this. Instead, maps abstract reality. A map is considered good if it leads you to where you want to go, or if it can reasonably provide you with a sense of where you are.

Maps are also distortions of reality. This raises the question of what kinds of distortions are appropriate for any given model. The answer to this question is that the appropriateness of the distortion is relative to what you want from the model. Sometimes a crude map—a map that is extremely simple, lacking in detail, and not especially precise in its scale—is adequate. At other times more elaborate maps are necessary. The point of this discussion is that although people sometimes think a model should be a perfect representation of the reality it describes, the necessary precision of a model is much more complicated. It is always relative to the use to which the model will be put.

With physical reality, surprisingly crude models or maps work amazingly well—not necessarily perfectly, but still quite well. A rough sketch can lead to buried treasure. Every day of the year millions of items move through the postal system using only a few numbers and letters of the alphabet to specify the points at which they are to be dropped. An address is a crude map, but it is sufficient for the job. You *expect* a letter to reach its destination if addressed (modeled) properly—and it usually does.

This raises the question of why the social sciences have encountered so much difficulty in providing the world with models that are accurate representations of the reality they seek to describe. To gain some understanding of the problem, we need to consider a few of the differences between *mechanical* and *probability* (or *probabilistic*) models.

The most effective models in the natural sciences are mechanical models. The least effective are probability models. There are myriad other models that come from outside natural science—from religion, art, fantasy, madness, opinion, and so on. These models are often asserted with mechanical rigidity and given an apparent mechanical quality. However, like a fantasy map, if you allow many of these models to determine your expectations, they can lead you into strange places.

Fantasy maps are generally rejected by scientists, technicians, and people who like to think of themselves as realists. In the world of social belief, however, a fantasy map can continue to work insofar as it has the odd ability to create the very terrain it maps. For example, social psychologists know that people tend to believe they are what other people tell them they are. You are given a map, so to speak, of yourself. A young boy might be told, for example, "You are brave and never cry, even when hurt." The child's map has the peculiar ability to become a part of the child's character. This is mentioned simply to point out that nonscientific models or maps also have a purpose and cannot be lightly dispensed with.[2]

The most generally useful and powerful of the various models that have come out of modern science, as well as Western culture generally, have been the classical mechanical models. These "clockwork" models have influenced virtually every facet of modern technology. So popular is the appeal of mechanical models that they are still relied on to provide us with an understanding of extremely complex events such as the dynamics of fluids or transformations taking place in the earth's biosphere.

Mechanical models have been sought in economics and the social sciences since the middle of the nineteenth century. However, these models have not been successful in fields such as sociology and psychology. Skinnerian behaviorism, perhaps, was one of the last powerfully influential mechanical approaches to human conduct. Psychologist B. F. Skinner wrote of the new Walden, a utopia based on mechanical learning through reinforcement. So far, though, the new Walden seems to be as elusive as it has ever been.[3]

While mechanical models are much sought after, it appears that as one moves more and more into the realm of the purely social—as opposed to the physical realm—events become "loaded" with large elements of chance (that is, unexplained variability). This weakens strictly mechanical descriptions of cause and effect relationships.

Using a probabilistic map is somewhat like trying to rely on an ordinary map that tells you that if you go in a particular direction you have only a degree of probability of reaching it. Most coefficients of correlation in the social sciences are less than 0.70. This suggests that fifty percent or more of what we are trying to understand has not been well defined or located. If, for example, total variance were represented by a distance of a thousand miles on a map, the social scientist's model would locate a point somewhere within half of that distance—a five hundred-mile segment.

Suppose you wanted to get from Mexico City to Chicago on the basis of your intuition. You really have little idea of where Chicago is, except that it is somewhere vaguely "to the north." If you rely on your intuition, you will have a range of error in reaching your destination that extends one thousand miles on either side of the city to the west and to the east. Then someone gives you a probability map that is correlated at 0.70 with your destination. Now you might end up five hundred miles on either side of the city. You are doing a lot better, but such a map would still be a limited device for practical use.

If the thousand-mile error represents your initial variability, then the five hundred-mile error represents your correlated map range of error. It is a lot better, but the heavy stochastic (probabilistic) element involved means that you still have to worry about whether you really are going to make it to your destination. Figure 15.1 shows the range before and after you use a correlation as a model of the terrain. (This example is purely illustrative and is offered as a cautionary tale for those who begin to think that they have resolved matters when they have attained a strong degree of correlation between variables. It is a start, but it still leaves a lot of questions unanswered.[4])

Figure 15.1. An Illustration of the Utility of a Correlation of $r = 0.70$

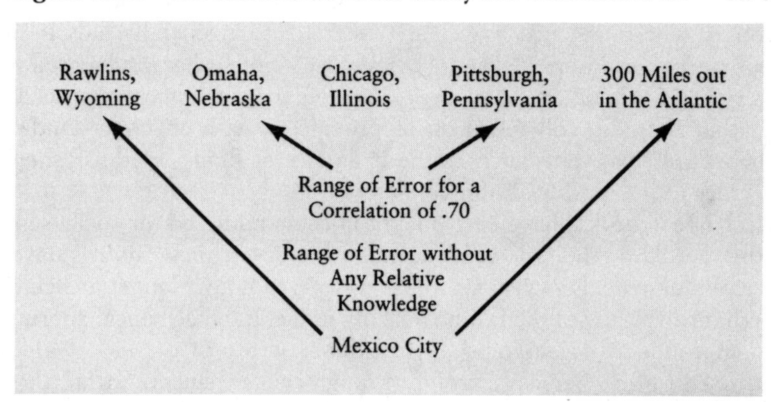

A correlation of 0.30, one that is often probabilistically significant, would generate a nine hundred-mile segment of uncertainty on either side of Chicago for our road map. Remember, if you did not have the map you would be even worse off. Without the correlation, you could land three hundred miles to the east in the Atlantic Ocean. With the correlation of 0.30 you probably will not land any more than two hundred miles out in the ocean. Correlations help.

However, you should be very careful not to get caught up in the belief that you are dealing with rigidly mechanical systems when you are dealing with correlated systems where the values of *r*, or other indexes of relationship, are even slightly below 1.00. Even an extremely high correlation of 0.90 leaves us with nearly 20 percent uncertainty.

In sum, probability maps are not completely reliable when it comes to providing us with a good sense of what to expect from our knowledge of a set of given conditions. They are, however, better than nothing.

THE EFFECTIVENESS OF MODELS

There is one last observation that needs to be mentioned in discussing the idea of models as they appear in the natural sciences and in the social sciences. Social scientists have long been aware of what they refer to as the self-fulfilling prophecy. When we have a map or a model of how gas reacts to temperature within a closed container, the gas is not affected by the model we have of it. When we create a map of human conditions, however, the map becomes a part of that which it is describing. (We alluded to this tendency in our earlier discussion of fantasy models.)

In economics, for example, model makers (that is, economists) hesitate to tell people that their models (maps) suggest that a total collapse of the market is about to occur because this expectation could in turn lead to a market collapse. You cannot be certain that a map, itself, will not actually create the conditions it is supposed to merely describe.

Model building in the social sciences is faced with serious problems. This does not mean that the attempt to create a model has been a waste of time. Much has been learned from statistical studies of human actions—at both the psychological and the sociological levels. One thing that has been discovered in *all* such studies is that social reality cannot be reduced to simple mechanical models. There is, without exception, a large amount of chance in any description of human activity. As the saying goes, nothing is certain except death and taxes—and there are even stories of people who manage to avoid paying taxes. Death, of course, is a bio-physical variable—subject to social forces, but physically inevitable.

This in itself is a major finding of profound significance, one that we have not seen too many social theorists really accepting or evaluating in terms of its deeper philosophical and moral significance. We do not want our priests or social advisors telling us things like if we are good we have a probability, significant at the 0.05 level, of going to heaven, or if we work very, very hard, we might succeed or we might not—it depends. Social policies and traditions are based on models of life that are often unreal, but are effective in getting people to act. Statistical studies are more realistic, but they always suggest that success, in any venture, is uncertain. This can have a chilling effect on social action.

Certainly the American belief that we can control our fate is not compatible with the findings of the modern social scientist. Social research findings show that no matter how you try to analyze social forces, a huge element of uncertainty always remains. Our finest analyses come up with correlations in the range of 0.70 and 0.80. This is an extremely high and rare result in social research. However, even when $r = 0.80$, a third of what we are trying to understand is still a matter of chance and uncertainty.

A NOTE ON PRE

It is generally the case that correlations in the behavioral sciences rarely exceed $r = 0.70$ and usually fall far below that level. This suggests that the predictability of social affairs is extremely chancy. To some extent this is true. Such things as communication, social identity, success, and other human activities have a high degree of inherent chance or probability involved in their outcomes.

At the same time, we have to look to another logic, that of proportional reduction in error (PRE), discussed in Chapter 13, to see that when we move away from attempts to be extremely precise—when we rely on "coarse" evaluations, predictability can be quite high. It is not at all unusual to attain predictability in excess of 99 percent efficiency. One can predict that all of the students coming into a classroom will be wearing some kind of clothing and be right nearly all of the time. This form of prediction is the basis for social relations and is sufficient. If we move toward greater precision with respect to our predictions, they seem to become more elusive.

A disturbing question comes out of this discussion of models. If precise probabilistic statistical models are unreliable as a basis for our expectations, then what else do we have to turn to? It is the old humanistic issue of whether we should be guided by reality or by dreams. Should we be led by the facts or by our prejudices? Should we rely on data or on stereotypes in our relations with each other? What sort of map is best for dealing with the complicated and labyrinthine pathways of human social affairs? Is there any "best" map? If there isn't, should we simply use whatever is at hand?

In the middle of the Nineteenth Century, the social sciences began to turn toward the vision (actually a kind of fantasy model) that there was a way to produce better social maps. At first the job seemed easy enough. Get information. Carry out research. Remain empirical. Seek precision. In sum, do all the things that any map maker knows are important. However, a century and a half has passed and the question still remains: is it possible to produce truly valid social maps? The position that we take in this book is that this question still merits serious debate.

Art and fantasy provide us with powerful models of social action. These models are never rigorous in the extent to which they measure and correlate significant variables. Nonetheless, the maps that artists provide us of the world and of ourselves can be extremely moving. Whether artful or scientific models are more appropriate for dealing with political, economic, and moral issues is not obvious. Perhaps the most reasonable view to take at the moment is that we should be aware of the relative strengths and weaknesses of both approaches.

Artistic models have as their major strength an ability to simplify moral and social issues and generate empathetic responses. Artistic models of social reality not only show us how to fall in love, they also motivate us to want to fall in love.[5] Artistic models can play with the problem of the extent to which an apparently trivial event ultimately determines the fate of a character in a drama. Artistic models not only show us a terrain, they make us want to venture forth into it—to fall in love, be successful, fight the villains, display honor and courage, help the oppressed, avoid greed, and so on.

The models of the social sciences, on the other hand, have as their strength a reliance on broad samples rather than idiosyncratic moments. They scrupulously try to avoid creating evidence that supports the basic theme of the research. As we have seen in previous chapters, artists are not the least bit hesitant to use artistic license to imagine events that substantiate the themes they are trying to promote.

Whenever we think of the arts and the social sciences we think of great drama on the one hand and solid statistical studies on the other. Cervantes' *Don Quixote* certainly ranks among the greatest works in world literature. It provides insights into the extent to which people are driven by fantasy models of reality, and it is an artistic mapping of the fantasy maps that shape our lives. (Recall that in the musical version of Cervantes' masterpiece, *The Man of La Mancha,* the good don enjoins us to "Dream the Impossible Dream.")

On the other hand, *The Statistical Abstract of the United States* consists of nearly 1,500 statistical tables. It reveals facts about the modern condition that cannot be revealed by fiction, by art, by journalism, or by anything else. The job can only be done the way *The Statistical Abstract* does it. In sum, there is room for both kinds of models (just as there is room for both classical music and rock and roll). To ignore the best that either form has to offer is to miss out on something.

The primary effort of intellectual work, whether in the arts or in the sciences, is to provide people with various maps. There was a time in the recent past when economists and sociologists talked about mapping the economic and social world in much the same way as astronomers go about mapping the heavens. One sociologist had his students describe in great detail everything they could possibly identify that would characterize a given, quite brief, interaction. The result was a map of such detail, complexity, and extraneous information that it wearied the reader with its ponderousness. Worse yet, it was confusing because it was too detailed. The question remains: what sort of map is best for dealing with social phenomena?

We (your authors) do not think that really good maps of human social affairs are possible—either through the arts or through statistics. For this reason we believe that human social relations will always be contentious. The social world is too intricate, abstract, and subtle to be reduced to simple and, at the same time, accurate maps. What is called for is a flexible approach. We should draw on the insights of artists, philosophers, moral leaders, and theorists, *as well as* the logic and data of statisticians and scientists. We should make our own estimates of how to set our course. Even as we do so, we should be aware that no matter how carefully we identify our position and our destination, we are likely to arrive at a point considerably different from the one we anticipated.

Research findings have shown that *probability is invariably an inherent part of any social exchange.* Thus, you should always expect the unexpected. Such advice can seem much too general. We would like more specific and more accurate maps. However, as we get more specific and more accurate about one problem, we also lose information by the trainload about others. There is no way out. Statistics is the development of a solid understanding of *both* what we know and what we do not know.

CONTROL

The most intricate and troubled aspects of statistical technique appear in the effort to get rid of, or **control,** extraneous influences that produce a kind of static or noise when we are trying to see how things are related. The classic example of an extraneous influence is the relationship between ice cream consumption and crime rates. It is a well-known statistic that in any given community, as ice cream consumption increases, the crime rate increases. Obviously something is interfering here. Ice cream as a locus of evil does not make sense. (Notice how our theoretical awareness forces us to reject certain causal implications of a strong correlation. Statistics should never be divorced from theory—and vice versa.) Everything is cleared up when we control the temperature. Ice cream is eaten in greater quantities in the summer, and crime rates also increase in the summer. When we take temperature into account, the correlation between ice cream consumption and crime disappears.

How do you take temperature "into account"? One way would be to gather ice cream and crime statistics for days that have the same temperature—let's say 95 degrees Fahrenheit. Obviously, temperature is no longer a variable condition. It is a constant. *Constants cannot be correlated with variables, by definition.* So, the influence of temperature has been removed from the scene.

In the remainder of this chapter, we will focus on four commonly used statistical control procedures. One of these is **cross-classification.** Another is **standardization.** The third is the use of **matched samples.** The fourth is relatively complicated, but the logic is straightforward. It is a variant form of cross-classification, but it goes under the name of **partialing.** The idea here is to "partial out" the effects of extraneous factors.

Partialing

Let's consider this last control technique, *partialing,* first. We will illustrate it as simply as possible with the ice cream and crime example. Hypothetical figures are used here to clarify the logic. In its more complex forms, this type of statistical control becomes extremely complicated and difficult to interpret. The point here is simply to get a sense of the logic that underlies the technique.

Suppose we have fourteen communities. We get data from the communities on the average number of quarts of ice cream consumed each day, the crime rate per 1,000 population, and the average temperature for a given year. Let's say our data look like those presented in Table 15.1.

Table 15.1. Fictional Data for Showing the Elemental Logic of Partialing

Community	*Ice Cream Consumed (quarts)*	*Crime Rate per 1,000*	*Temperature (°F)*
A	10	15	30
B	15	25	30
C	20	20	40
D	25	30	40
E	30	25	50
F	35	35	50
G	40	30	60
H	45	40	60
I	50	35	70
J	55	45	70
K	60	40	80
L	65	50	80
M	70	45	90
N	75	55	90

Obviously there is a strong correlation between ice cream consumption and crime rates. There is also a strong correlation with temperature. In this example, everything is correlated with everything. You must rely on a broader understanding of the variables involved to make sense of what is happening.

When we correlate the crime rate with temperature we get a scattergram like that in Figure 15.2. For any value of temperature we have a predicted or estimated average value of crime, designated *E* in the scattergram. Notice that this estimate is the average for any level of temperature, and there is still some variability around the average. Crime values still vary, but they do not vary as much when we know the value for the temperature as when we are ignorant of this value. This is called *residual variance,* and it is the variance that is left over after we have obtained a correlation. If the correlation were perfect, there would be no residual variance.

Now the question is, can we find something that will correlate with the residual variance? When we try to correlate ice cream consumption with the residual values, we get a scattergram like that in Figure 15.3.

Ice cream consumption does not correlate with the residuals. In other words, after you correlate crime with temperature, what you have left over

Figure 15.2. Scattergram Using Data in Table 15.1

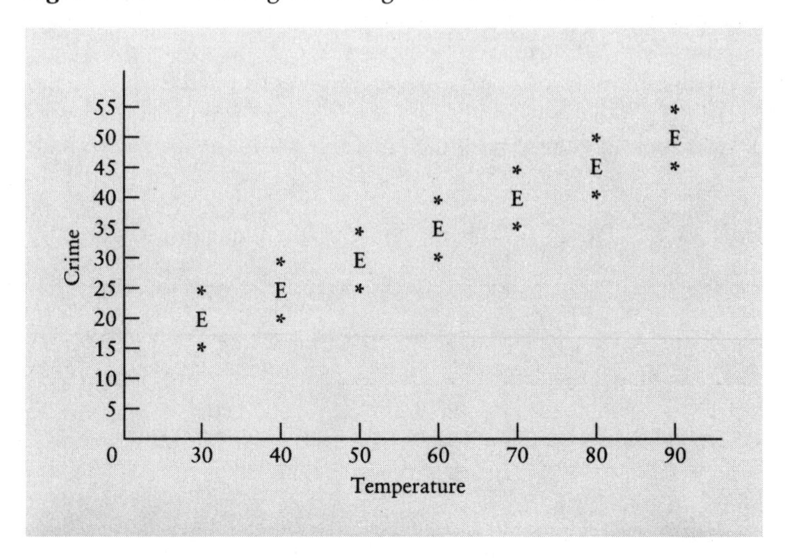

Figure 15.3. Residuals in Figure 15.2 Correlated with Ice Cream Consumption

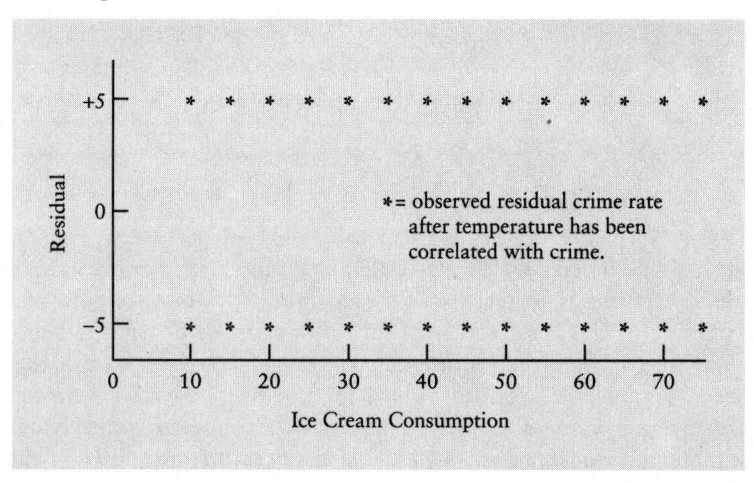

does not correlate with the temperature. It might correlate with something else. The influence of temperature has been taken into account by correlating it with the crime rate and then dealing with the variance that is left over (the residual variance).

This is the primary logic of a procedure known as **partial correlation.** By partialing out the effects of one variable, we try to determine if other variables are still capable of maintaining a relationship with the dependent variable. Partial correlations are subject to the problems that beset any kind of correlation analysis—bias from extreme cases, sampling variability, and, of course, the need for good interval or ratio scales.

Cross-Classification

Cross-classification is the simplest, most direct, and also probably the most commonly used form of statistical control. Many of the tables presented in *The Statistical Abstracts of the United States* control for factors such as race or region by cross-classifying data.

With cross-classification we begin to see that statistical control is really an extended development of the idea of relationships. Relationships are an attempt to bring variability under control. When a dependent variable is freely varying, our best guess as to the value of an individual observation is a function of the extent to which the variable, in fact, varies. When we know that there is an independent variable related to the dependent variable, we can use that knowledge to make a more controlled estimate of an individual value. Cross-classification provides a statement of relationships when it involves two variables, and it moves into the realm of multi-variable statistical control when it involves more than two variables.

Table 387 of the 1990 *Statistical Abstracts* offers the data shown in Table 15.2.

Table 15.2. Characteristics of U.S. Readers in Percentages, 1983

	Book Reader	*Nonbook Reader*	*Nonreader*	*Total*
All persons	50	44	6	100
Male	42	52	6	100
Female	57	37	6	100

There is a modest association between gender and book reading, such that women are more likely to be book readers than men. What might influence this relationship? Perhaps women spend more time at home and have a greater opportunity to read books. Perhaps women read fiction while men read technical reports. Perhaps men are more interested in reading about sports, which are reported on in newspapers, while women are more interested in literature. Perhaps women are simply brighter. All sorts of possibilities come to mind.

If we say that women read more books than men, an argumentative person might reply, "Yes, but men don't have as much time for book reading." Or, "Yes, but that's because men would rather be out playing than indoors reading." When someone says, "Yes, but . . .", it usually means the argument is about to move from a simple two-variable form into a multi-variable form.

Suppose we were able to collect further data on the reading habits of the American reading public. Suppose further that we found the information presented in Table 15.3.

Table 15.3. Characteristics of U.S. Readers, in Percentages, Cross-classified by Time Spent at Work Outside Place of Residence, 1993 (Data are fictional.)

	Time Spent at Work Outside Residence							
	Less than 10 Hours/Week				*40 or More Hours/Week*			
	Book Reader	*Nonbook Reader*	*Non-reader*	*Total*	*Book Reader*	*Nonbook Reader*	*Non-reader*	*Total*
Men	65	30	5	100	35	55	10	100
Women	65	30	5	100	35	55	10	100

When time spent outside the residence is taken into account, the association between gender and reading habits disappears. (Keep in mind that the data in Table 15.3 are purely fictional and illustrative.) This suggests that it is available leisure time, rather than gender, that influences reading.

Cross-classification has the advantage of being simple. Its logic is straightforward. If you examine a relationship while holding some third factor constant (in this case time spent outside the residence), then that factor cannot be influential because it is a constant. You cannot relate a constant to a variable condition.

The problem with cross-classification becomes evident even in the simple illustration in Table 15.3. The problem is that the more variables you add to the list of cross-classifications or the more detailed you make any cross-classification, the more complicated your table becomes, until it is quite difficult to make sense of it at all.

Standardization

Entire courses are offered on *standardization,* which is a statistical technique used across a broad range of disciplines, from the natural sciences to economics, demography, and the social sciences more generally.[6]

The logic of standardization again has to do with transforming a variable condition into a constant. In this instance, the task is done by making calculations that determine how conditions would appear among a variety of events *if* each of the events had some standard or common value. This might sound ominous, but we are all familiar with and make daily use of one of the most common of standardization procedures—the simple percentage. The percentage tells us what things would look like if all of the cases we are interested in came from groups with a standard value of one hundred.

Suppose George makes thirteen hits during the season. Helen makes fifty. Gregory makes seventy. Leona makes twenty-nine. Which hitter is the best? It depends, you say (another one of those ordinary phrases that warns us that we are about to complicate things a little). Depends on what? It depends on how many times each person was at bat. We find that George batted forty times, Helen ninety-three times, Gregory two hundred and ten times, and Leona forty-three times. Their averages (there's no getting away from that word when you are describing things) are reported as:

Batting Average	
George	.325
Gregory	.333
Helen	.538
Leona	.674

Batting averages are reported with a base equal to one. If the base were one hundred, then George would have a 32.5 percent average and Leona would have a 67.4 percent average. Common bases for this kind of simple standardization are 1, 10, 100, 1,000, 10,000, and 100,000. The principle is always the same.

Batting averages are interesting. When major league baseball is presented on television, the commentators tell us that such and such player is batting .236 against left-handed pitchers and .294 against right-handed pitchers. They might then add that the hitter bats .217 when no one is on base and bats .309 when someone is on base. Here is a place where the problem of control and standardization sometimes leads commentators to lapse into odd parodies of statistical analysis.[7]

The simple percentage is also another example of an average, but one that applies to populations where we *pretend* that the total set of observations in the population is always equal to one hundred. The ordinary percentage, simple though it is, is a place where the average, variability, standardization, and probability all fold in on one another in one of the most commonly used statistical indexes of all.

Once again, it is good to keep in mind that the percentage is a simple construction. It is a kind of fiction. It tells us how things would look *if* everything came from a population with a value of one hundred. The extent to which the percentage can distort things becomes increasingly evident when the real population from which it is drawn is relatively small.

Consider the course evaluation table included in Chapter 4 (page 73), which reported the number of students from each class who filled out the evaluation. One class had only one student, and that student filled out the evaluation—in percentage terms, a one hundred percent response rate. A good rule of thumb is not to calculate percentages for populations of less than one hundred. However, this rule is commonly violated.

Distortion works in the other direction as well. If you have large values and want to hide them in some manner, you can often do it by presenting them as relative values, such as percentages, rates per 1,000, or some other equivalent. For example, Americans spend large absolute sums of money, in the billions, on pets each year. However, in relative terms, the amount spent is minuscule, amounting to something like one percent of one percent of the gross national income.

Absolute values and relative values are both proper measures of a quality.

However, they produce results that look quite different. The appearance of such results depends on the base. A thousand dollar fine for parking is severe, in absolute terms, only if you are poor. If you have wealth in the billions, it is nothing. The law is equitable: it fines the poor in exactly the same manner as it fines the rich. This example, once again, reveals the place of statistics in ordinary affairs, because fines are almost always absolute rather than relative. Here is a place where a simple statistical decision, to use absolute or relative values, has profound moral, legal, and judicial implications. Absolute fines are much more punitive for the poor. There should be a movement for relative fines. Such fines would standardize the punitive intent of fining. It would also bring more money into the communal coffers. (However, don't hold your breath waiting for people to act on this statistically grounded injustice.)

Another straightforward example of the logic of standardization can be seen in the use of standard dollars. Any attempt to compare the prices of goods and services in, let us say, 1940 with those of 1990 will be influenced by the fact of inflation.

Table 15.4 provides a list of some dollar costs for various items for two time periods in American history. The third column presents costs for 1948 if one were using 1991 dollars to buy things.

Table 15.4. An Illustration of Standardization: Taking Inflation into Account

	Cost of Selected Items		
	1948	*1991*	*1948 costs in 1991 dollars*
Three-bedroom home	$7,700.00	$94,300.00	$92,400.00
Average income	3,187.00	34,213.00	38,244.00
Price of new Ford	1,150.00	11,835.00	13,800.00
Gasoline, 1 gallon	0.26	1.20	3.12
Bread, 1 pound	0.14	0.71	1.68
Milk, 1 gallon	0.88	2.72	10.56
Postage stamp	0.03	0.29	0.30

If we set the 1948 dollar as having twelve times the purchasing power of the 1991 dollar, *on the average,* then we can make some interesting comparisons between the two periods. We can take inflation into account by making the purchasing power of the dollar constant in 1991 dollars. By doing so, we see that housing was cheaper in 1948 and incomes were relatively higher. An automobile cost relatively more. Milk was astonishingly expensive in 1948 or amazingly cheap in 1991, depending on how you look at it. In constant dollars, a gallon of milk would have cost over ten dollars in 1948, while in 1991 it cost less than three dollars.

Notice the nature of the "what if" game that is being played here. In this instance, standardization asks, what if you established an *average* inflation rate and used it to establish a dollar value with a constant purchasing power? However, different parts of the economy inflate at different rates. Medical costs and baseball players' salaries skyrocketed in the past decade while the costs of gasoline and farm products inflated at a much slower pace. Extreme values can distort averages. As with the cases described in previous chapters—Nevada with its extreme divorce rate, and Washington, D.C. with its extremely high proportions of lawyers and attorneys—an important question arises: should we leave hyperinflationary figures in the distribution, knowing that they skew values, or should we not violate the ideal of using all data and keep them in?

In the example above we casually assigned a constant inflationary rate of twelve to the 1948 dollar to standardize it with the 1991 dollar. Achieving a more careful standardization requires going back to the most basic statistic of them all—the average—and trying to find which value, over the spectrum of economic activities, best represents what is meant by an average inflationary rate. This is a technical problem of some complexity, and we do not need to deal with it in detail here. The important thing is to recognize that the standard dollar is an average. How that average is calculated will affect comparisons.

Before leaving standardization, let's consider a typical standardization procedure that is commonly used in a variety of contexts. Suppose we want to compare two countries in terms of fertility, which we will measure as the number of live births per 1,000 women per year. One community has a birth rate of 22 births per thousand women each year. Another has 27 per thousand per year.

However, this comparison is not fair in a sense because the second community has a larger percentage of young women than the other community. Standardization, in this instance, asks the question, what would the rates look like if the two communities had the same female age distributions? Table 15.5

Table 15.5. Fictive Data for Illustrating Standardization by Equalizing Categories within Two Populations

Oakdale			
Age	***No. Women***	***No. Births***	***Birth Rate***
15–30	2,500	75	30
30–39	7,000	175	25
40+	6,000	90	15
TOTAL	15,500	340	21.93

Elmsville			
Age	***No. Women***	***No. Births***	***Birth Rate***
15–30	15,000	160	32
30–39	6,000	150	25
40+	1,000	15	15
TOTAL	12,000	325	27.08

presents a simplified form of this kind of standardization. Here the birth rates are weighted by the single set of relative proportions of women in the different age groups.

When we look at the unadjusted or unstandardized birth rates for these two fictional, illustrative communities, Elmsville has a larger birth rate. However, Elmsville is a young town and Oakdale has a much older female population. Perhaps the difference is simply due to the different age distributions for the female populations in the two communities.

Because the rates are made the same for the age categories in this illustration, it should be obvious that if we make Elmsville's female population have the same distribution as Oakdale, we are going to get the same overall birth rate. This is what happens when we standardize fertility. What is important here is to see the basic technique that is used to achieve statistical control. We ask the question, what would Oakdale's fertility rate look like if it had the same age composition as Elmsville? To standardize, all we need to do is pretend Oakdale has a similar age composition to that of Elmsville, using the same rates for Oakdale that were in the original table (see Table 15.6).

Table 15.6. Oakdale's Fertility Statistics for a Female Age Distribution Like That of Elmsville

		Oakdale	
Age	*No. Women*	*No. Births*	*Birth Rate**
15–30	5,000	150	30
30–39	6,000	150	25
40+	1,000	15	15
Total	12,000	315	26.25

*Note that age-specific birth rates stay the same.

The relatively low fertility rate for Oakdale was a result of its age composition. When age was held constant by pretending Oakdale had the same age distribution among its women as Elmsville, the fertility rate went from 22 to 26—a value similar to Elmsville's rate. In standardization of this kind you use hypothetical weightings to achieve similarities between groups for some given characteristic—in this case the age distribution of the communities. Notice, once again, the "what if" nature of this statistical argument. *What if* Oakdale's women were distributed, by age, the way the women of Elmsville are? If that were true, what would the fertility rate for the total community look like?

The thing to keep in mind with standardization is that it relies on this game of "what if." What if all cases for each group consisted of precisely 100 events? What if the age composition of the United States were the same as that of Southeast Asia? What if you were buying today's goods with a 1940 dollar? Or, what if you were buying 1940 goods with today's dollar? Standardization gives us a chance to make comparisons that would otherwise be obscured. It plays with reality. It is also a kind of model. Standardization tells us what we might have expected *if* such and such had been the case.

Statistical Matching

The logic of *statistical matching* argues that if you begin with two groups that are perfectly identical and then subject one to an experimental condition while withholding the experimental condition from the other group, any differences that appear between the two groups can be attributed to the experimental condition. *All* extraneous influences have been canceled out.

Biological studies are perhaps the most famous examples of the use of matched experimental populations. Experimenters have gone to great lengths to produce cloned laboratory rats, mice, or other creatures. Clones are nearly exact duplicates of each other. If you administer a carcinogenic agent to one group of clones but not another (maintaining constant living conditions for both groups) and then discover that the first group develops tumors, you have strong evidence that the carcinogenic agent was the single responsible agent. No extraneous factors have been allowed to intrude. The environments, the food, and the animals are all carefully regulated and matched.

In psychological and sociological studies that deal with human subjects, attempts at matching can only be approximate. We can compare two similar nations, for example, before and after one of them institutes a new policy. However, the length of time required to observe the effect of a national policy, along with the approximate nature of such matching, make such statistical studies far less precise than the use of matching in, for example, biological research.

SUMMARY

Our discussion of mechanical and probability models, along with our brief, elementary discussion of control, suggest that social statistics are relatively imprecise. This is true. However, there is no set of models, no logic, no mathematical technique that can enhance precision much beyond the level it has already reached. Statistical analysis (along with good old ordinary life experience) teaches us that social reality is not perfectly predictable or controllable.

This does not mean that we should abandon statistics. It also does not mean that we should pour more energy into a quest for exact precision where it simply does not exist. This *does* mean that we should proceed with caution, no matter what our basis of understanding might be—statistical or literary. It means that when it comes to understanding our social selves, we should be mentally flexible. We should know when and where a good data base is worth more than an elegant poem—but we should also realize that there are times when a good poem is worth more than a carload of statistics. Such knowledge does not come from a study of techniques, though one cannot get by without some understanding of statistical technique. It comes, if it comes at all, from constantly reviewing the current state of knowledge. It comes from experience.

Thinking Things Through

1. Models are commonly complex typifications though they can also be quite simple. The primary thing they do is to generate expectations. Describe, as briefly as you can, two different models of the condition Americans refer to as "success." How do different cultural models of success produce expectations or responses that differ from your own personal experiences? Where do these models come from? (Notice that when people say, "Where'd you get *that idea,*" they are often asking you to define some kind of model.)
2. As you get older, your medical expenses increase. Like most generalizations about human affairs, this is only partly true. There are always exceptions. How would you take these exceptions into account? How would you go about doing this in a study of the relation between medical expenses and aging? How would you standardize for sex differences in medical expenses?
3. Why are psychologists running into so much trouble in their efforts to find a standardized measure of intelligence?
4. In the United States, there is a major controversy over whether high school students should be required to take a standardized knowledge test before being allowed to graduate. There is also talk of standardized tests being used to determine whether or not universities and colleges are being educationally productive. Suppose you have gone on to acquire a Ph.D. in the social sciences, and you have gained fame as a wise person. A federal committee asks for your professional opinion with regard to making these tests a part of national educational policy. What would you say in a three-page report to the committee?
5. Games are sociologically interesting because they are highly standardized social activities. They are also of interest because people are, generally, fascinated by games. Do you think standardization has something to do with this? Why are games so highly standardized? Why not let people play games however they wish? For example, you could make up whatever number of strikes you think is reasonable before being counted out in baseball. Is standardization a disguised form of authoritarian control?
6. While standardization has been discussed in this chapter as a specific statistical technique, it actually permeates nearly all statistical operations. For example, we've considered *standard* deviations and efforts to obtain *standard* measures of association. Standardization is also important in ordinary argumentation and rational thinking. For example, if you think members of one ethnic group have a greater aptitude for success in college, you might want to standardize for the extent to which the family environment sustains a commitment to intellectual activities. Give an example of how problems of standardization might be revealed in an ordinary conversation.

Notes

1. This aspect of models is definitely worth thinking about. What it suggests is that since all human communication relies on models of some kind, and models are necessarily distorted, there is, as a consequence, an inherent element of distortion in *all* human communications. This is particularly the case when we are communicating about social qualities.
2. In this discussion we are approaching the topic of the social creation of reality. This philosophy is rarely, if ever, examined from a statistical perspective. However, like any other human intellectual effort, the concept of the social creation of reality relies on the logical syntax that defines statistical procedures, and it can gain greater logical rigor by taking that syntax into account.
3. Skinnerians are fond of thinking of people as elaborate machines. This school of thought is a tribute to the power of mechanical metaphors. Perhaps people are machines—highly complicated ones. If so, then eventually psychology should be able to attain the same mechanical precision as electronics or ballistics. It is the argument of this book, however, that all the studies done by all the social and psychological sciences over the past century have not come anywhere close to findings with mechanistic precision. Reading the work of B. F. Skinner offers insight into just how potent the mechanical metaphor has been in Western thinking throughout the Twentieth Century. See, for example, B. F. Skinner, *Beyond Freedom and Dignity* (New York: Knopf, 1971), *Science and Human Behavior* (New York: Free Press, 1965), and *Walden Two* (New York: Macmillan, 1948).
4. The mathematics behind this illustration are much more complex than we need to go into. Actually, hitting or missing the objective is a randomized normal event. Extreme misses would be somewhat rarer in the correlated case, but there would still be a lot of misses.
5. Many Americans learned how to kiss not by reading scientific sex manuals or statistics on kissing but by watching professional film actors do it on the screen. We might not be inclined to think of love scenes in films as models, but they are. We are then led to wonder how cinema, in general, has affected human actions as a popular presenter of models that not only describe a broad spectrum of actions, but do so with the added effects of dramatic intensity.
6. In a sense nearly all statistical work involves standardization in some form. Correlation coefficients attempt to find standardized ways to refer to relationships; the standard deviation is a standardized approach to variability; science, whatever its topic, seeks standard measures. What is interesting is to speculate on the question of why standardization is such a highly sought-after goal.
7. This nearly absurd reporting of statistical details is mentioned to remind you that people are not always antagonistic to statistics. It simply depends on the context. In this instance the baseball fan is entranced by figures that, when subjected to deeper thought, are found to be of extremely dubious value.

16

CLASSIFYING AND COUNTING: FUNDAMENTAL PROBLEMS IN REASONING

"I hate definitions."

—Benjamin Disraeli

"When you can measure what you are speaking about, and express it in numbers, you know something about it; but when you cannot measure it, when you cannot express it in numbers, your knowledge is of a meager and unsatisfactory kind: it may be the beginning of knowledge, but you have scarcely, in your thoughts, advanced to the stage of science."

—Lord Kelvin

"The best criticism is to do the job yourself—and do it better. Anything else is just steam"

—Donald Bogue

STATISTICS AND STORIES

Our discussions of statistical forms in this book have been only an "introduction to an introduction." The modern study of statistics is so complex that no one has a total command of the subject. There is always more to discover. It is like learning Japanese. You are not going to become an accomplished speaker of Japanese after a semester spent seriously studying one or even a dozen books on the subject. It is much the same with statistics. An understanding of statistical methods calls for reviewing more than a few books. In addition, a few years spent doing statistical research is rewarding. The more you practice statistical thinking, the more you begin to see that there is always something new to learn—and that, in an informal way, you can practice statistical reasoning anywhere at any time. You begin by thinking that an arithmetic mean is easy enough to understand and simple enough to determine. Eventually you discover it is possible for a set of figures not to have a meaningful typical value, and that supposedly simple things are more complicated than you've been led to believe.

It is time to bring our discussions to an end. We have elected to close this introduction to an introduction by returning to a subject that was discussed at some length in Chapter 2: the matter of categorization. The reasons for this are that, first of all, categorization is a much more serious and difficult problem than the simple logical operations of statistical analysis. Second, most statistical texts presume that the problem of categorization has been attended to and go on from there, leaving a lot of students unaware of just how relevant categorization is, and how incredibly complicated it can be. Third, the topic of categorization reveals why stories or literature as social descriptions or models are complementary rather than antagonistic to statistical studies. There is absolutely no way the story can be avoided as a descriptive form. But once we gain a comprehension of the *limitations* of stories as descriptive forms, we gain a better awareness of why statistical forms are also necessary. We also begin to see that each form—the story or the statistical table or index—is always an intellectual compromise based on arbitrary decisions and, therefore, always subject to controversy and criticism.

When you are merely illustrating a statistical technique, it does not make much difference what kinds of examples you use. You can go ahead and pretend the problem of categorization has been solved and concentrate on the logic of a particular technique. The discussions in this book make use of casually established or even fictional categories on a number of occasions. There is nothing wrong with this when the task is simply to illustrate a particular technique or the logic underlying some index. In serious work, however, the success of the entire enterprise depends on how well you have ordered your observations into classes, categories, or scales. If this has been done improperly, the rest of the statistical procedure is a waste of time. Junk categories produce junk statistics.

Numbers are powerful devices for communicating. They provide us with a sense that the maps we are using are good ones. When they appear on the scene, they tend to dominate the stage. Whether we like them or not, numbers are convincing in an argument—and one of the reasons for using the statistical method is to prevail in arguments. We are impressed by numbers because we are intuitively aware that precise classifications move in the direction of numerical forms. We then conclude that something with a numerical form is precise. This, of course, can be a highly fallacious conclusion.

Our intuition is correct when we comprehend that precise classifications have quantitative forms. This is, in part, the power of modern natural science: its categories are expressed as numbers. It is not valid, however, to conclude that whenever something is expressed in terms of numbers, it naturally follows that precision has been attained. Precision is not simply a matter of numerical expression. It is a matter of the extent to which a given classification device—a name, a class, a term, a quality—includes only that to which it refers while excluding that to which it does not refer. In other words, within-category variance should equal zero in a truly precise instance.

The U.S. Bureau of Labor Statistics publishes labor force data. Until quite recently, these data did not classify private household workers as a part of the labor force, raising the following question: is the work done in the private household, usually by housewives, a part of labor? If it is, should it be included as a part of the study of labor in the United States? If it is not, what differentiates it from labor? If you argue that the labor force is made up of people who get paid, then you are not measuring labor but something quite different. There might also be the implicit suggestion that those who get the most pay do the most work—and that is patently false. Incidentally, it is still the case that even if you get part-time pay, you are included among the fully employed in employment statistics. Something is wrong with this. What is labor? How do you count it?

It should be evident to you by now that a great variety of economic, psychological, and sociological terms are aggravatingly loose in their meanings. They are loaded with ambiguity. Simply counting ambiguous entities does not transform the initial ambiguity into greater precision. On the other hand, so powerful are the advantages of having a count, that the problems of ambiguity sometimes simply have to be swept aside (instead of becoming a source of philosophical hair-splitting).

Even when you have categories that are not particularly ambiguous, problems remain. The simple population counts obtained by the U.S. Census Bureau are instructive. You can count people. You can define a person as any entity that is alive and born of a human mother. This is a relatively unambiguous definition. It permits accurate counting to take place. It is a physical definition. Physical events are much easier to define than truly social ones, and we can assign meaningful numerical values to them. Still, the problem remains. If we count the number of people living in the United States, have we counted the same thing as we might count if we counted the number of people living in China? Or in Ethiopia? Or in Greenland? Even in a limited, physical sense, these populations differ. The population of the United States, for example, is physically and socioeconomically different from the population of China in a wide variety of ways.

If this sounds like niggling, we have to think about what we mean by *precision* in our descriptions of the world. The meaning that physicists give to the word *precision* is barely comprehended by the social scientist, whose tolerance for imprecision, even in the most carefully designed studies, is extremely high. Rigorous physical categories or measurements provide us with opportunities for accurate counting and for obtaining a precise meaning from that count. Social measurements do not. For example, if we tell you that the temperature outside is 95 degrees Fahrenheit, you do not need to ask any further questions about temperature. If we tell you that Edna is a leader, you have only a most general notion of what we are talking about. The statement calls for further clarification. The category leader, by itself, does not carry much information. It calls for a story about Edna in order to gain access to what was meant by the original statement.

Population counts, as physical measures, have advantages and disadvantages. When you say a room contains fifty-seven people, you have made a precise statement. You have come up with a population count—a statistic. You have also made an extremely general statement. You gain precision, in this instance, by abandoning information. This problem exists in any kind of attempt to deal with the world of experience. Once you know a room contains fifty-seven people, the next question is, what kind of people? The further you press this question, the more information you obtain. The more information you obtain, the more you move away from a numerical-statistical description of events into a literary form. You begin to tell the story of the room.

THE INTERPLAY BETWEEN LITERATURE AND STATISTICS

The distinction between literature and statistics, contrary to the opinion fostered by many statisticians, is not a sharp one. There are, of course, statisticians who would like to make statistics a pure and isolated discipline, remote from the fictions and the fantasies of literature. There are also people, perhaps too many, who look on statisticians as barbarians.

The shift from stories to statistics and back again is a matter of how you deal with the categories that concern you. If you want to get good counts of things, you have to lose information. If you want information, you have to think about slacking off on the business of counting. The advantages that come from using numbers are paid for by a loss of detail and individuality. As detail and individuality are brought into focus, our counts of things become less definite and, eventually, impossible.

This issue is introduced here because there are people (such as sociologist George Lundberg, the author of *Can Science Save Us?*) who have argued that literature should be abandoned in favor of science. There are others (the poet E. E. Cummings might be one) who have argued that scientists do more harm than good—that poetry and songs are the way to go.

The real world is more complicated than either of these positions allows. Neither statistics nor literature can ever be a totally dominant form of expression. We have statistics and we have literature, and the two forms are integral to describing our world. They are different solutions to the same problem: how to communicate effectively. Effective communication requires accurate or appropriate description. The two forms are a response to contradictory descriptive and communication demands: the demand for precision on the one hand and, on the other, the demand for information.

The use of the term *information* is awkward, but we cannot think of a better term. One of the reasons people find statistics exasperating, perhaps, is because it is not informative in the same way that a story is informative. It is one thing, for example, to know that a baseball hitter has an average of .323, and it is something else to watch the performances from which that number is derived. The average informs us, but it obviously does not give us any profound sense of the events it comes from. It is a relatively pure distillation of information. (In modern sociological methods, the issue being considered here would probably be referred to as the distinction between *qualitative* and *quantitative* research. Neither approach can really prove dominant because each approach seeks to resolve contradictory communicative demands—the demand for breadth on the one hand and the demand for detail or information on the other.)

Or, to press the point further with a commonplace example, the baseball player Babe Ruth exists within two types of literature. There is, on the one hand, the statistical counts that define him as one of the great baseball heroes of America. Babe Ruth cannot be comprehended without the statistics that define his athletic life. Babe Ruth is also a story. If you knew only the story, you would have a sense of Babe Ruth. If you knew the statistics, you would also have a sense of Babe Ruth. To have a relatively complete sense of Babe Ruth, however, requires an awareness of *both* the statistical and the literary forms that now are all that is left of his life. The two broad forms—literary and statistical, quantitative and qualitative—are not antithetical but complementary. We need both.

In the social sciences the interplay between literature or the story and precise classification is a close one. The reason for this is that unlike the physicist or biologist, the social scientist is primarily concerned with human relations. Human relations are not easily reduced to single-variable analyses. This problem can be illustrated by asking whether the term *woman* can best be seen as (a) a logical, statistical classification, (b) an instance of a single variable called *gender* with male and female as its polar opposites, or (c) an experience that can only take the form of a story about a specific woman. If it is a classification, then what operation best performs the task of defining *woman?* (Such a definition would include all members of the category *woman* and exclude all nonmembers.) If it is a story, then how do you capture the most relevant details and information? The same is true with any other social category: Native American, African American, businessperson, student, criminal, leader, doctor, cowhand, or anything else.

If you try to define *woman* through stories, compiling in detail the biographies of the billions and billions of women who have ever lived over the span of human history, you will have a pile of information of such density and magnitude that it is beyond the imagination of even the greatest literary genius. On the other hand, if you reduce the concept of *woman* to some simple definition that permits counting, you have abstracted out of an infinitude of unique experiences a kind of statistical fiction. You draw on a simple classification to bring order and breadth into your observations. The story does one thing. Statistical observations, as we know them today, do another. There is a place for statistics and a place for literature—each is necessary to deal with the limitations of the other.

OPERATIONAL DEFINITIONS

In order to obtain a count of any kind, you must know, first, what it is you are counting. You need an operation that specifies the nature of the object to be included in the count, as well as the objects that should not be included in the count. Some kind of *operational definition* has to take place before a count can be made. In the social sciences, as in science generally, it is important that the operation be made explicit so that others can use it to check the count you have made.

The difference between statistical observations and stories can be seen quickly in the fact that operational definitions, especially in the social sciences and in psychology, are convenient simplifications. The operation for defining a delinquent might be anyone between the ages of twelve and nineteen who is convicted for criminal activity. The operation for defining a drug user might be anyone who has indulged in the use of an illegal drug at least once within the last three months.

Obviously, such definitions are arbitrary and contain within them a wide range of actions that can only be sorted out by going into the details of individual cases. A girl might be defined as a delinquent for putting a lipstick in her purse while shopping at a local mall. A young man might be placed in the same category for stealing a car and raping the driver. The range contained by the operational definition is such that there is still a great deal of variability within classifications. Variability, we now know, is a major problem. A good definition restricts within-group variability; a bad one allows a high degree of variability.

Each year, on several thousand college campuses around the country, in tens of thousands of classrooms, operational definitions are put into effect that deal with the problem of grading. Grading is a counting problem.[1] For each class we have to count the members of, in the typical case, five classes: the class *A*, the class *B*, and so on to the class *F*. By what operation is such a count made? Students pay close attention to this operation and constantly question it. On this operation depend hundreds of thousands of academic careers. Students raise questions like the following:

> "Will the mid-term count as much as the final?"
>
> "Do we have to know the stuff in Chapter 8?"
>
> "How much does the term paper count?"
>
> "Can I get extra credit for reading an additional book?"
>
> "Do I have to know names?"
>
> "Will we be quizzed over the guest lecturer's material?"
>
> "Do you take off points for guessing?"
>
> "Is the quiz going to be essay or objective?"

These questions deal with operational definitions. Trying to get a definition of what differentiates an A from a B, let us say, invariably has a heavy element of arbitrariness in it and is, therefore, subject to controversy and criticism. Some teachers give A's to the top ten percent of the scorers on a test. Others have absolute standards—the whole class can make an A. Some reward effort and good attendance; others do not care about such matters. It is a mixed bag. At the bottom is the question of what precisely is meant by excellence. It is not an easy question, and different people deal with it differently. That's the big problem with operational definitions: with respect to defining social characteristics, they are quite arbitrary in character.

Grading distributions, obviously, vary as the operation defining the grades varies. This is the Achilles' heel of the operational definition. It permits variability. Since something should always be whatever it is, any definition of it should be consistent with any other. However, this ideal is difficult to attain in the social sciences. Perhaps that is why we find sports so satisfying. It is a place where the operations for defining excellence have been established in an astonishingly consistent manner. Of course, sports also provide us with an extremely broad and useful data base—one that offers an opportunity to make the best use of statistical techniques.[2] Once again we see the significance of standardization, now applied to the problem of categories. Where the natural sciences have relatively standardized categories, the social sciences do not.

RELIABILITY

When it is possible to achieve a high degree of standardization, then operational definitions provide consistent results time after time after time. A number of operations in the physical sciences are of this nature. Measuring temperature provides consistent or reliable results. Temperature is what a temperature gauge measures. Moreover, if you have a good temperature gauge, you will get reliable measurements. What is *reliability?* It is a relationship. A measurement is reliable if measurements of the same thing taken at one time or place are the same as those taken at another time or place. Reliability for the fictive values in Table 16.1 is perfect. The results at time 2 correspond perfectly with the measures taken at time 1.

Grades, as measures, are interesting because their reliability usually is on the order of correlations of $r = +0.60$. More rarely, reliability of grades is as high as $r = +0.75$ or, perhaps even $r = +0.80$. At best, grades allow for a

Table 16.1. Measurement Reliability (Data are fictional.)

	Time 1	*Time 2*
A	33	33
B	54	54
C	61	61
D	49	49
E	05	05

considerable degree of measurement error. A really careful or precise count is obviously impossible. Within a single class students generally distribute themselves in a similar fashion on each test that is taken. However, a lot of shifting is likely to take place in the center of the distributions. Super students tend to ace each exam. The students with serious problems tend to keep on getting F's. It is within the lower C range and the upper B range that there is considerable movement. The stability of these measures, to the extent it exists, generally comes from the extremes in the distribution, not the center. When we see this happening, we know that we are being forced to rely on crude and imprecise measuring devices.

There is also the problem of between-class variability. A student who gets a good grade in history might get a low grade in statistics, or it might go the other way. This can be caused by the content of the courses, by variability in the specific talents of the students, or by the operations used for grading in each class. The operation in one class might rely heavily on recall while another leans on problem solving. The statistics teacher asks for lists of limitations of particular techniques; the history professor asks you to consider what might have happened if Abraham Lincoln had not been shot. One teacher might rely on essay tests, another on objective tests. Careful counting requires good operations and the standardization of those operations. Grading is a crude operational procedure. It relies heavily on tradition and academic ritual for acceptance.

The problems involved in grading students are not commonly discussed as problems in operational definitions and in counting, but that is what they are. Great effort has been expended throughout the Twentieth Century to find standardized tests for intelligence, scholarly ability, knowledge, and so forth. The difficulties these tests have run into are instructive. They produce different distributions when administered to the same groups at different times. That is, a person who scores, let us say, an 80 on one administration of the test might score a 95 on another administration. The extent to which test results correlate with similar test results reveals a large margin of error—residual variance is commonly at least thirty percent and often as great as fifty or eighty percent of the total variance.

We will not go into the additional problems such tests have encountered with respect to problems of control. These problems are popularly referred to as those of attaining tests that are free of *culture bias,* or *culture-free* tests.

All these concerns have to do with what statisticians and methodologists refer to as the problem of reliability. The problem of reliability raises the question of how reliable a test or an operational definition is in the kind of counting it generates.

VALIDITY

The central problem with operational definitions is the problem of **validity,** and here Twentieth Century social scientists have come up with a slogan that merits deep consideration: *Intelligence is whatever an intelligence test measures.* In other words, X is whatever you have decided is an appropriate operation for counting X, so long as another person agrees and can employ the same operation for further testing of the count. If there is agreement, then the test or operational definition has *face validity*.

Actually, this is about all you can do. It is a terrible solution to a problem that is perhaps unsolvable. If nothing else, the operational definition allows people to get on with the business of counting things in the world around them. However, in the realm of social analysis, it is a solution that generates about as many problems as it resolves. The question is simple enough: how do you set up a way of getting an accurate count of students, wives, delinquents, Arabs, Catholics, cheats, social isolates, smart people, leaders, successful people, and so on? Each of these categories is, in one sense, easy to operationalize and, in another, nearly impossible. If a classification is viewed as a simple model, then it is a model that provides us with a map, or expectation, for counting. What we expect when we include someone in a given category might be quite different from what we find if we examine the individual case in more particular detail.

Take counting husbands. The operation usually used for this count is whether the men in a given household are married. But a man can be married and not really be acting as a husband. On the other hand, a man can be unmarried and be acting very much like a husband. Now, this might sound like more niggling. But keep in mind that in this case, you are not counting husbands, you are counting married people. There is a difference. However, for practical work, we have to take things for granted rather than get into categorical hair splitting. But we are not, in science, supposed to take things for granted if we can possibly help it.

The operational definition runs into some of the problems alluded to earlier when we considered the problems involved in sampling a structured complex (see Chapter 10). What is meant by a marriage, by minority group status, by being a woman, or by living as a student? Each of these categories refers to complicated events involving idiosyncratic details that are particular to each case. The operational definition tries to find a single or at least a relatively limited number of qualities—a selected *sampling* of salient qualities—that summarizes the event being defined. As we saw in the discussion of sampling, no systematic procedure currently exists for sampling structural complexes. Therefore, no truly systematic device exists for defining such complexes. Within this observation lies the basis for a lot of conflict and disputes.

This is as good a place as any to bring our discussions of statistical forms to a close, because in considering the problem of the operational definition, we find the elementary logic of the average, variability, sampling, probability, and relationships at work simultaneously. An operational definition is an attempt to state the primary characteristics of the quality being defined. It is an attempt, in other words, to express typicality. A definition of wife should get at what is typical of wives; a definition of African American should get at what is typical of African Americans; a definition of student should get at what is typical of students. With social categories, however, there is always within-group variability in each instance, and our statement of typicality is affected by that variability. In a sense, a categorical classification is a presumed sampling of a population.

With respect to social events, the things being classified are not simple homogeneous sets of events, but rather complex structured actions. Social categories have a probability of being accurate in individual cases rather than being universally accurate. The relationship between the definition and the event is not a close one but a relatively loose one.

It is true that being married typifies all husbands; it is a perfect modal quality. Moreover, it is a modal quality that includes *all* members of the class "husbands." However, despite the invariant quality of the modal value, we know that other factors are involved in being a husband, or in being a member of any other social category. To sample the population of husbands by using the fact that they are formally registered as married provides a perfect operation in terms of an invariant quality. Unlike the operation that defines weight, however, the population being sampled is consistent only in terms of one relevant aspect of the category "husband." The definition is formally precise, but the question remains as to whether it is substantively precise. This question continues to jeopardize the value of the operational definition of social concepts, because it suggests that the operational definition does not really absorb all the varied aspects of the role implied in the concept.

In other words, you can use the term *husband* as a kind of average or typicality to describe a particular person. However, the operational definition does not permit you to go any further than to say that the person you are describing is formally married. You are not permitted any other guesses about the person's life or role. Two problems arise. One is that social terms are loaded with rhetorical and experiential meanings that are not eliminated by the operational definition. It is probably naive to think that a study in which the researchers have a well-defined sense of their operations will be understood in the same fashion by the general public when the findings are reported. More to the point, researchers themselves rarely have a disciplined sense of the rigorous limits any operational definition imposes on qualities as complex as a human personality or a social role.

If you try to get around such problems by staying with a rigorous operational definition and then relating its values to qualities believed to be a part of that event, then you are going to run into a definition that is probabilistic in nature rather than absolute. Husbands, for example, might or might not be living with their wives. This quality (of living with wives) is no longer a rigorous definitional detail but is, instead, a probabilistic detail.

In sum, no matter which way the social scientist turns, the operational definition, when used to deal with social categories, does not produce the same degree of precision with respect to its referent as do categories used in either mathematical logic or the physical sciences. The problem is inherent in the nature of the populations being defined. The natural scientist commonly defines a population that has relatively little or even zero variance across all occurrences. The social scientist, when trying to define a category such as "student," is dealing with an event that is extremely variable. Worse yet, each individual case is a complex structural process or action; "students" are really verbs, not nouns.

In the social sciences, further problems arise because the use of different operational definitions generally produces different results. Both the U.S. Bureau of Commerce and the U.S. Census Bureau retain an antiquated definition of employment, one that includes people who are barely existing on part-time incomes, while neglecting to count those who have given up looking for work as unemployed. This antiquated definition is retained for a good reason, however. If the operational definition were changed, then comparisons with earlier figures on unemployment would be made invalid.

In the natural sciences a great deal of effort is expended on the standardization of operational definitions (the measuring devices used to make careful counts). An entire agency, the National Bureau of Standards, is dedicated to the task of making sure that researchers are provided with standard operations. In the social sciences there is relatively little, if any, standardization. Attempts have been made to find a standard, noncontroversial intelligence test, but to the best of our knowledge, none has succeeded.

This is a serious problem because research results can vary considerably depending on the operations used in counting. A good operational definition, at the very least, calls for an intimate and thoughtful understanding of the condition being defined. Casual operational definitions are likely to lead to casual statistical studies.

As a social scientist, you are always faced with the problem that if you want to know what any individual case in a population is really like, you have to turn to the story that reveals structural details that the operational definition has glossed over. In sum, the importance of the story cannot be ignored.

SUMMARY

The statistical method, throughout most of its range, is essentially a two-note concerto. The two fundamental notes are measures of typicality and measures of variability. These essential forms appear within and are a part of the other basic forms: probability, sampling, relationships, control and categorization, models, and counting. For example, in order to count, you have to have good categories. Good categories typify the object or quality being classified. If there is within-category variability, then your categories lose precision.

Whether you are engaged in some mundane activity involving a daily routine or caught up in an elaborate research program that involves the crunching of gigabytes of data, you will, in some manner or other, be dealing with typicality and variability. These two fundamental concerns are an important part of many aspects of your life. To have at least a basic understanding of the problems involved in assessing typicality and variability is the first step toward a broader awareness of how you think about things. For this reason, if no other, an elemental knowledge of statistical forms is a vital part of the development of your own self-awareness.

In order to get good counts, we must have good categories underlying our counting procedures. In the social sciences categorization problems are dealt with by means of operational definitions. The main problems facing the social scientist who must use operational definitions have to do with the fact that such definitions are (a) highly arbitrary, and (b) unable to get rid of within-category variability.

Thinking Things Through

1. You are asked to carry out some research in which the object is to explore success and failure among college students. You obtain a sample of 1,200 people who are enrolled in various classes. Does formal enrollment really identify an individual as a student? What do you think the term student truly means? Does your conception of a "real" student agree with what others think? What problems arise with respect to doing precise counting when you are confronted with categories such as this?
2. Can you come up with a perfectly valid and reliable system for grading people in school? If current grading procedures (operations) are imprecise, what are the implications of this imprecision? For example, there are probably some people who miss out on membership in the Phi Beta Kappa honor society because an instructor made an error in evaluating their talent. How might this error affect these students' futures?
3. If something as simple and obvious as grading cannot be given a standardized operation that reduces categorical errors to zero, what are the implications for studies that deal with complex issues such as crime, marital happiness, ethnic sensitivity, intelligence, or executive ability? (Hint: At one level it can be shown that all Americans engage in crime of some kind nearly every day. Is everyone, then, a criminal? If not, how do you define criminality in such a way that you eliminate this absurd conclusion? To what extent is the definition necessarily arbitrary? Do politics play a role in the definition?)
4. Write a short but thoughtful essay on the relative strengths and weaknesses of literature and statistics as ways of trying to describe our social lives.
5. The terms *liberal* and *conservative* are highly emotive and significant social categories for large numbers of Americans. Suppose you have been assigned the task of counting the number of liberals and conservatives in the United States. How would you go about this task? You are also given the additional assignment of determining how members of these two populations compare with respect to income, education, and attitudes toward capital punishment. List, without necessarily resolving, the various problems you would expect to encounter as your project got underway.

Notes

1. Even more interesting is the modern approach to teaching evaluation. What operation defines the variable nature of the teaching process? Is a teacher who gets a rating of 2.96 necessarily a better teacher than one who gets a rating of 2.85?

Incidentally, a consideration of the criteria that are often used in such evaluations can give you a better sense of what is meant by our argument in this book that social categorization is forever haunted by the problem of superficiality. What, really, is a good teacher? Is a drill sergeant in the Marines a better teacher than, let us say, a professor of medieval history? Does subject matter make a teacher good or bad? If not, why do teachers of some subjects almost invariably get higher evaluations than teachers of other subjects? How would you standardize for subject matter?

2. The limitations of statistics with respect to predicting sports outcomes should be instructive to social scientists who think social affairs can be studied with precision. Formal statisticians do little better than "shoot-from-the-hip" sports analysts in predicting winners. In fairness to formal statistics, however, we should stress that good sports analysts have a well-developed—but *informal*—statistical sense.

GLOSSARY

absolute values If we have 15 A's and 5 B's, then the *absolute* number of A's is 15. Relative to 100, the number would be 75, or 75 percent. Relative to 1,000 the number would be 750. When absolute values are extremely large, they can often be made to look small by expressing them as relative values. For example, an absolute number of deaths from AIDS of 30 million people sounds shockingly high—and it is. However, relative to the total planetary population of 5.5 billion, it is about one-half of one percent—a number that can seem small. See *relative values.*

addition rule What is the probability (Py) you will have a car accident (Pa) this year *or* not have one (Pn)? If $Pa = 0.25$, then $Pn = 1 - 0.25$, or 0.75. The question asks you to determine the probability of experiencing *either* of two events. Since one or the other will occur, you *add* the individual probabilities. Therefore $Py = 0.25 + 0.75 = 1.00$. See *multiplication rule.*

alpha If you are going to conclude that something is not happening simply by accident or chance, you must define, at the outset, what you mean by chance. Behavioral science researchers commonly define something as not being a matter of chance if it has a probability of occurrence equal to or less than 0.05. This is the cut-off point, called the alpha level, that allows for rejection of the idea that chance has accounted for a given observation. Notice that the 0.05 level for alpha is arbitrary. Alpha should be determined according to specific concerns surrounding the decision-making process.

ANOVA ANOVA stands for the analysis of variance. Essentially, ANOVA examines how much it is possible to reduce the variance in a given scaled dependent variable after it is sorted into different nominal categories. For example, height is a ratio scale that displays variability or variance. If we classified a sample of people into the nominal categories "leaders" and "followers," examining height for each nominally identified group, we would have a relationship if variability was smaller within these categories than it was before we sorted our sample.

asymmetry If the relationship of X to Y is different in degree from the relationship of Y to X, the relationship is asymmetrical. This is a common occurrence with nominal data.

average The term *average* is used in ordinary discourse to refer to a variety of conceptions of typicality. These can range from batting averages (basically simple percentages) to extremely complex stereotypes grounded in folklore, such as the average professor or an average old person.

between-group variability When an interval or ratio scale variable differs greatly from one nominal category to another, there is considerable between-group variability. In this case within-group variability is necessarily lowered, and the nominal classification can be said to account, statistically, for at least some of the variance in the dependent variable. See *within-group variability.*

bimodal distributions Bimodality occurs when frequency distributions display two well-defined modal values. Such distributions make estimates of typicality difficult. As a consequence, other statistical procedures, such as estimating the extent of relationship between two variables, can also be more difficult.

careful counting We offer the term *careful counting* to alert readers to the extent to which casual counting takes place. Careful counting cannot take place unless the events being counted have first been carefully defined. Problems with respect to defining social elements makes counting them in a careful or precise manner especially difficult. Standard texts in social statistics commonly gloss over this issue and assume that careful counting is characteristic of any count. This, of course, can be misleading.

categorization In this book we use the term *categorization* to refer to the problem of classifying some set of events in such a way that the symbol used for classification includes all members of the set and excludes all nonmembers of the set. This is the beginning and, usually, the end point of statistical research. Without good categorization, precise counting simply cannot be done.

central limit theorem If an infinite number of samples are drawn from a given population and the mean of the statistical values obtained from these samples is calculated, the mean will approach the population statistic or parameter in value. (In effect, this theorem says that an infinitely large sample should provide an accurate estimate of any statistic for the population. Like any good mathematical theorem, this is difficult to argue against.)

central tendency An attempt to estimate or summarize the values in a data set by relying on a single value. For example, given the set 6, 8, 8, 8, 9, 10, 11, 12, 12, the values 8 (the mode), 9 (the median), and 9.33 (the arithmetic mean) are estimates of central tendency. A measure of central tendency does not necessarily define an individual, or group of individuals, within the set. See *typification.*

chi-square A popular statistical procedure for determining whether nominally sorted values are significantly associated. The basic logic consists of comparing a set of values that show how the data would appear under a state of zero association with how they actually appear. If there is a significant difference, then a relationship is presumed to exist. The logic of chi-square needs to be thought about carefully, because even a small degree of association can be shown to be probabilistically significant if large samples are involved. See *hypothesis testing, proportional reduction in error (PRE).*

combinations If you are dealt six cards from an ordinary deck, you expect some to be red and some black. However, it is possible to get a variety of combinations of black (Bk) and red (Rd) cards as the cards are being dealt. There is only one way to get all red cards (Rd, Rd, Rd, Rd, Rd, Rd). There are twenty different deals, or combinations of cards, that produce combinations of three red and three black cards. Here are two of these combinations:

Rd, Bk, Rd, Bk, Bk, Rd
Rd, Bk, Rd, Rd, Bk, Bk

common sense *Common sense* is mentioned here to underscore the tendency of people to reach "hard" or definitive conclusions about human social actions and statuses. Carefully researched social investigations almost invariably reveal that human affairs are not "hard," deterministic matters but rather "soft," probabilistic ones. Common sense is a subjective term, generally referring to a body of knowledge with which one is in general agreement.

compound interest formula This formula provides answers to questions such as the following: If inflation continues at 3.2 percent per annum and a loaf of bread now costs $1.25, how much will it cost in one hundred years? Notice the big *if* in this question. (Answer: $38.99—nearly forty dollars a loaf.)

confidence In statistics the term *confidence* is used largely to refer to how certain we can be of the reliability of generalized descriptions based on a sample. The generalization "All humans are mortal" is based on a sample. We can, however, put great confidence in it. (A Christian might reject this generalization, not on sampling grounds, but because of a belief in the immortality of Christ and eternal life for those who are saved.)

confidence intervals Limits that define the extent to which a generalization based on a sample is *probably* reliable. For example, a voting poll presented on the television news might say that 40 percent of voters were in favor of a new bond issue, and this figure is accurate by plus or minus 3 percent. Any value between 37 and 43 percent, then, is probably correct in this instance. The values 37 percent and 43 percent are the confidence intervals—that is, the intervals within which you can have confidence in the sample's findings either 95 or 99 percent of the time, depending on alpha. Unfortunately, the setting of alpha is never mentioned in the media. You can generally safely assume that alpha is set at the 0.05 level to make the survey results look better. See *alpha*.

constants A set of values for which any estimate of central tendency also defines the value of any individual in the set.

control The elimination of confusing or extraneous factors by various techniques. A common example of controlling an extraneous element, such as the varying number of cases within which some event is occurring, is the percentage. It sets the number of cases in any situation at a consistent value of one hundred. Without such controls, fair comparisons cannot be made. See *standardization*.

correlation coefficient The correlation coefficient, ideally, is the square root of the proportion of the variance in a dependent variable that is accounted for by an independent variable or set of variables. It should, in most but not necessarily all instances, be squared. If we find a correlation of +0.14 between a test of aptitude for being a firefighter and actual ability, the aptitude test accounts for roughly two percent of the variance between firefighters with respect to their competence. See *proportional reduction in error (PRE)*.

criterion variables A psychological test might be used to predict whether you would be an effective police officer. Being a good or bad police officer is an example of a criterion variable, and the test is a predictor variable. See *dependent variable*.

cross-classification One way to eliminate a disruptive factor in a study is to examine how events take place for given or fixed levels of that factor. Educational levels of individuals in a given sample might interfere with a study of how people react to medical advertisements. Cross-classification of the sample by educational level removes at least some of the effects of education insofar as it is now possible to examine the problem for, let us say, people with only a grade school education—a group within which the effect of education has been reduced to such an extent it is no longer considered a variable factor.

curvilinear relationships A relationship between X and Y that is defined by any line other than that derived from $Y = a + bX$. For example, as you add salt to food, the taste continues to improve until it "turns around" and begins to get worse. Most social research relies on the assumption of simple linear relationships.

degrees of freedom A nontechnical definition of degrees of freedom might suggest that it is the number of spaces in any array of values that can be filled with optional values before the remaining spaces must be filled with fixed values according to marginal totals. For example, in the following array, the marginal total is 223. There are six spaces. We are free to put differing values in five of the spaces.

$$[-] + [-] + [-] + [-] + [-] + [-] = 223$$

After we do this, however, the value that goes in the last space is fixed.

Free	*Fixed*
1 + 2 + 3 + 5 + 21 +	[191] = 223
10 + 55 + 66 + 44 + 97 +	[1] = 223

In this example we have five degrees (spaces) of freedom. Ascertaining the correct number of degrees of freedom is necessary for determining appropriate probabilities for obtained values of statistical procedures such as chi-square and ANOVA.

dependent variable A dependent variable is a variable you would like to account for in some manner. If you can reduce its variability by associating it with some plausible independent factor, you have statistically accounted for the dependent variable—to the extent that variability was reduced. See *independent variable.*

descriptive statistics Descriptive statistics involve any statistical effort that is not concerned with generalizing beyond a specified sample. So powerful is the human inclination to generalize that this species of statistical information is relatively rare. Many studies that are passed off as inductive research should have settled for descriptive status. See *inductive statistics.*

deviation, mean absolute See *mean absolute deviation.*

directionality Directionality is especially important when testing relationships. For example, if we predict that women will be inferior to men in performing a given task, even a modest suggestion by data that they are superior immediately leads to a rejection of the speculation. If no direction were specified before data collection, then a modest difference would not sustain rejection of the argument that women differ from men.

direction of relationship When you increase the value of X and the value of Y also increases, a relationship is positive. For example, there is a positive relationship between amount of cigarette smoking and respiratory illness. If, when you increase the value of X, the value of Y decreases, you have a negative relationship. For example, the older a person becomes, the lower the probability that he or she will become engaged in a violent crime.

equal interval scales Measures or forms of categorization that provide equidistant intervals between varying degrees of some condition. For example, an IQ of 140 should be exactly five units higher than one of 135 and five units lower than one of

145. The question is, how certain can we be of the equidistance of such a scale? If we *can* be certain, such scales provide more information than either ordinal or nominal scales.

expected values The values we anticipate on the basis of some model, theory, or speculation, as opposed to the values we actually observe.

Gaussian distribution See *normal curve.*

homoscedasticity We know that when a distribution is affected by extreme values, it is more difficult to determine the usefulness of any measure of central tendency obtained from that distribution. In a similar fashion, when two or more distributions are used for correlation or ANOVA studies, skewedness affects the validity of any conclusions that might be reached. Therefore, correlation and ANOVA require an assumption of *homoscedasticity,* or homogeneity of variance throughout the distributions involved in the study.

hypothesis testing Hypothesis testing is the reigning research paradigm in the behavioral sciences. It relies on the determination of the extent to which there is a difference between what you have observed and what you might have expected if the difference in parameters for two (or more) populations is zero. This is the so-called *null hypothesis.* The weakness of this approach rests primarily on the fact that as sample size (N) is increased, you can reject even trivial or meaningless observed differences as having happened just by accident. It is a most serious weakness. See *null hypothesis.*

independent variable When you try to account for something, you commonly do it by associating it with an explanatory variable or set of variables. Such variables, in statistics, are formally referred to as *independent variables.* For example, if we try to explain the difference between the rich and the poor by saying that the poor are lazy, degree of laziness is the independent variable. See *dependent variable.*

inductive statistics Statistical work directed toward generalizing to larger populations. A study of a few high school students, often biased and limited, can quickly become the basis for statements that refer to *all* high school students. The good statistician is meticulous about the rules that permit valid generalization. However, even good statisticians often fail in this effort because we are forced to generalize, in the realm of human affairs, on the basis of inadequate information.

lambda A means of assessing proportional reduction in error (PRE). Consider the following illustrative table:

	Men	*Women*	*Total*
Good Spellers	52	93	145
Poor Spellers	96	14	110
Total	148	107	255

In a state of ignorance about the men and women categories, we would guess that most people are good spellers and, in each individual case, we would be wrong 110 times out of the total of 255 cases. If we knew about the categorization by gender, we would guess poor spellers and be wrong 52 times with the men; with the women we would guess good spellers and be wrong 14 times, for a total of 66 errors. Lambda is the total errors before knowledge minus the total errors after knowledge ($110 - 66 = 44$), divided by the errors we made in a state of ignorance ($44/110 = 0.40$). In this case we reduced our error by 40 percent. Lambda gives misleading results in some instances, telling us there is no relationship when one does in fact exist. It also provides asymmetrical values for associations. See *asymmetry, proportional reduction in error.*

logical forms The best term the authors of this book could think of for the logical matrix underlying the ways we go about describing the complex world in which we live. The main point we are trying to make in using such a term is that the forms used by the statistician (such as averages, deviations, sampling, relationships, categorization, models, probability, counting, and standardization) appear in any human effort at description, including such nonstatistical domains as poetry or drama. The problem of standardizing the meanings of specifically social terms, for example, poses as big a problem in literary theory as it does in the methodology of the social sciences.

marginal totals Marginal totals are usually the sums of values that make up the cases contained in the rows and columns of a table. In the following table the values 13, 16, 20, and 9 are marginal totals. The overall total, 29, is the sum of the marginals.

			Total
	8	5	13
	12	4	16
Total	20	9	29

matched samples Matched samples provide one way to standardize groups for comparative purposes. The best example of this is the use of cloned mice in biological studies. The truly precise matching of human social populations is impossible, however, so such matches can only be approximations in social research.

mean The single value that represents a set of values by determining what each individual in the set would get *if* all the values in the set were shared equally. For example, in the set 5, 5, 20, 20, 20, the sum of values is 70. If the total were shared by the five members of the set, each would get 70/5, or 14. Notice that no individual value in the set conforms to this value.

mean absolute deviation Take the difference between each value in a set and the mean for the set. Disregard the sign of the difference, sum all absolute differences, and then divide by the number of values in the set (*N*). You now have an average difference value. This results in the average amount of deviation within the set by arbitrarily disregarding whether individuals were above or below the mean. Since it is based on all values in the set, it is more reliable than the range as an estimate of variability. See *range, standard deviation.*

mechanical models We use the term in a general way to refer to statistical or other descriptions with extremely high predictive abilities as opposed to models where predictive reliability is relatively uncertain, or in which the probability (or stochastic) component is very low. See *probability models.*

median A measure of central tendency that uses the middle value in an ordered array of values as the best single representative of the array as a whole. For example, in the array 6, 17, 42, 112, 1,340, the value 42 is the median value. Notice that if the extreme value were reduced from 1,340 to 134, the median would not change.

mode A measure of central tendency that relies on the most common value in a set as the best single representative of the set. For example, in the set 2, 2, 2, 2, 2, 5, 13, 67, the number 2 is most common and therefore typifies the set. Note that distributions with several modes make it difficult to find a value that represents the set as a whole.

model A contrived reconstruction of some real process or condition that enables us to expect certain consequences either by manipulating the model or by referring to its description of the reality it attempts to define. For example, a map is a model of a territory. If we follow its directions, we expect to arrive at a certain destination. If it is a good model (map), we reach our destination. If it is a poor model (map), we get lost. Working airplane models, such as those used by aeronautical engineers, are often smaller versions of planes that are intended for production. By manipulating the models, the engineers can expect certain consequences to follow if similar modifications are made in the larger production aircraft. In social research, statistical models result, invariably, in descriptions with a high probability component. This restricts the practical applications of the research. The term *model* is quite similar to what people refer to, more generally, as a description.

multiplication rule Solving probability problems calls for knowing what is being asked. If you are asked to find the probability of drawing an ace and then, with replacement, drawing another ace, the solution is $1/4 \times 1/4$, or $1/16$, or 0.0525, or about 6 times out of 100 attempts, on the average. The individual probabilities are multiplied. (The major difficulty with this is that ambiguities of communication often make it easy to misinterpret what is being asked.) See *addition rule.*

nominal categories Nominal categories are extremely crude categorizations consisting of no further measurement than naming or labeling those cases that are thought to belong to a given set. College students, for example, are labeled *students.* But if we define a student as a person dominated by a love of knowledge and a burning desire to master an intellectual discipline, then certainly many people taking courses for other reasons would fall outside this nominal set. Nominal categorization points up the significance of categorization insofar as statisticians, universally, agree that nominal data inhibit the use of sophisticated techniques of analysis. Note that nearly all ordinary discourse relies on nominal classification. To that extent it is inherently imprecise.

normal curve (Gaussian distribution) Basically, any curve or distribution in which the mode lies on the arithmetic mean and the mean divides the distribution into two equal halves; moreover, the farther you move from the mean, the fewer the number of cases in the distribution. The normal curve is associated with the standard deviation in that the proportion of observed values, above or below a given value, is precisely defined by how far these values are from the mean. For example, in a perfect normal curve, sixteen percent of the total distribution lies above one standard deviation to the right of the mean and sixteen percent lies below one standard deviation to the left of the mean. The normal curve appears in many natural contexts.

null hypothesis The hypothesis of no difference between groups. If you can reject this notion, then you must accept the possibility that there really is a difference. This simple idea has one major problem: there is, nearly always, *some* kind of difference between groups. Therefore for most practical purposes one can always, in principle, reject the null hypothesis. See *hypothesis testing.*

operational definitions Categories established in terms of the means or operations devised for observing the referent implied by the category. For example, we might define a juvenile delinquent as anyone within specified age limits who is apprehended and convicted of a felony. This definition is formal and simple. It also provides the operation that enables identification of individuals as delinquent. It has the disadvantages of excluding those who commit felonies and are not apprehended and including those who are innocently convicted. It also has the disadvantage of including an extremely wide range of actions—from being caught with a bit of marijuana to being a serial murder-rapist. It has the advantage of being simple and permitting counting. The traditional example of an operational definition is the assertion that intelligence is whatever intelligence tests measure. In effect, the operational definition claims that the operation used for counting also defines the reality being counted.

ordinal categories or scales Ordinal categories permit ranking events from the highest to the lowest along some criterion of classification. Each year, for example, college football teams are ranked from the first team down to the twenty-fifth team. This is an ordinal ranking system. Notice that this ranking does no more than rate teams from number 1 to number 25. It does not tell you how much better team number 1 is than team number 2; it might be a lot better or only slightly better.

outliers Outliers are extreme values that can distort given statistical estimates. For example, the mean wealth of the entire American population is biased by the few extremely wealthy multi-billionaires who skew the distribution. A good statistician is always on the alert for distortions created by outliers.

paradigm Within the context of statistical literature, a paradigm is a general procedure for carrying out intellectual work or for assessing the merit of a particular effort. The methodological paradigm that guides the literary novelist, for example, is different from that which directs an experimentalist. The precise meaning of this term is difficult to determine; various writers use it differently. Thomas Kuhn, historian and philosopher of science, used the term to refer to modern paradigms of scientific research. In this case Kuhn was referring to the spirit behind entire generations of scientific effort.

parameter A true value for a given population. Because populations that concern us are commonly extremely large, we have to estimate the true value, or parameter, by using samples. If a population is small or manageable, we can obtain the parameter directly from data relevant to that population. It should be noted that the standard statistical techniques for estimating population values from samples are coming under serious critical review by the present generation of statistical analysts. See *descriptive statistics, inductive statistics.*

partial correlation Partial correlations attempt to get rid of an extraneous influence by taking into account how it is related to the dependent and independent variables under consideration. For example, the correlation between crime and ice cream consumption might be $r = +0.70$. But after we partial out temperature effects, the correlation drops to zero.

partialing Partialing is a general term for removing, by statistical (as opposed to experimental) procedures, the disruptive influences of any intrusive variables or factors.

Pearsonian correlation coefficient (r) Devised by Karl Pearson, r provides a proportional estimate of the extent to which an explanatory variable accounts for variability in some problematic variable. Where variables are freely independent, r should be squared. If, for example, amount of time spent studying is correlated $r = +0.43$ with grades for 297 college students, the studying accounts for roughly 18 percent of the variance in grades. Since r refers to variance, it can only be used with data that allow variance to be determined.

probability Probability, in ordinary statistical analysis, is basically an assessment of the proportion of times something falls, *on the average,* between the two extremes of being perfectly certain of occurring or perfectly certain of not occurring in a specified time period.

probabilty models In standard discussions of statistical issues, probability models are devices used for solving probability problems. For example, if we toss a coin millions and millions of times we expect, *on the average,* that heads will appear half the time. This is an extremely simple probability model. Note that for any individual case we might get heads or we might not. We expect to get heads half the time if we toss the coin more than once, preferably many times. Probability models cannot, by their logical character, provide certain expectations for the individual event. We have broadened the implications of this term slightly to include statistical findings that contain a large probability component. Such findings are less certain than those with very low probability components. (An exception occurs in quantum mechanics, a field that rests on probable events but that also has precise solutions to many of its problems.) See *mechanical models.*

proportional reduction in error (PRE) This concept provides a distinctive approach to the question of how we can evaluate the extent to which two or more factors are related. PRE begins by asking how many errors you would make in guessing an individual value based on how you have assessed proportions. For example, if you know the drop-out rate at Siwash University is 30 percent, then if you predict, at random, that any student will not drop out, you will be right 70 percent of the time—an extremely high level of predictability in social research. You will be wrong 30 percent of the time. Any kind of association to improve predictability will have to improve on the already established level of 70 percent predictive efficiency. This concept deserves a lot of thought because it suggests that predictability is much simpler for nominal than for more refined categories. But such a conclusion is not a cause for celebration. Nominal social categories are highly ambiguous—that is, they contain a great deal of within-category variability. We are therefore left with a situation in which we can achieve greater predictability as we become less certain of just exactly what it is we are trying to predict. See *correlation coefficient.*

range The difference between the lowest and highest values in a distribution—a crude and unreliable measure of variability. See *standard deviation, mean absolute deviation.*

ratio scales Ratio scales have equal intervals and an empirically established zero point. They permit comparisons on the basis of ratios, such as the following: The consumption of energy by a typical American is fifty times greater than that of a typical Ugandan. Notice that we are referring to a physical variable, energy, that has a knowable zero value. Social and social psychological events are commonly resistant to established zero points. For example, can we establish a zero point for, let us say, scientific knowledge? Since even an infant reacts with shock when it is dropped through the air, we would have to say it has a kind of scientific knowledge of the world. If you wish to argue with this, we would suggest that the matter is at least controversial; real zero points, however, are noncontroversial.

relationships The attempt to reduce variability in a dependent variable by determining the extent to which it becomes less variable when another factor is taken into account. For example, some men commit acts of domestic abuse, but others do not. If we find that all men who were abused as children abuse their spouses while nonabused men do not, then we have accounted for domestic abuse through a relationship. One of the most common and best indexes of relationship in statistics is the Pearsonian correlation coefficient (r).

relative values A numerical count for a given set of events relative to some standard value such as 1, 10, 100, 1,000, 10,000, or some other fixed reference point. For example, in the set A, A, X, X, X, X, X, X, X, X, the relative value of A is two per every ten members of the set. See *absolute values.*

reliability The ability of a scale, sample, or other observational device to generate consistent findings. A rubber yardstick would give different measures of height for a particular object. It would not, therefore, be as reliable as a metal yardstick.

residual variance Suppose we obtain a correlation, r, of -0.57. We know to square this to obtain the proportion of variance in the dependent variable that has been accounted for. We have accounted for 32 percent of the variance. The 68 percent of the total variance remaining is the residual, unexplained, or unaccounted-for variance. See *correlation coefficient, variance.*

sampling The use of part of a set to determine some quality characteristic of the entire set. For example, suppose you want to have your eyes examined. The optometrist asks you to look at a few letters. Then a diagnosis is made. The diagnosis is made on the basis of a brief sampling (a part of the set) of your vision (the total set). If you do not like samples, you might ask the optometrist to examine your vision twenty-four hours a day for the rest of your life—an obvious waste of time, and also a good example of why we constantly rely on samples.

sampling variability The degree of variability between values obtained from samples as opposed to the variability obtained between individual cases. If sampling variability is high, information obtained from such samples will have a high probability of misreading population values. See *standard error.*

scattergram A visual depiction of a relationship between variables measured at least at the interval scale level. A scattergram is a good device for checking out the possibility that outliers are distorting indexical values. See *outliers.*

semi-interquartile range An attempt to avoid the capricious estimates of variability that can come from relying on the truly extreme values in a set. The semi-interquartile range relies on the difference between the value for the 75th percentile of a distribution and the value for the 25th percentile.

social categories Social research, by definition, is forced to make use of a variety of social concepts. Social categories have been singled out in this book because they are extremely resistant to definitions that are noncontroversial. Furthermore, it is difficult to deal with them objectively. They are inherently rhetorical. The consequence is that it is difficult to attain statistical studies in the social sciences that are noncontroversial.

spurious correlations Spurious correlations are usually suspiciously high correlations obtained through error in calculation, faulty understanding of technique, and/or bad categorization. For example, the correlation between the percentage of Marines who graduate from various boot camps and the percentage who do not is $r = -1.00$—a perfect relationship, but a spurious or meaningless one.

standard deviation (*SD*) The square root of the mean for all squared deviations from the mean in a set. The important thing to see with the *SD* is that it is an average of all deviations from the mean for all observations in the set. See *mean absolute deviation, range.*

standard error (*SE*) The standard deviation for variable values obtained from a hypothetically infinite set of samples of the same size drawn from the same universe. If there is a lot of variability between samples, *SE* will be a large value. If there is no variability (the case with physical constants, for example), then *SE* = zero. See *sampling variability.*

standardization Procedures directed toward the end of making comparisons fair, or uncontaminated by extraneous influences. The quest for standardized intelligence tests provides an example of concern over the extent to which such factors as education, experience, ethnic background, test bias, and so on can contaminate test results. Standardization involves any procedure that attempts to rid the testing process of these extraneous influences. See *control, statistical control.*

statistical control The use of statistical means to take into account factors that might disturb an accurate description of an event or process; the attempt to achieve fair comparisons by data manipulation as opposed to experimental control. See *standardization.*

statistical reasoning Decision-making based on descriptions that rely heavily on accurate counting and relatively large numbers of observed cases.

statistical tables Formal summarizations of numerical information, generally in rows and columns. A well-designed statistical table is, to a considerable extent, self-explanatory.

sum of squares The sum of the squared differences between the mean of a distribution and each observed value within the distribution. For example, if the observed value is 2 and the mean is 5, then the difference between the observed case and the mean is −3. The square of −3 is 9. Each such difference in the distribution is squared, and the sum of these squared differences is called the sum of squares.

tabular cells The boxes in which values are located in tables.

total variance See *variance.*

true limits (of intervals) Suppose an interval is from 5 to 9. A score of 4.4 would be allocated to a lower interval. A score greater than 4.5 would be allocated to the 5–9 interval. Therefore, the true lower limit for the interval 5–9 is 4.5. A score greater than 9.5 would be placed in the next higher interval. The true range for the interval, then, would be from 4.5 to 9.5, or five units. Notice that all of this depends on a consistent rounding convention. (Here any number that takes the form 5.5, 6.5, 3.5, 4.5, etc., is rounded up to the next whole number.)

Type 1 error Concluding that something other than chance is responsible for a given observation when the observation was, as a matter of fact, a chance happening. For example, suppose you feel that someone snubbed you by ignoring you when passing you in the hall. Actually, this person simply happened, by accident, not to see you. No snub was intended. You committed what statisticians call a Type 1 error.

Type 2 error Concluding that chance is responsible for an observation when something other than chance was at work. For example, suppose you think your car had a flat tire because it accidentally ran over a nail. Actually, an enemy punctured your tire when you were not looking. Chance was not at work. You committed what statisticians call a Type 2 error.

typification A more ordinary or nontechnical term for what statisticians refer to as *central tendency.* See *central tendency.*

unexplained variance See *residual variance.*

validity The extent to which a category or measure actually refers to the entity or process it is supposed to refer to. For example, are intelligence tests valid measures of whatever it is we mean by intelligence? This is, perhaps, one of the most challenging problems facing the statistical researcher when dealing with uniquely human issues. Validity can generally be challenged by those who are interested in doing so.

value Values, in statistics, refer to numerical quantities rather than values of a moral, religious, or cultural nature.

variability Some things are identical within the limits of measuring error. Hydrogen atoms offer an example. Other things are not identical—they always differ from each other. There is little more to variability than the simple arithmetic of difference. However, variability confronts us with profound intellectual problems. For example, how do we account for the astonishing variability that exists in the area of personal wealth among Americans? Literally millions of pages of analysis have been devoted to this issue, and the matter has yet to be resolved, at an intellectual level, in a noncontroversial manner. See *constants.*

variance Technically, variance is the square of the standard deviation, or SD^2. It is the average of squared deviation from the mean. More generally, it is another measure of the extent to which values in a distribution vary from each other.

weighted mean Suppose we have three groups. The mean height of group A is 66 inches, for group B it is 70 inches, and for group C it is 74 inches. The mean for the groups, overall, is 70 inches. We discover that group A is eight times as large, in number of cases, as C; group B is twice as large. We can use the population size of the groups as weights and obtain a more accurate assessment of mean height. So, $[(8 \times 66) + (2 \times 70) + (74)]/11 = 67.45$ inches, a much lower mean value because of the large contribution of group A. The important thing is to know when unweighted means are disguised as more accurate estimates in order to serve some special interest.

within-group variability An examination of variance in a set of observations after the observations have been grouped, generally within nominal categories. For example, how is variance in intelligence affected by being partitioned into groups such as males and females? If within-group variability is reduced to zero, then a set of nominal categories can be said to explain, in the statistical sense, a set of scalar data—if various assumptions have been met. If the within-group variability is the same as it was before the scalar data were divided into nominal cases, then no statistical explanation has been attained. See *between-group variability.*

z-score A standardized score for determining how common or uncommon a given observation might be within a particular set. For example, if the mean for a set is 100 and the standard deviation for the same set is 20, then a particular value within the set that was equal to 140 would be two standard deviations above the mean. The *z*-score is equal to 2.0. (This assumes that the set of observations is normally distributed.)

zero point That value in a measurement of a variable that states the variable does not exist. The zero point is common for physical scales. For example, there are gravity zones with very high levels of gravitational force (black holes, for instance) and other areas where there is zero gravity (inside a space ship orbiting earth). One might argue that a small degree of gravitational attraction exists within the space ship—small, but measurable. For practical purposes, however, we can refer to it as zero gravity. With social variables, finding a zero point is commonly an arbitrary matter. The most common scales with established zero points are percentages because it is possible to have a zero percentage (for example, zero percent correct items in a test). However, getting zero percent correct on a test does not necessarily mean that an individual has zero knowledge of the subject of the test.

ABOUT THE AUTHORS

R. P. Cuzzort (Ph.D., Minnesota) is Emeritus Professor of Sociology at the University of Colorado at Boulder. Much of his early work focused on demography and statistical analysis. He was a coauthor of Dudley Duncan's *Statistical Geography: Problems in Analyzing Areal Data* (Free Press, 1961) along with Beverly Duncan. Professor Cuzzort then turned his attention to sociological theory, becoming concerned with the creative applicability of theory. His recent work *Using Social Thought: The Nuclear Issue and Other Concerns* (Mayfield, 1989) was selected by the Association for Humanist Sociology as its book of the year in 1989. His early work *Humanity and Modern Sociological Thought* (Holt, Rinehart, and Winston) is now in its fifth edition.

James S. Vrettos is a graduate of Columbia University and teaches sociology, criminology, and statistics at the City University of New York–John Jay College of Criminal Justice. Among his recent articles are "The Medicalization of Deviance and Criminality" and "Deviance and Social Structure: An Analysis of Christmas in July" (about the 1977 New York City blackout riots).

INDEX